PHYSICS
AS A LIBERAL ART

PHILOSOPHIÆ

NATURALIS

PRINCIPIA

MATHEMATICA.

Autore *JS. NEWTON,* *Trin. Coll. Cantab. Soc.* Matheseos
Professore *Lucasiano,* & Societatis Regalis Sodali.

IMPRIMATUR·
S. PEPYS, *Reg. Soc.* PRÆSES.
Julii 5. 1686.

LONDINI,

Jussu *Societatis Regiæ* ac Typis *Josephi Streater.* Prostat apud
plures Bibliopolas. *Anno* MDCLXXXVII.

PHYSICS
AS A LIBERAL ART

JAMES S. TREFIL
Professor of Physics
University of Virginia, Charlottesville

Pergamon Press
New York   Toronto   Oxford   Sydney   Frankfurt   Paris

U.S.A.   Pergamon Press Inc., Maxwell House, Fairview Park, Elmsford, New York 10523, U.S.A.

U.K.   Pergamon Press Ltd., Headington Hill Hall Oxford OX3 0BW, England

CANADA   Pergamon of Canada Ltd. 75 The East Mall, Toronto, Canada, M82 5W3

AUSTRALIA   Pergamon Press (Aust.) Pty. Ltd., 19a Boundary Street, Rushcutters Bay, N.S.W. 2011, Australia

FRANCE   Pergamon Press SARL, 24 rue des Ecoles, 75240 Paris, Cedex 05, France

WEST GERMANY   Pergamon Press GmbH, 6242 Kronberg-Taunus, Pferdstrasse 1, West Germany

---

First edition 1978

Library of Congress Cataloging in Publication Data

Trefil, J.   S
     Physics as a liberal art.

     Bibliography: p.
     Includes index.
     1. Physics. 2. Science—History. I. Title.
QC23.T8   1978   530'.09   77-6729
ISBN 0-08-019863-5

Picture Credits

Photographs were supplied by courtesy of the following Atomic Industrial Forum; Bettmann Archive; Brookhaven National Laboratory; Burndy Library; California Institute of Technology, Archives; Comsat; Consolidated Edison; ERDA; Fermilab; Hale Observatories; IBM; Mary Evans; NASA; Niels Bohr Library, American Institute of Physics; UPI.

Printed in the United States of America

TO JEANNE AND DOMINIQUE

CONTENTS

AUTHOR'S PREFACE

    One of the great joys of a liberal education is the chance that it gives you to study things, just for the fun of it, that will in all likelihood be of no practical use to you in the future. Certainly, when I look back on the courses I took as an undergraduate, the ones that stand out most are not those needed for my profession, as good as they were, but, those courses with names like "Art Appreciation," and "Opera Appreciation." These courses added nothing to my skills as a physicist and never earned me a dollar, but I know that I would be a much poorer person if I hadn't taken them.

    The idea of the "Appreciation" course is simply to open a new world—a new way of looking at things—to each student. These courses do not teach skills, they teach ideas and concepts. My courses did not turn me into an artist or a musician. What they did was give me some understanding of the art or music of a particular period so that when I encountered it, I had a deeper understanding of it than I would otherwise have had. This is the kind of enrichment of life that is supposed to be the legacy of a liberal education.

    I began to wonder why courses like this were so seldom found in the offerings of science departments. After some investigation, I discovered that the reason had to do with two rather rigidly held set of prejudices—one set for scientists and another for humanists. Unfortunately, the vast majority of students, holding neither set of prejudices themselves, seemed to be denied the opportunity to learn about science because of them.

On the one hand, we have the humanist prejudice. Hasn't it ever struck you as odd that a man who has never read Shakespeare would be considered uneducated, while a man who had never studied Newton or Einstein would not? This sort of attitude is not so uncommon among people who have studied the traditional literary subjects. In some way, science is excluded from "culture," even though the unique contribution of western civilization to human knowledge has been the scientific outlook.

On the other side of the coin, when I discussed the ideas for this course with my colleagues, I frequently encountered remarks like "Well, if you teach it that way, you won't really be teaching physics, will you?" In other words, there is a strong feeling in the sciences that unless someone is willing to put in the years of study needed to learn a science in its fullest sense, it just isn't worth teaching him anything. The idea of a "Science Appreciation" course with the same goals as an "Art Appreciation" course is incompatible with this way of thinking.

In a way, this book is an attempt to find a way between these two conflicting viewpoints, one saying that science isn't worth studying and the other saying that it isn't worth teaching. An attempt is made to see science in general, and the science of physics in particular, as a part of the development of western civilization. The connection between the cultural background of a scientist and the kind of scientific outlook that emerges is discussed in several examples, so that when we get to modern science, we can see it as a part of modern life.

The idea of the book is not to teach someone how to work physics problems or to make anyone into a miniature physicist. Instead, I have tried to explain the central ideas of modern science in simple terms, with an eye toward giving the student some idea of how science operates and how a scientist sees the world. If, after going through the book, the student can pick up the "Science" section of a newsmagazine and have a little deeper understanding of what he reads than he would ordinarily have had, then I will consider the experiment in teaching contained herein a success.

I would like to thank my son Jim (who has matured considerably since the incident described on p. 6) for help in proofreading the manuscript.

James Trefil
Charlottesville, Va.

GENERAL
BIBLIOGRAPHY

In addition to the texts listed at the end of each chapter, there are a few books that the student can use for general reference for almost all of the material presented. These are:

Greider, Ken. *Invitation to Physics*. New York: Harcourt, Brace Jovanovich, 1973. This book follows a general historical approach, and is quite readable. A number of specialized topics are presented, and a very interesting set of biographical sketches of important scientists is given.

Hewitt, Paul G. *Conceptual Physics*. New York: Little, Brown, 1974. Although this text does not follow a historical outline, it treats so many topics so well that it is worth the student's time to attempt to cope with the differences in order of presentation of subject matter.

March, Robert H. *Physics for Poets*. New York: McGraw-Hill, 1970. One of the first (and still one of the most readable) textbooks on physics for non-scientists.

CHAPTER I

HOW WE KNOW THINGS

"The Savage is to the ages what the child is to years."
Percy Bysshe Shelley,
"A Defence of Poetry"

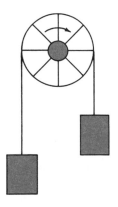

The 200-inch reflecting telescope at Mount Palomar, California—one of the most important instruments in astronomical research in the world.

A. What is Physics?

Physics can be defined simply as the study of matter and motion. As such, it is the science that is most basically concerned with uncovering the laws that govern the action of the material world. The laws discovered by the physicist often become the bases for new sciences, but the physicist himself is not usually involved in the application of these laws.

For example, we will study the way in which Galileo and Newton discovered the basic laws that govern the motion of falling bodies. Today, this knowledge allows us to send probes to the farthest reaches of the solar system and to put men on the surface of the moon—things that Newton could only have dreamed about. In the last century, physicists discovered the laws that govern the motion of fluids. Today, scientists are applying these laws to problems as diverse as the flow of protoplasm in a cell and the flow of traffic in a city.

Thus, when we learn about the way that scientific discoveries were made in the past, we will be learning about things that affect us constantly in our daily lives; when we learn about the kind of research that is going on in the sciences today, we will be getting some insight into things that will be affecting our lives in the future.

Of course, physics is only one science among many. In general, we speak of three kinds of "sciences"—physical, biological, and social. The physical sciences (physics, chemistry, and astronomy) deal with material, inanimate systems. The biological sciences deal with living systems, while the social sciences deal with individual or group behaviors.

Nevertheless, physics plays a rather special role among the sciences. As the science that is concerned with understanding the laws of material objects, it is often intimately involved in the basic workings of the other physical sciences. For example, the laws of physics are used by astronomers to calculate the motions of stars and planets. The laws of the atom and of heat transfer are used by chemists to deal with reactions which they study. Almost all of modern engineering science is based on applications of the laws of physics.

In addition to providing the basis for the physical sciences, it may very well turn out that the laws of physics provide the basis for understanding the biological sciences as well. One of the most exciting developments in modern times

has been the birth of the field of biophysics, which is concerned with the application of the laws governing atoms
and molecules to the large molecules found in living organisms.

In what follows, then, we will be tracing the slow
acquisition of knowledge about the material world by scientists in many different cultures. The goal is always the same:
to understand the world in which we live. The name that
men have applied to the process has varied—philosophy,
astronomy, and physics, to cite a few. But, no matter what
name is used, the quest remains the same.

B. Knowledge

Before we start to talk about how nature works and
how the world around us is put together, we should pause
to ask how it is that human beings acquire knowledge about
their surroundings. The general area of study dealing with
how we know things is called epistemology. It is one of the
oldest fields in philosophy. We are going to concentrate on
two different aspects of the acquisition of knowledge—the
cultural and individual.

By cultural knowledge, we shall mean the accumulated
knowledge about the physical world to which a particular
culture has access. For example, in 20th-century America,
we have access to knowledge about nuclear physics, about
the germ theory of disease, and about the laws of falling
bodies. You can add many more items to this list. In 16th-
century Italy, on the other hand, only the laws governing
falling bodies would have been part of the general store of
knowledge, while in classical Greece, even this would not
have been known.

Obviously, the acquisition of knowledge by the human
race has been a long and arduous process, and each group of
people has built upon the knowledge of previous groups. It
is also obvious that we can talk about the growth of knowledge available to people at any given time in terms of a general
growth of human knowledge spanning cultural and geographical boundaries. Thus, by the acquisition of knowledge "cul-
turally," we shall refer to this steady growth of knowledge
over a long period of time which is the result of the work of
a large number of people.

There is another sense in which we can talk about the
acquisition of knowledge. When a child is born, he knows

Albert Einstein (1879-
1955) discussing the
theory of relativity.

3

very little about the world around him. By the time he reaches adulthood, he knows as much about it as other members of his culture. Some place along the line he acquired knowledge as an individual. We shall discuss this process later, but it is to this type of learning process that we apply the name "individual acquisition of knowledge." One of the themes that we shall try to develop is to see how these two different processes—cultural and individual acquisition of knowledge—parallel each other, and how we can draw striking analogies between the state of development of a culture and the state of development of an individual.

Before we begin discussing the acquisition of knowledge, however, it should be made clear that throughout this text we shall be talking only about *scientific* knowledge. This is knowledge that is gained by use of the logical, rational, deductive capability of the human mind. It is not, however, the only kind of knowledge that human beings can acquire, nor is it the only kind of truth to which human beings can aspire. For example, we have all had the experience of having an intuition or hunch, when we "knew" something as surely as we ever knew anything in our lives, without being able to prove it. When we love someone, we don't ask for a logical proof of the fact. A mystical or religious truth, likewise, cannot be "proven" in the scientific sense, but it can be as true as the law of gravity.

The point is that the scientific development of man, with which we shall be concerned here, is only one part of his development, just as rationality is only one part of a complete person. Since this is a book about physics and the development of scientific thought, we shall be concentrating exclusively on rational and deductive truths. But this should not blind us to the fact that other kinds of truths exist whose study is as interesting and important as the one we are beginning here.

C. Cultural Knowledge: The Technical and Philosophical Traditions

Let us begin by stating a few generalities about cultural knowledge. The next several chapters will be devoted to a discussion of different cultures and their scientific development, but a few introductory remarks here will give us a frame of reference for that discussion.

From the earliest times, men have been interested in understanding the world around them. There are many reasons for this. It is important to know when to plant crops, which means that something about soil and weather has to be learned. Telling time and understanding the seasons, in turn, involves some sort of crude astronomy. Developing tools and weapons involves acquiring some practical knowledge of metallurgy and the properties of materials. The list could be extended indefinitely, but the point is clear. In order to survive—to feed, clothe, and shelter himself and his children—man has had to learn how to use things that he finds in nature, and this means that he has had to study and learn about nature. This pragmatic, practical kind of endeavor we shall call the technological tradition. It will be dealt with at length later on. For now, you should bear in mind when you pass a construction site or see an airplane or a car that you are looking at the fruits of man's labor in this tradition.

Leonardo da Vinci (1452–1519) drawing of giant crane.

At the same time, there is in man a yearning for something more than just the material needs of life. He wants to know not just the fact that he should plant in the spring, but why spring always follows winter. He may need to be able to use the stars to navigate his ship, but he wants to know what the stars are, why they are there, and what man's place is in the total scheme of things. There are many aspects to this quest—too many, in fact, to mention here. However, we shall call the body of thought that concerns itself with providing answers to these deeper questions (at least as far as they are concerned with the physical world) the tradition of natural philosophy. Some examples of this tradition might be the myths explaining the creation of the earth or the Greek theories about the nature of matter.

We shall try to show how the development of science, which took place in Europe after the Renaissance, can be thought of as a unique blending of these two traditions. While every civilization has both technical and philosophical knowledge, only in Europe were they combined in such a way as to produce the scientific and technological revolution in which we are still living and which, for better or worse, has fundamentally altered the course of human history.

5

D. Development of Knowledge in Individuals

It is obvious that between the time an individual is born and the time he dies, he learns a great deal about the world. For centuries, philosophers have concerned themselves with the process by which this knowledge is gained. Some, like Plato, have argued that the knowledge is already "there" in the human mind, and simply needs to be brought out. Others have developed different theories. Yet, until this century, no one thought to answer this question by studying those directly concerned with the acquisition of knowledge—children. It remained for Jean Piaget, a Swiss psychologist, to realize that by observing children as they learn one can trace the process by which full-blown concepts are acquired by adults.

It is one of the marks of genius that when a new, revolutionary idea is enunciated, one's first reaction is to say "of course—that is the way to do it." It seems obvious to us today that the proper way to learn about the development of concepts in human beings is to trace these concepts from the earliest age when they appear to adulthood. Yet for thousands of years philosophers argued about these concepts and never thought to test their ideas against observation.

Jean Piaget was born in 1896; he became interested in natural science at a very early age. When he was 10, he published an article in a scientific journal, and was reportedly offered a museum directorship because of it. (The offer was withdrawn when his age was learned.) He began working with children in Geneva in the 1920s, and, although he quickly began developing the theories that we shall discuss later, his work was largely ignored until quite recently. He was the classic case of an "isolated genius" who developed his ideas without a great deal of interaction or encouragement from other members of his profession.

The basic theme that Piaget discovered in the development of children can be characterized as a continual shrinking of the ego—a steady realization that the external world is composed of objects that are not like human beings and that must be thought about from a non-egocentric point of view.

The first step in such a process is the child's realization (at a very early age) that there is a difference between "me" and "not-me." This fact was emphasized for me soon after the birth of my first child. I came into the nursery one day

6

and saw him lying in the crib, looking at his upraised hand. He would move the hand and then laugh uproariously. This went on for a long time. What delighted him, in our terminology, was the discovery that an object out there—that pinkish thing—could be controlled by his own mind. He also knew (or learned very soon) that other objects were not like this. He could not, for example, affect the walls of the room the way he could affect his hand. This realization—that there are objects in the universe that are not part of oneself—is the first step in the shrinking of the ego.

Piaget concerned himself primarily with what followed this first step. He found that the ego-shrinking continued for many years and was accompanied by an increasing sophistication and ability to distinguish between things in the world. His method of study was to take a single concept (such as causality, or space, or velocity, or life) and see how it developed over the years. He would then construct elaborate and subtly differentiated lists of stages for the development of each of these concepts. In the example that we shall consider in Table 1.1—the development of the idea of causality—17 distinct stages are differentiated by Piaget.

For our purposes, however, we can simplify (perhaps even oversimplify) and say that the development of concepts in children can generally be thought of as proceeding through three stages: (1) personal, (2) transitional, and (3) rational-deductive.

Table 1.1 The Stages of the Development of the Concept of Casuality According to Jean Piaget*

1. *Motivational.*
   Things happen because objects want them to.

2. *Finalistic.*
   Things are part of a divine plan.

3. *Phenomenistic.*
   Things that are contiguous in space and time are connected.

*In general, stages 1-6 correspond to what we have called "personal," stages 7-9 to "transitional," and stages 10-17 to "rational-deductive."

4. *Participation.*
    Like things can influence each other.

5. *Magic.*
    Things can be controlled by gestures, thoughts, etc.;
    This is similar to stage 4.

6. *Moral Causality.*
    Things happen because they must. This is similar to
    stage 1, but with the objects obliged to act as they
    do.

---

7. *Artificialistic.*
    Events occur because of human acts; this is comple-
    mentary to (and occurs at the same time as) stage 6.

8. *Animistic.*
    Things become what they are through a process of
    growth after creation.

9. *Dynamic.*
    Things happen because of a "life force" in objects.

---

10. *The Surrounding Medium.*
    Things happen because of the reaction on the object
    of its surroundings; this is the first physical expla-
    nation.

11. *Mechanical.*
    Explanation through physical contact.

12. *Generation.*
    Things in the world are born out of each other; this
    is animistic, but admits that substances may change.

13. *Substantial Identification.*
    Things are made up of other, simpler things (e.g.,
    sun made from clouds rather than made by men
    and then growing).

14. *Condensation.*
    Like stage 13, but with the idea of closeness or
    packing of constituent materials included.

15. *Atomistic.*
    Everything is made of small particles of material (don't confuse this with the modern idea of atoms).

16. *Spatial.*
    Explaining things by laws of perspective, relative volumes (not always found.)

17. *Logical Deduction.*
    Explanation by deduction from general laws.

One way of thinking about the personal stage of development is to understand that, although the child realizes that external objects are not a part of himself, he nevertheless believes that they are *like* himself in that they have a personality, a will, and feelings. In this stage, then, the child will see events happening because they are willed to happen by objects whose personality is much like his own.

If a child in this stage sees a balloon rising in the air, he will say that it rises "because it wants to fly" or "because it wants to see everything from up there" or "because it was cold here and it wants to go away."

The important point about this stage is that there is no question of the balloon obeying some natural law. There is no guarantee that another balloon will rise. After all, it might not "want" to. This idea of seeing every event as a personal, individual happening is the main characteristic of the first stage that we shall want to remember later. The personal stage lasts roughly from ages 1 through 6. The second stage (which we are calling transitional) is shorter, lasting roughly up to age 9. In this stage the child realizes that external objects do not have personalities, but he nevertheless endows them with a "life force" or "strength," and attributes their actions to this force. In the example of the balloon mentioned above, the child would not say that the balloon rose because it wanted to, but rather that it rose "because it was full of gas, and strong, so it flies." He might say (as Greek scientists did) that objects fall toward the earth "because it is their nature to move toward the center of the universe." Because of the idea of a "life force," we shall sometimes refer to this stage of development as the "animistic" stage, rather than the "transitional" stage.

9

The final stage of development is completed by late adolescence and is characterized by the concept of general laws that are obeyed by physical objects. This, we shall come to see later, is also characteristic of the scientific attitude toward the world.

In the balloon example, a person in this stage might say something like "the balloon rose because it was filled with a gas that was lighter than air, and lighter objects will float." There is no question here of the balloon possessing a will or a vital force; it is simply a question of an impersonal object obeying a natural law.

As should be obvious from Table 1.1, there are subtle distinctions within each of the three stages. In addition, this attempt at categorization suffers from much the same fault as any other scheme that tries to capture the infinite variety of real life in a simple system—life is much more complex than the scheme appears to make it. For example, in a group of children of a given age, it is quite possible that we would find some children in each of the three stages listed above. Nevertheless, these broad categories will serve as a useful framework for our discussion. The main point that we want to make is that each individual human being reaches adulthood by passing through a series of fairly well-defined stages. Even though each person proceeds through these changes at his own speed, and may skip some of the substages listed in Table 1.1, each person will eventually go through all of the main categories.

What we see, then, is a steady progression in the individual from the personal to the impersonal—from a view in which each event is the result of a conscious act of will on the part of some object to a view in which objects, now devoid of personality, behave according to well-defined and discoverable laws. Another way of stating this transition is to say that the child develops from a view in which external objects are "thou" (i.e., are personalities much like himself) to one in which the objects are "it."

In the chapters that follow, we shall see that this same transition from "thou" to "it" was made gradually by the human race in its cultural development as well. We shall see how the world view of the Egyptians and Babylonians could easily be characterized as being a "thou" view of events, and that post-Renaissance Europe was involved in developing an "it" view. The transitional stage in this scheme of things occurred with the Greeks.

We do not wish to imply, in setting up a scheme of this sort, that there is necessarily some deep metaphysical connection between individual and cultural development, although such a connection could well exist. Rather, we are trying to set up a framework of ideas into which we can fit the things we are going to learn about the ideas of many different cultures and peoples. You may, if you like, simply regard the developmental stages of the individual as useful analogies when we discuss the development of cultures.

E. A Short Preview

In the succeeding chapters, we shall be studying the slow process by which men acquired knowledge about the world around them. As we have already said, there will be analogies in this process to the individual acquisition of knowledge. However, the main point of this story is not so much in analogies to other things but in the sheer fascination of being able to see the best minds of the human race grappling with unanswered questions.

The first "scientists" in the modern sense were astronomers, so we shall concentrate on how men came to know about the heavens and the earth's place in the universe. We shall see how this knowledge, in turn, led to important questions about how material objects behaved. The study of this question, which we call the Science of Mechanics, was what eventually led Isaac Newton to formulate the scientific method, and to use this method to show that the same laws governing objects on the earth are responsible for the motion of the planets.

Once the scientific method was available, it was widely applied. We shall discuss what was learned about electricity and magnetism and phenomena concerned with heat (the study of these latter phenomena is thermodynamics). These three sciences—mechanics, electricity and magnetism—and thermodynamics form the basis for what is called Classical Physics. By the end of the 19th century, scientists felt that they were very close to knowing all of the answers to questions about the physical world.

The period that followed marked one of the great intellectual revolutions of history. In a few short years, two

11

totally new scientific disciplines were born. One, called relativity, dealt with objects moving near the speed of light; the other, called quantum mechanics, dealt with objects that were very small (of the size of atoms or less). This revolution is still not finished. When the last chapter of this book is completed, the reader will discover that, although we have learned a great deal about the world in the last 3,000 years, we still have a great deal to discover.

SUGGESTED READING

Frankfort, Henri. *Before Philosophy.* Baltimore: Penguin Books, 1946. This book contains an excellent discussion of the "thou-it" relation in primitive cultures.

Mailer, Norman. *Of a Fire on the Moon.* New York: Signet, 1971. The first chapter of this book ("The Psychology of Machines") contains a modern attempt to personify a complex object.

Piaget, Jean. *The Child's Conception of the World.* New York: Harcourt Brace, 1929.

Piaget, Jean. *The Child's Conception of Physical Causality.* Totowa, N.J.: Littlefield, Adams, 1966. Like most of Piaget's own writings, these books are rather heavy going. It is worth reading, if for nothing else than the direct quotations from the children he worked with.

QUESTIONS AND DISCUSSION IDEAS

1. Make a list of items that might be classed as cultural knowledge for the following groups:
   a. 20th-century urban Americans
   b. 18th-century American pioneers
   c. present-day Australian aborigines
   d. medieval Europeans
2. Could there be cultural growth without writing? If not, why not; if so, how much?
3. Make as complete a list as you can of ways of "knowing" things other than those associated with rational thought.
4. In the text, we discussed the idea that the laws of physics might someday be found to explain biological systems. Discuss the idea that the laws of physics might someday be found to explain individual psychology as well.

5. A group of children watch the following experiment: A brick, painted red, and a block of wood of identical shape, painted blue, are dropped in a tub of water. The children are then asked to explain what they saw. Categorize the stage of development of the child who gives the following answers:
   a. The amount of water displaced by the brick weighs less than the brick, so the brick sinks.
   b. The brick sinks because it is red, and red things go to the bottom, where it is warmer.
   c. The wood floats because it is blue, and blue things are stronger than red ones.
   d. The wood floats because boats are made of wood; if it didn't, boats would sink and people would drown.

6. A man throws a rock into the air. A number of children see this and then are asked to explain why the rock continued to move after it left the man's hand. Give an answer typical of each of the three main stages of development we have discussed.

7. Give an approximate age for each child in your answer to Problem 6. Can you give more exact answers in general? Why or why not?

8. Piaget listed 17 stages in the development of the idea of causality, and felt that there were no more. Discuss the question of whether or not there might be further stages of development for the human being.

9. List some modern activities that might come under the heading "technical tradition." Which of these might have been part of the technical tradition 500 years ago?

10. List some modern activities that might come under the heading "philosophical tradition." Which of these might have been part of the philosophical tradition 500 years ago?

11. Make a guess at what the stages of development might be for the concept of motion. Look up the actual stages and see how close you came to what Piaget found.

CHAPTER
II

PHILOSOPHY AND TECHNOLOGY IN TWO ANCIENT CIVILIZATIONS

And on the pedestal these words appear:
"My name is Ozymandias, king of kings:
Look on my works, ye Mighty, and despair!"
Nothing beside remains. Round the decay
Of that colossal wreck, boundless and bare
The lone and level sands stretch far away.

Percy Bysshe Shelley,
"Ozymandias"

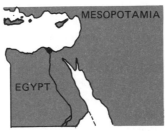

A. Introduction

The best way to learn about technology, with its concern for the practical aspects of life, and natural philosophy, which concerns itself with more spiritual questions, is to study societies in which the two were more separated and distinct than they are in our own. In the cultures that existed in Egypt and Mesopotamia, there was a steady growth of the ability of men to control their environment. We would call this a growth in their technological ability. In addition, there was a coherent, well-thought-out natural philosophy (which we usually refer to as a religion or mythology), which supplied each member of the society with a very clear and precise statement of what the world meant and what his place in it was. We shall see, however, that there was little connection between these two areas of thought.

We shall also see that there is a strong analogy between the way that people in these societies viewed the world and the first stage of development (discussed in the last chapter) in which a child sees every event as a distinct "thou." This fact had a strong influence on the kind of scientific questions that could be asked and on the kinds of science that could develop in these cultures.

B. Egypt: Philosophy and Technology

The first civilization we shall discuss is the Egyptian kingdom that grew up on the banks of the Nile around 2500 B.C. (exact dates are not really important for our purposes). There were two central geographical facts governing the Egyptian outlook on life. The first of these, of course, was the Nile river, which, with its yearly flood, nourished the fertile strip of land along its banks and made human life possible. The importance of the Nile is made especially clear to a visitor because it is often possible to stand with one foot in the desert and one in growing plants.

The second important geographical fact governing the Egyptian outlook concerned the splendid isolation of Egypt. On either side of the narrow strip of tillable land there stretched a wide expanse of desert ringed with mountains, which served as a natural barrier against invasion for much of Egyptian history. The kind of view of the world that developed under these circumstances is summarized very

16

well by the noted historian Otto Neugebauer, who in his book *The Exact Sciences in Antiquity*, puts it this way:

> Of all the civilizations of antiquity, the Egyptian seems to me to have been the most pleasant. The excellent protection which desert and sea provide for the Nile valley prevented the excessive development of the spirit of heroism which must often have made life in Greece hell on earth. There is probably no other country in the ancient world where cultivated life could be maintained through so many centuries in peace and security.

When an Egyptian wondered about the questions that we associate with the philosophical tradition—"What does it all mean?" "Where do I fit in?"—there was a well-developed set of beliefs to answer them. To the Egyptian, the universe was the kind shown in Fig. 2.1. The earth was a flat plate with mountains around the edges and Egypt in the middle. The plate floated in the primordial waters (which were the sources of the Nile). There was the sky above, and a counter sky below. The stars and planets (insofar as they were considered at all) were simply fixtures in the sky. The daily cycle of the sun was thought to be a journey of the

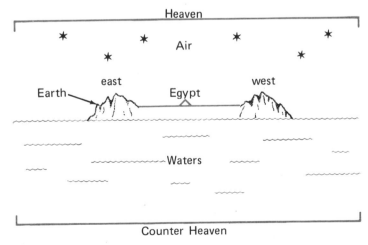

Figure 2.1.   The Universe as seen by an Egyptian.

17

Sun God through the heavens. At night, he entered the underworld and fought a battle with the forces of darkness, emerging victorious each morning.

There were many explanations about how the world came to be. The elements of the most common version of the creation myth are quite common in other civilizations. In the beginning, there was only the primordial waters. From this water sprang gods, and the Sky God separated the waters into earth and sky. The study of Egyptian mythology is a truly fascinating subject, and the interested reader is referred to the texts mentioned at the end of this chapter.

For our purposes, we make special note of several aspects of the Egyptian world view. In the first place, we find that events tended to be regarded as special and personal, rather than general. For example, each time the sun rose, it was because a very personal god had won a battle during the night. The rising of the sun was not the result of a general law (as it is for us) but a singular and special event each time it occurred. It was, in other words, a "thou" rather than an "it."

In the second place, the mythology contains many examples of nonrational, "mythopoeic" thought. For example, to the question "What holds the sky up?" you might get one of several answers. You might be told that it is held up by four pillars, or that it is the belly of a cow (the sky goddess), or that it is held up by the sky god (as in the creation myth). If you were to press this point, an individual would tell you that *all* of these answers were true at the same time. This is hard for us to understand, but in the mythical mode of thought the idea that something is either one thing or another simply doesn't hold. Thus, the natural philosophy of the Egyptians did not necessarily involve what we would now call rational answers to questions about the world. This does not mean, however, that it did not fulfill its function, which was to provide answers to difficult questions.

Finally, we note that while this world view does, indeed, provide answers to the type of questions that are asked in the philosophical tradition, these answers do not have much to do with helping one with problems of everyday life. For example, knowing that the Nile has its sources in primordial waters under the earth does not help in predicting when it will flood or in surveying the fields after the waters have receded. Thus, it became necessary to develop, completely independently, a way of dealing with these practical questions. This set of methods we shall call the Egyptian technology.

There can be no doubt that the Egyptians had developed their technical skills to a high degree. The pyramids and other buildings that still stand today give testimony to that. We also know that they carried on an extensive commerce and had highly developed medical skills. We shall look at just one aspect of Egyptian technology which is typical of all of the others: the development of mathematics.

There are many practical reasons why a knowledge of mathematics might be necessary. For example, suppose you wanted to divide the contents of a granary among a man's heirs, so that the eldest child receives twice as much as the others. How much does each one get? Or suppose that after the annual flood, you wanted to reestablish the boundary markers so that farmers could begin to plant their crops. You would need some knowledge of geometry to do this. Thus, very early in their development, Egyptian scribes began to work on techniques for handling these types of problems. Most of what we know about Egyptian mathematics is contained in the "Rhind Papyrus," which was deciphered in 1877.

The system that was finally developed certainly did the job for which it was designed, but it was very different from what we now call mathematics. It was essentially a set of "rules of thumb" arrived at through trial and error, rather than a set of general laws from which consequences could be derived by rational deduction.

The Egyptian number system worked in this way: a single stroke (I) stood for one, a hoop (∩) for 10, a coil (ȣ) for 100, and so forth. To write the number 37, for example, one wrote three 10s followed by seven ones, or

∩∩∩IIIIIII

The two mathematical operations that the Egyptians could carry out—multiplication and division—were based on very simple procedures. For example, if an Egyptian wished to work the problem

4 X 5 = 20,

he would start by constructing a table of multiplications by four; that is,

$$1 \times 4 = 4,$$

$$2 \times 4 = 8,$$

$$4 \times 4 = 2 \times 8 = 16.$$

He would then note that the first and last row could be combined to give

$$1 \times 4 + 4 \times 4 = 5 \times 4 = 4 + 16 = 20,$$

which is the desired answer. The reader can gain some facility with the technique (which is actually used in some modern digital computers) by working some of the examples at the end of the chapter.

Division was simply the inverse of this process. For example, to divide 20 by 5, we simply construct a table of multiplication by 5, or

$$1 \times 5 = 5,$$

$$2 \times 5 = 10,$$

$$4 \times 5 = 2 \times 10 = 20.$$

The last line, of course, contains the answer. Egyptian scribes compiled long lists of such tables, so that these mathematical operations were reduced to simple bookkeeping.

If this were all there was to arithmetic, grade school would have been a lot easier for all of us. Unfortunately, in performing a division, it does not always happen that everything works out evenly, as in the above example. When it doesn't, we encounter fractions, a subject to which the Egyptians devoted considerable study.

They developed a system (for reasons that are not clear) in which every fraction was expressed as a sum of other fractions, each with a numerator equal to one. For example, we would write

$$\frac{13}{16} = \frac{1}{2} + \frac{1}{4} + \frac{1}{16},$$

which the Egyptian would render as

$$\overset{\bullet}{2}, \overset{\bullet}{4}, \overset{\bullet}{16}.$$

20

The pyramid at Giza, reflected in flooded land, illustrating the technical abilities of the ancient Egyptians.

Again, long tables were compiled by trial and error, in which vast numbers of fractions were written this way. One could, for example, consult such a table to learn that

$$\frac{2}{99} = \dot{6}6, \ 1\dot{9}8.$$

By the use of these tables, it was possible to do simple calculations involving fractions, but this is as far as the arithmetical skills went. When you think about it, this illustrates a very important point. Can you imagine trying to use this cumbersome kind of arithmetic to figure the interest of a bank balance or the return on a stock (not to mention designing an airplane or automobile)? Clearly, with the type of mathematics available to the Egyptians, advanced concepts in many fields (including science) were simply not available. We shall see this theme over and over again as we trace the development of science. Before important new breakthroughs can be made, it is necessary that the mathematical techniques needed for that new understanding be available. For example, a real understanding of the solar system could not be obtained without the development of the calculus (which we shall discuss later). The fact that, unlike the Babylonians, the Egyptians never developed astronomy may be related to the way in which Egyptian mathematics was developed. This is best illustrated by looking at geometry, an area that was critically important in the problem of resurveying fields after the annual Nile floods. The Egyptians knew, again by trial and error, that the area of a rectangle was the product of height times width. They had also discovered an approximate formula for the area of a circle. In the Rhind Papyrus, this is given in the following words:

"A field round of khet 9 (i.e., diameter 9). What is the amount of its area? Take thou away one ninth of it, namely 1. Remainder is 8. Make thou the multiplication 8 X 8. Becomes it, 64. The amount of it, this is the area."

In modern language, the Egyptian would give the area of the circle of diameter d as

$$A = (\frac{8}{9}d)^2,$$

while the correct formula, we know, is

$$A = \pi r^2 = \pi(\frac{d}{2})^2.$$

We can get some idea of how good the Egyptian approximation to the area of a circle was by noting that an Egyptian would write the area of a circle of diameter d as

$$A = 0.790 \, d^2,$$

while we would write the area of the same circle as

$$A = 0.785 \, d^2.$$

Written this way, the two results seem very similar indeed. In fact, if we worked out the correct answer to the problem posed in the Rhind Papyrus, we would get 63.6 square units, as opposed to the 64 square units the Egyptian author got. The two answers differ by less than one percent!

How did the Egyptians arrive at such a good approximation to the area of the circle? We can't know for certain, of course, but the most likely thing is that some very bright young scribe in a temple somewhere discovered, by trial and error, that the approximate form given above gave answers very close to those that were known to be true. It is doubtful that accuracy of more than one percent was needed in surveying fields or calculating the volume of cylindrical granaries. The important thing from our point of view, however, is that this is purely a rule of thumb procedure. The Egyptians had no general rule for finding areas, but had developed,.by trial and error, formulae for finding the areas of certain kinds of shapes.

To drive this point home, ask yourself how the author of the Rhind Papyrus would have approached the problem of finding the area of an ellipse. Since he had no general rule for finding areas, he would have had to start from scratch—with trial and error again. In fact each new shape would have to be handled as a separate problem, to be dealt with as if no other formulae for area were known. Unlike modern mathematics, where areas are derived from a general law governing all plane figures, in Egyptian mathematics each figure was treated individually, and there was no general law of areas.

Thus, the "thou" aspects of the Egyptian world view are reflected in their technology. Their great achievements in engineering and other technological fields rested on the ability of the artisan to cope with individual problems as they presented themselves, rather than on any type of deductive reasoning.

C. Technical Digression: The Phases of the Moon

Before we turn to a discussion of the Babylonians and their contributions to the science of astronomy, it is necessary to remind ourselves of some of the features of the night sky, particularly the moon, as it appears to observers on the earth. Throughout this discussion, we shall discuss the motion of the moon as *we* now know it to be. It must be borne in mind that this is most emphatically *not* the way it would be perceived by a Babylonian astronomer.

Let us first consider the earth by itself. Suppose we had a point of observation in a rocket ship directly above the north pole. Looking down, we would see the earth turning underneath us. If we looked at the United States, we would see San Francisco and New York (see Fig. 2.2). They would be turning in the direction indicated (the easiest way to re-

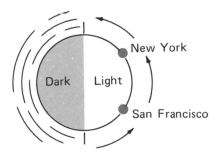

Figure 2.2.

member this is to recall that the sun comes up in New York before it does in San Francisco. Looking down on the earth, we would see the city of New York go from light into dark and, about 12 hours later, from dark into light again.

Now let us ask how this set of events would appear to a man standing in New York. While (from our rocket ship) we see the sun standing still and the world turning, the man in New York believes that *he* is standing still and the sun is moving. What we would call the movement of New York from light into dark, he would call the sun setting. Furthermore, it would appear to him that the sun had moved down toward San Francisco (i.e., had set in the west). In the morning, although we would say that he had moved back into the light, he would see the sun coming up in the east.

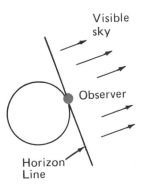

Figure 2.3.

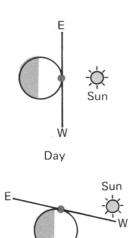

Day

Sunset

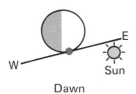

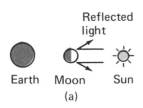

Dawn

Figure 2.4.

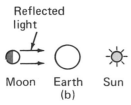

Earth   Moon   Sun
(a)

Moon   Earth   Sun
(b)

Figure 2.5.

Another way of thinking about this sequence of events is to consider what a man can see when he stands at any point on the surface of the earth (see Fig. 2.3). Obviously, if he looks straight down, he will see the ground under his feet; if he looks straight up, he will see the sky. If he looks straight out along a line tangent to the earth at the point where he is standing, he will be looking at his horizon. (To see this, note that if he is looking in this direction initially, and then looks down a little, he will be looking at the ground, while if he looks up a little, he will be looking at the sky). The only parts of the sky he can see are those above this horizon line. (This region is labeled "visible sky" in Fig. 2.3.)

Let us now watch from our vantage point above the north pole as the man in New York goes from day to night to day again (see Fig. 2.4). During the day, the sun lies in the "visible sky" region, above the horizon line. At sunset, the horizon line passes through the sun. We see the man in New York pass from light into dark, but he sees the sun set below his horizon line. During the night, the "visible sky" does not include the sun; so it is dark. At dawn, the other end of the horizon line sweeps across the sun, and the man in New York sees a sunrise.

With his way of looking at sunset and sunrise, let us now turn our attention to the problem of the appearance of the moon to an observer on the earth. We know that the moon generates no light of its own, so that what we see is actually light from the sun reflected from the face of the moon. If we again go to our vantage point above the north pole, we know that when the earth, moon, and sun are as in Fig. 2.5a, no reflected light can reach the earth, so it will be the time of the month when there is no moon at all. Similarly, when they are as in Fig. 2.5b, observers on the earth will see the reflection from the entire face of the moon—it will be full moon.

A common cause of confusion in thinking about the full moon is the question of why the moon is not in the earth's shadow, and therefore invisible. The answer is that the sun, moon, and earth's shadow do not lie exactly in a plane, such as the plane of this page, so that the moon will normally be a little above or a little below the earth's shadow. (When this doesn't happen, and the moon does pass into the earth's shadow, we have an eclipse of the moon!)

24

Let us proceed by asking how the moon appears to our observer in New York. For the sake of argument, assume that the moon is in the position shown in Fig. 2.6a. Then when it is noon in New York the moon is below the horizon and not visible. A few hours later, Fig. 2.6b, the horizon line crosses the moon, which means that the moon comes up on the eastern horizon. Note that at this time the sun is still above the western horizon, so that it is still daylight. (Have you ever seen the moon during the day?)

A few hours later, Fig. 2.6c, the horizon line sweeps across the sun, and the man in New York sees the sunset. At sunset, he observes the moon high in the eastern sky, as illustrated in the figure. The moon is in "first quarter," since the man can see only a part of the lighted half of the moon. In the early morning hours (see Fig. 2.6d), the western horizon line will sweep across the moon, and the man will see the moon set in the west. Between sunset and moonrise, he will see the moon move across the sky toward the west. We see, then, that the fact that the earth rotates means that someone standing on the earth will see the sun and the moon go across the sky in circles.

One special configuration of the earth-moon-sun system will be of interest to us in our study of the Babylonians, and that is the new moon. We mentioned above that when the moon is between the earth and the sun, we can see nothing. Soon afterwards, however (see Fig. 2.7), when the moon has proceeded a little way around its orbit, it will again be possible to see just a small sliver of the illuminated part of its face.

If we ask where this visible sliver will be at sunset, a glance at Fig. 2.7 should convince us that just at sunset, when the horizon line is sweeping across the sun, and the observer on the earth is passing from light into dark (i.e., when he sees the sun set in the west), the moon should be just above the horizon line. This means that an observer will see just a sliver of the moon on the western horizon at sunset. The significance of this particular event will become obvious when we discuss Babylonian astronomy in the next section.

It is also very instructive to follow through the first few days of the cycle of the lunar month. On the first day of the month, as discussed above, the moon appears just on the

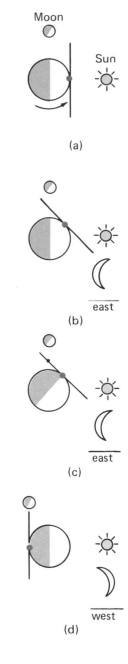

Figure 2.6.

25

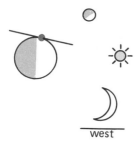

west

Figure 2.7.

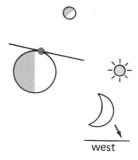

west

Figure 2.8.

western horizon at sunset. Soon after the sun goes down, the moon sets as well. (Can you prove this to yourself by thinking of the rotation of the earth and horizon lines?)

The next evening, the earth, moon, and sun are in the position shown in Fig. 2.8, the moon having traveled about 1/28th of the way around its orbit. Using the figure, we see that the moon will be farther from the horizon line at sunset, and hence will appear to be higher in the sky. It will also appear fuller, since more of its lighted face is visible. As the evening progresses, the rotation of the earth will carry the horizon line past this new position as well, and we will see the moon set (but later than it did the night before).

This progression continues throughout the month. Each night the moon is farther from the horizon line at sunset, and it appears to travel across the sky to the west and then to set. Working out what happens in the last half of the month is left to the problems at the end of the chapter.

The recurring phases of the moon constitute a kind of "clock" which early civilizations could use to record the passage of time. In fact, any repetitive phenomenon—the moon, the swinging of a pendulum, the alternation of the phase of an electric current—can be used to mark the passage of time. The first appearance of the moon after a night of darkness is a dramatic event, and has the obvious religious and mystical implications that go along with rebirth. It should not be too surprising, then, that the Babylonians adopted the phases of the moon for their calendar.

With this background on the appearance of the heavens, we can go on to discuss the Babylonian civilization, with a special emphasis on their astronomy. In a real sense, they were the first astronomers, and the attempt to use the lunar clock described above led them to new and interesting ways of working with the world.

D. Mesopotamia: Philosophy and Technology

The civilization that flourished in the valley between the Tigris and Euphrates rivers from roughly 2500 B.C. to about 300 B.C. provides another example of the dual institutions of philosophy and technology in early societies. (Actually, there was not one "culture" but many which dominated the region in turn over this period of time.) This area is probably more familiar to the reader than Egypt, because the Hebrews recorded many of their contacts with Mesopotamian people in the Old Testament. Most of us have heard of the Babylonian lawgiver Hammurabi, for example, and of the Assyrian kings who led military campaigns against Israel. Some may have heard of the Sumerian *Epic of Gilgamesh*, which is the earliest known example of written literature. The readings suggested at the end of the chapter should prove very interesting to students who would like to pursue this subject in more detail.

Like the Egyptians, the Babylonians had a fully thought-out picture of what the world was and of their place in it. To the Babylonian, the world was pictured as a flat disk surrounded by water, which was in turn surrounded by the outer mountains (see Fig. 2.9). The sky was a covering over the entire system.

Within this universe, a hierarchy of Gods existed. Each city, each clan, and each family had a personal deity whose importance waxed and waned with the successes of his worshipers. The creation of the world (chronicled in a text called the *Enuma Elish*) was thought to have taken place when the primeval chaos (represented by a female monster called Tiamat) was killed in battle by Marduk, the Storm God, who then cut her in two, forming the earth with one half and the sky with the other. Men were created from the blood of Tiamat's mate, and their function was to relieve the gods of the need to work.

Once again, we see the development of a world picture in which events take place in a personal, individual way,

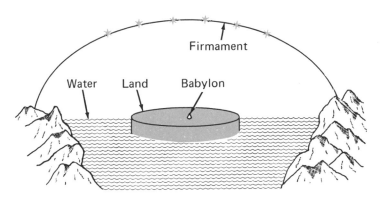

Figure 2.9.

rather than as the result of general laws. This view permeated the entire culture and was not restricted to the creation stories alone. For example, in the biblical account of the flood story, God sends the flood as punishment for man's violation of his laws. We would say that the flood occurred as a consequence of a general law—the law that punishment follows transgression. In the *Epic of Gilgamesh,* however, the reasons for the flood are quite different. There we read that it happened because

"In those days the world teemed, the people multiplied, the world bellowed like a wild bull, and the great Gods were aroused by the clamor. Enlil heard the clamor and he said to the Gods in council, 'The uproar of mankind is intolerable and sleep is no longer possible by reason of the Babel.' So the Gods in their hearts were moved to let loose the deluge. . ."

Once the floods started, however, the Gods were not able to control it:

"even the Gods were terrified of the flood, they fled to the highest heaven. . . they crouched against the wall, cowering like curs."

The Babylonian definitely did not live in a world governed by natural laws of the type we are used to!

Another example of the personal element in the Babylonian religion is the story of the development of disease among men. The God Enki and the Goddess Ninmah, after drinking some beer, have an argument. She claims that she can make well any person that he creates with a defect. The contest starts; he creates crippled men, and she cures them. Finally, he creates a man that she cannot cure. Nothing can be done, and disease has been loosed upon the earth.

So while the Babylonian's philosophy very definitely told him what his place in the world was (a servant to the gods), it did not supply him with any way of improving his material life. As was the case with the Egyptians, this was left to the men who worked in the technical tradition.

This tradition was developed in Mesopotamia over a period of a thousand years. Tracing its development is beyond the scope of this text. Like the Egyptians, the Babylonians developed a number system for use in commerce and engineering. One of the first problems we find treated in their writing is the problem of measuring out land tables. It was shown that two pieces of land having the same perimeter need not have the same area. Apparently, Babylonian real estate agents sometimes could not resist the temptation to sell small lots with large perimeters at premium prices.

The number system was a decimal system based on the number 60. To understand what this means, consider what we mean when we write the number 73. What we really mean is

$$7 \times 10 + 3 \times 1 \ (= 73).$$

In a sexagesimal system, however, this number would be written

$$1 \times 60 + 13 \times 1 \ (= 73),$$

which the Babylonians would have written

$$1, 13.$$

It is much easier to carry out calculations in a system which uses a repeating base number. Our own decimal system is based on the number 10, and even though we would find the Babylonian system cumbersome, it was a decided improvement over the Egyptian system discussed previously.

A large part of the Babylonian skill in arithmetic arose because of their need to create an accurate calendar in the context of their lunar timekeeping system. We tend to take our calendar for granted, but a little thought will convince you that it really is an essential part of life. For example, we have to know how many days there will be in each month to make out payrolls. We have to know when to plant and when to harvest. On a more philosophical level, we have to know when to celebrate the various religious holidays (which, in early civilizations, are usually connected with agriculture).

The reason that our calendar is so simple, and the reason that April is always in the spring and September is in the fall, is that we use a calendar based on the motion of the earth around the sun. The earth completes one orbit in 365 1/4 days, and the seasons are determined by the location of the earth on that orbit.

Consider what would happen, however, if for some reason we wanted to mark time by the revolutions of the moon, rather than the sun. We know that the moon makes one circuit around the earth in about 28 days; to the Babylonians, this process would appear as a regularly recurring change in the phases of the moon. Thus, they would mark the first of the month as that day on which the first new crescent was visible in the western sky at sundown. Operationally, this is an easy way to define a month.

Unfortunately, the month defined in this way does not fit easily into a year defined in terms of the movement of the earth around the sun. In fact,

$$19 \text{ solar years } = 235 \text{ lunar months.}$$

It is not hard to see that in such a system the month called "April" would not always occur at the same season. Since it is very important to know when to plant crops, however, a system was devised in which a 19-year cycle was used. Twelve of these years had 12 lunar months, but interspersed with them were seven years in which an extra month was added. In this way the lunar calendar never got too far out of line with the actual seasons. This system of using the lunar, rather than the solar, calendar is no longer used in our times, but it persists in some very traditional areas, such as setting the dates of religious festivals. (Haven't you ever wondered why Easter doesn't fall on the same day every year?)

The life of an astronomer in Babylon was not easy. In order to do his job, he had to have at his disposal mathematical tools much better than those available to his Egyptian counterpart. One of the main motivations for the development of the hexadecimal number system, and the ability to perform calculations that went along with it, must have been the need to make precise predictions about the positions of the stars in order to update the calendar.

The instruments used by the Babylonians to make observations were crude by modern standards. They measured the position of stars by sighting along a rod—a procedure something like aiming a rifle. Over the centuries, however, they accumulated a mass of astronomical data which was used (and represented the best attainable) until the breakthrough in astronomical instrumentation brought about by Tycho Brahe in the 16th century.

Most science textbooks end their discussion of the contributions of the Babylonians to astronomy at this point. However, there is another aspect of Babylonian astronomy which ought to be discussed, and that is astrology. Today, astrology is viewed (by scientists, at least) as a somewhat disreputable activity, not fit for a serious scientist to be involved in. This view was not always held, however, and the advances in astronomy were linked to astrology well into modern times.

The basic idea of astrology as practiced by the Babylonians is that there is a relation between the position of the sun, moon, and planets in the sky to events that take place on earth. In particular, it was thought that a knowledge of certain aspects of astronomy would give the astrologer the ability to predict future events, and therefore to advise kings and other leaders on courses of action.

In order to use astrology, however, it was necessary to "cast a horoscope." Casting a horoscope for an individual involves locating each of the heavenly bodies as they were at the time of that person's birth. Clearly, if an astrologer is to carry out his trade, it is absolutely essential that reliable measurements of planetary and lunar positions be available. Thus, the skills of the astronomer were useful to those in positions of power, and this usefulness clearly played a role in the development of Babylonian science.

SUGGESTED READING

Bok, B. J., and Jerome, L. E. *Objections to Astrology.* Buffalo, N.Y.: Prometheus Books, 1975. A modern (negative) discussion of astrology by a well-known astronomer. If you read this, you should also look at the book review by Robert March in *Physics Today,* March 1976.

Frankfort, Henri. *Before Philosophy.* Baltimore: Penguin Books, 1946. Contains a readable and engrossing description of Egyptian and Babylonian culture. Provides excellent background material for these subjects.

Heidel, Alexander. *The Babylonian Genesis.* Chicago: University of Chicago Press, 1951. Translation of early Babylonian epic poems, with commentary.

Neugebauer, Otto. *The Exact Sciences in Antiquity.* Princeton, N.J.: Princeton University Press, 1952. Good descriptions of Egyptian and Babylonian mathematics, with a heavy emphasis on how scholars deduced what is now known.

Peet, T. Eric. *The Rhind Mathematical Papyrus.* Liverpool, England: University of Liverpool Press, 1928; Kraus reprint, 1970. A translation of the papyrus, with the problems worked out, and a useful commentary. It is fascinating to browse through a book like this and see the way the Egyptians thought!

Sandars, N. K. *The Epic of Gilgamesh.* Baltimore: Penguin Books, 1960. A translation of the first known work of literature. Very readable (particularly the flood story discussed in the text).

Scott, J. F. *A History of Mathematics.* London, England: Taylor and Francis Ltd., 1958. The first chapter gives a concise summary of Egyptian and Babylonian mathematics.

QUESTIONS AND DISCUSSION IDEAS

1. Do the following multiplications using the Egyptian technique (be sure to write the answer in Egyptian numerals):
   a. 3 X 16
   b. 7 X 8

2. Write the following fractions in the Egyptian manner (be sure to write the answer in Egyptian numerals):
   a. 4/9
   b. 17/24
   c. 19/32

3. Give some examples of modern activities which the Egyptians could not have developed, given their mathematical abilities.

4. Give some examples of modern activities which the Egyptians *could* have developed, given their mathematical abilities.

5. How would an Egyptian or Babylonian philosopher have answered the question "Why did it rain today?"

6. In Egyptian mythology, for every God there is a "counter God," just as there is a sky and counter sky. What aspects of Egyptian geography do you think might have led to this idea that symmetry is an important part of nature?

7. Using a technique similar to our discussion of the apparent motion of the sun and the moon, discuss the apparent motion of a fixed star. Why are some constellations visible in the evening only at certain seasons?

8. Write the following numbers in the Babylonian hexadecimal system:
   a. 87
   b. 260
   c. 150
9. In the text, we discussed the use of Babylonian arithmetic in real estate. Put yourself in the position of a
Babylonian scribe, and find (by trial and error) an
example in which two tracts of land have the same
perimeter (distance around the outside), but different
areas.
10. We speak of the "signs of the Zodiac." What is the
Zodiac, and what are the signs?
11. Conduct one of the following "experiments":
   a. Go into newspaper files and find the predictions
      for a year made by several professional astrologers.
      Compare their success in predicting events with
      that of news columnists (who also often make
      predictions).
   b. Obtain horoscope predictions for a particular day
      from some source. After the day has passed, find
      a group of people and ask them to say which de-
      scription most nearly matches their day (but don't
      tell them which description goes with which sign).
      Compare their answers with their birthdays.
12. Complete the discussion of the lunar month started in
the text. In particular, draw a diagram for the earth-
moon-sun system for the following cases:
   a. first quarter
   b. full moon
   c. third quarter
   d. dark of the moon
13. Draw a diagram of the earth-moon-sun system for each
of the following cases:
   a. moon directly overhead at midnight
   b. moon rises at 10 P.M.
   c. moon on the western horizon at noon
14. What was the phase of the moon last night?

CHAPTER III

THE BIRTH OF HUMAN REASON

" 'Don't be an ass, Heraclitus. You could step into the same river twice, if you walked downstream at the same rate as the river.' He was amazed. . .''

Severn Darden,
"The Philosophy Lecture"

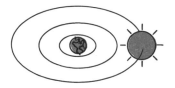

A. The Ionians: The Technical Tradition Becomes Philosophic

As different as they were, the Egyptian and Babylonian civilizations had two important things in common. First, they were essentially static. Changes in the culture and in scientific thought took place so slowly as to be virtually unnoticeable to a single individual. For example, the only major innovation in Babylonian mathematics from about 1800 B.C. to 700 B.C. was the introduction of the "zero" into the number system.

The full impact of this sort of stable, static type of life can be appreciated if you think of what the western world was like 1000 years ago, think of all of the changes that have taken place since then, and then imagine what it must have been like to live in a culture in which very little change took place over the same time span. It must have been very hard to conceive of new ideas in such an atmosphere.

The second important feature of the civilizations we have been studying is that a very real gap existed between the technical and philosophical areas of thought. Each area had its own ways of dealing with the world, and each had its own area of competence on which the other would not encroach.

Given the general static nature of things, there would probably have been very little motivation for an individual to challenge the existing way of doing things and strike out in new intellectual directions. It is not surprising, therefore, that the first new direction of thought— a direction that ultimately would lead to the development of modern science—came from a new area and a new people who were not weighed down by centuries of tradition.

The western coast of Turkey (now called Anatolia) was settled by Greeks in about 1000 B.C. This area was called Ionia, and for a long time it dominated Greek thought and development. Our own perspective on Greece tends to be centered on Athens, so that few names from Ionia are familiar to us. One king, Croesus, has come down to us in legend, but important intellectual developments were taking place here around 600 B.C. which far overshadow any military or commercial happenings.

36

Ionia is situated more or less midway between Egypt and Babylon and very early developed a strong mercantile culture, with each Ionian city founding colonies along the coasts of the Mediterranean and Black Sea. Because it was a new, vigorous kind of place, without the dead weight of tradition, new ways of looking at the world could be accepted. At the same time, people living there were in contact with the older civilizations, and could see for themselves that there was nothing special about one set of beliefs or traditions among the many with which they were in contact.

In the period around 600 B.C., the main city in Ionia— the "Big Apple"—was called Miletus. This was a center of commerce and technology, with a large merchant fleet and a very high level of technology. For example, the engineers of the Persian armies were sent there to study, much as engineers from other countries now come to the United States. I like to think of Miletus as something like modern New York City or Renaissance Florence. It was a place where people came and went, where ideas flowered, and where a large part of the world's business was carried on.

Greek vase, showing marriage procession.

It was a place where men were expected to be active in civic affairs and to engage in everyday technical undertakings, whether this meant running a fleet of ships or developing new navigational instruments. This concern with the techniques of the world gave rise to a pragmatic way of looking at the nature of things which is very familiar to modern Americans. We are used to asking "does it work?" and feel intellectually at home with men who do the same.

The most interesting of the Milesians for us is a man named Thales, who is usually called Thales of Miletus. This man has a strong claim to be known as the "first scientist," and yet we actually know very little about him. We don't really know if he was Greek or Phoenician. None of his original writings survive and therefore we know of him only through what later writers said about him.

As is usual in cases where facts are few, stories abound. From the legends, we can piece together a picture of Thales that is typical of what we would expect of a prominant Milesian. He is thought to have been alive in about 585 B.C. He was apparently an important political figure in Ionia, and is supposed to have tried to promote a scheme to get the various Greek city states to form an alliance against the Persian menace (the effort was unsuccessful, but it was a

37

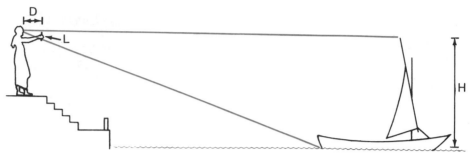

Figure 3.1.

good idea—the area was later conquered by the Persians). He was an engineer, and apparently traveled both to Egypt and Babylon. He is supposed to have invented an instrument that would give a man on shore the ability to estimate the distance a ship was out at sea. This instrument is pretty much the same as the instruments still used in coastal navigation.

Its operation is very simple, and is illustrated in Fig. 3.1. A man stands on shore and sights the top and bottom of the ship on a stick of length L a known distance D from his eye. Then, knowing the height h of the ship, he can find the distance to the ship by simple geometry.

Thales is also reputed to have predicted an eclipse of the sun in 585 B.C., and to have stopped a battle by doing so. This incident is the one on which his reputation among the ancients was based, and is probably the basis of his inclusion in the list of "The Seven Wise Men."

These accomplishments, even if they all actually existed, would hardly earn Thales a higher place in our attention than any of hundreds of other men in pre-Socratic Greek history. What did he do that was so remarkable?

What he did was to ask an old question and give it a new answer. The question he asked was: "How is the world made?" The answer he gave, according to Aristotle, was: "The first and basic principle of all things is water."

Now this hardly seems like a giant breakthrough to us today. But let's think for a moment about how Thales probably came to make this statement, and perhaps it won't seem so obscure.

When we look at the world around us, we see that there are three types of matter (in modern terminology, three phases of matter). There are solids, like wood or stone, there are liquids, like water, and there are gases, like air or steam. A man who worked with materials (such as an engineer) could hardly fail to be aware of this fact.

He could also hardly fail to be aware of the fact that given the proper circumstances a given chunk of material could change from one of these states to another. Metals could be melted and solidified in a foundry, for example, and fluids could be boiled and converted to vapor.

When you think about it, there is one common substance that can exist in all three phases. That substance is water. It exists as ice (a solid), as steam (a gas), and, normally, as a liquid. Thus, it should not surprise us that someone looking at the world might guess that everything he saw was simply a different form of this common substance.

What should surprise us is the tremendous difference between this sort of statement and the kind of thinking about the world that had existed up to this point. In the first place, the question "How is the world made?" would normally be considered a philosophical question, not a technical one. It was the type of question that, in Egypt or Babylon, might have been asked by a priest. It certainly would not have been asked by a technician.

And yet this man, Thales of Miletus, not only asked a philosophical question, he answered it with the methods of the engineer! With him, we have the first attempt by men to bridge the gap between the world of technique and the world of thought. It does not matter very much that his answer to the question was wrong. What matters is that for the first time we have an attempt to give a natural, non-religious explanation of the world, whereas before him such explanations would always have been couched in the language of religion and revealed truth. This point can be grasped most fully by comparing the way Thales discussed the creation of the world from water and the way the world was created from water in the Babylonian creation epic, where creation took place not by a natural process such as freezing and boiling, but by the will of the gods.

Another important aspect to this new mode of thought is that whereas events and things were viewed previously as particular and personal, we now have the beginnings of the idea that things in the world are not so different from each other after all. Not only is one rock pretty much like

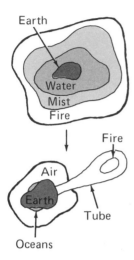

Figure 3.2. The evolution of the world according to Anaximander.

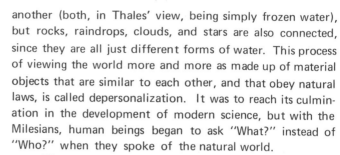

another (both, in Thales' view, being simply frozen water), but rocks, raindrops, clouds, and stars are also connected, since they are all just different forms of water. This process of viewing the world more and more as made up of material objects that are similar to each other, and that obey natural laws, is called depersonalization. It was to reach its culmination in the development of modern science, but with the Milesians, human beings began to ask "What?" instead of "Who?" when they spoke of the natural world.

The pragmatic, naturalistic way of thinking that was pioneered by Thales was continued by a large number of Ionian philosophers. We won't discuss them all, but a few should be mentioned for the sake of completeness.

Anaximander was a Milesian, probably a student of Thales. He improved on Thales' work. For example, where Thales had held that the earth must float on water, Anaximander held that it must hang in space, since it is equidistant from everything else. This is a very modern way of looking at things. He also replaced the single element—water—with four elements—the familiar earth, air, fire, and water.

Anaximander also developed a cosmogeny. In his picture of the development of the world, everything started out in a stratified form, with earth (the heaviest) at the center, then water, mist, and finally fire around the outside (see Fig. 3.2). Then the fire heated the water, making it evaporate and allowing the land to appear. When the pressure of the vapor got too high, the outer circle of fire ruptured and formed into circles of fire surrounded by air. These circles are the stars. We see them through holes in the mist which, when they become clogged, shut off our view and are responsible for eclipses.

The important thing about this rather detailed cosmogeny is that every event in it is related to something that could be observed. There are no supernatural events, nor are the original materials of the universe any different from those we encounter in everyday life. When we compare this to the creation stories of earlier peoples, we can see the process of depersonalization very clearly.

B. Pythagoras: The Magic of Numbers

The name of Pythagoras is probably more familiar to the reader than the name of the other Greek philosophers we

Dust cloud

40

have discussed up to this point. Yet for all of this familiarity, we know surprisingly little about the man. We believe that he was born on the island of Samos in about 580 B.C., and that he traveled extensively—possibly even as far as India. In his later years, he left his native home for political reasons, and founded a school in the Greek city of Crotona, in southern Italy.

The term "school" is used loosely here, since it was regarded by the members as a primarily religious institution, probably closer in spirit to a modern commune than to a school. The Pythagoreans had a very mystical view of the world, and numbers played a very important role in that view.

Pythagoras himself does not fit into our conceptual scheme very well. In many ways he was a man far ahead of his time. Although a religious leader, and therefore presumably very much in the philosophical tradition, he carried out experiments to test his ideas—an activity we would normally associate with the technical. His major contribution to science was the introduction of the idea that an important relationship exists between numbers (and therefore mathematics) and the physical world. This idea (in a considerably altered form) is one of the main ingredients of the scientific method as we now know it.

Legend has it that Pythagoras came to this conclusion one day when he was passing a blacksmith shop. He noticed that the different hammers made different musical tones when they struck the metal being worked—the heavier the hammer, the lower the tone. He is supposed to have asked the smiths to strike their hammers in a particular sequence, thereby producing a melody. History does not record what that melody was nor, for that matter, how the smiths felt about the whole procedure. For Pythagoras, however, the principle that something quantitative (in this case the weight of the hammer) could be related to musical tones had been firmly established.

From this conviction to further experiments with musical instruments is a short step. He discovered that when strings are plucked, the length of the string is related to the tone produced. Furthermore, harmonious tones are produced by strings whose lengths are in the ratios of whole numbers. For example, if the lengths of two strings are in the ratio 1:2, we have an octave, the ratio 3:2 produces the fifth, and so forth.

41

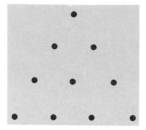

Figure 3.3. The Tetractys

From the statement that "Music is Number," the generalization to the statement "All is Number" was an easy one. This statement became the watchword of the Pythagorean school, and led to a kind of mystical numerology which it is very hard for modern minds to follow. Elaborate numerical schemes were devised. The even numbers were male, the odd numbers were female. Numbers associated with the Tetractys, the triangular arrangements of dots shown in Fig. 3.3, were considered to be associated with the Gods. A cosmology was devised in which there were nine "planets" (Mercury, Venus, Mars, Saturn, Jupiter, Earth, Moon, Sun, and a nonobservable "counter-Earth") circling around the "central fire" (see Fig. 3.4). This number was arrived at because this left, when the sphere of the stars was included, a total of ten heavenly bodies, and ten was a perfect number to the Pythagoreans.

This sort of thinking seems very naive to us today, and it smacks of the type of thing we usually associate with cranks. However, in the first flush of discovery, it is not surprising that the Pythagoreans extended an idea that contains some truth to areas where it was never meant to apply.

The school of Pythagoras encountered two major crises. The first is related to the theorem bearing his name. As he stated this theorem, it said: "The sum of the squares erected on the sides of a right triangle is equal to the square erected on the hypotenuse." The way of stating the theorem emphasizes the fact that, to the Pythagoreans, a number was a very real thing—in this case being given by the area of a square of a given length.

To the Greek mind, symmetry was a very important

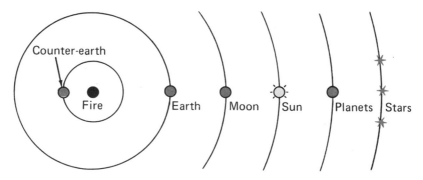

Figure 3.4. The Pythagorean Cosmology.

concept. Clearly, the most "perfect" right triangle would be one with two of the sides equal. But, as we know, the ratio of the length of the hypotenuse to the length of a side in such a triangle is $\sqrt{2}$: 1. This ratio was totally incomprehensible to the Pythagoreans, who conceived of numbers as real things, and who felt that all harmonies in nature had to be expressed as the ratio of whole numbers, as were the musical tones. No matter how hard they tried, they simply could not express $\sqrt{2}$ in this way. So great was their consternation that the knowledge of this dilemma was revealed only to members of the inner circle of the school, who were pledged never to reveal it to outsiders. It is probably the first example of a "top secret" classification. It is not hard, however, to see how minds so attuned to the mystical play with numbers would find numbers like $\sqrt{2}$ to be nonsensical—in fact, irrational. (Haven't you ever wondered where that term came from?)

The second great crisis of the Pythagorean school came when the citizens of Crotona, grown increasingly uneasy about the "wizards" and other weird types that were flocking to their town, formed a mob and burned the school down, killing or banishing all of its members.

Pythagorean thought remained a potent force in Greek philosophy for a long time, but, aside from the geometrical theorem we have discussed and the general idea that numbers have some relation to the physical world, little of Pythagorean influence remains in modern science.

C. The Problem of Change and the Atom

One of the central problems that occupied Greek thinkers was the question of change. They could not see how the fact of change in the world could be reconciled with the demands of reason. The pursuit of this problem led eventually to the theory of atoms, one of the major scientific ideas we associate with the Greeks. However, the *way* in which this theory came about illustrates the important differences between Greek and modern scientific inquiry, for the Greeks arrived at the idea of atoms as a sort of compromise between conflicting schools of philosophy without any appeal to experience or observation. Later, when we study the modern ideas about atoms, we shall see a very different sort of development.

43

We can begin the story of change as we did the story of the development of new ways of thinking in the technical tradition. Heraclitus, of Ephesus, a city in Ionia, lived in about 500 B.C. Unlike the Milesians, we do have some surviving fragments of his writing, in addition to the accounts of his life from other sources. He was born into an aristocratic family, but preferred a quiet life as a philosopher to the position in the political life of the city that he could have had by right. His writing betrays his aristocratic origins, however, being set into very terse and somewhat obscure parables.

Like all of the men we shall study from now on, his philosophy dealt with much more than simply the questions related to the development of science. He thought about ethics and about the age-old problems of good and evil. From the point of view of our development, however, his most important contribution was his thinking about the way the world operates.

In a nutshell, Heraclitus saw the world as "process" (to use a modern term). That is, he saw reality as a constantly shifting, changing interplay between opposing factors. His most famous statement of this idea is: "You cannot step into the same river twice."

Of course, we do not know what the factors were that brought Heraclitus to this sort of conclusion, but it is tempting to think that his Ionian background had something to do with it. After all, even though he was an aristocrat (as was Thales), he would have been aware of the industrial life in his city. More important, he would have absorbed the Ionian ideas of learning about the world by observation. One of the most important things that we see around us is constant change—the weather, growth, death, and so forth. It is not too hard to imagine how a man brought up in an atmosphere in which the technical tradition was so strong would make this simple observational fact one of the cornerstones of his philosophical system.

In any case, with Heraclitus, we have one way of looking at change in the world. Change is observed to occur, is accepted as a fact, and is even elevated into a position as a central fact of existence. This way of thinking is something that 20th-century men find congenial, and we would probably not be bothered much by it.

44

To another school of Greeks, however, this easy accept-
ance of the appearance of the world was not to be tolerated.
We have to remember that this period represents the first
flush of man's love affair with reason, and like most affairs,
it tended to provoke excessive behavior at the beginning.
The most important men who followed the new modes of
thought were Parmenides and his student, Zeno. They
lived in the Town of Elea, and are sometimes referred to as
"the Eleatic school."

The main emphasis of their thought was a rejection of
everyday experience, especially the experience of the senses,
and a total reliance upon reason to find the truth. Listen to
Parmenides when he says: "Do not let custom, born of
everyday experience, tempt your eyes to be aimless, your
ear and tongue to be echoes. Let reason be your judge."

This preference for reason over experience reaches its
peak with Plato, and we shall see much more of it later. For
the problem of change, however, it poses rather formidable
difficulties. For if we are to know the truth through our
reason, then that truth must not change, because the truths
of reason must be immutable. How, then, can we deal with
a world in which our experience tells us that change is constantly occurring?

Zeno posed this problem in its ultimate form in his
famous "proof" that motion is impossible. He argued that
it is impossible to cross a room (for example) because first
we would have to go halfway across, then halfway across
the remaining half, and so on. Since there are an infinite
number of such divisions, it would take an infinite time to
get anywhere; hence, motion is impossible. We have the following version of this argument in Zeno's own words:

"If anything is moving, it must be moving either in
the place where it is or in the place where it is not. However, it cannot move in the place where it is (for the place
in which it is at any moment is of the same size as itself and
hence allows it no room to move in), and it cannot move in
the place in which it is not. Hence, movement is impossible."

Thus, after the Eleatic school, we had two very definite
and distinct ways of looking at the world—the Ionian way,
which accepted change and relied on observation, and the
Eleatic way, which rejected change and relied on reason.
This was the climate of thought in which the first ideas of
atomism arose. This Atomic school was founded in about

530 B.C. by Leucippus, but his student, Democritus, is usually associated with the concept of the atom.

Democritus argued that if we began cutting up a piece of matter with the sharpest knife imaginable, we would eventually come to a piece that could not be cut any further. This he referred to as the "atom" (from the Greek word for indivisible). He then argued that these atoms were unchanging and eternal, but that the *relation* between atoms was in a state of constant flux.

This compromise seemed to have something for everyone. It had unchanging elements (the atoms) and changing elements (the relations between them). When we observed a change, such as the melting of ice, we would then say that what we were seeing was the change in the relation between the atoms in the material (in modern terms, the breaking of the bonds between water molecules in the ice), but that the true reality of the atoms themselves was unaffected by the apparent change.

Unfortunately, this compromise solution, as clever as it was, did not gain wide favor. Plato never mentioned it, and Aristotle regarded it as a sort of novelty not to be taken seriously. It is only with the modern development of the atomic theory that we look back at Greek atomism as the precursor of things to come.

The way that atomism arose illustrates the difference between the Greek and modern way of arriving at knowledge about the world. When we study John Dalton and modern atomic theory, we shall see how he introduced the idea of the atom in order to explain observations of chemical reactions. To the Greeks, this sort of approach would have been incomprehensible. Greek atomism came out of a dispute between philosophical systems and had very little to do with actual detailed observation and experiment. Even the debate about the place of experience in human knowledge was carried out by means of reason alone, without appeal to observation.

D. The Athenians: Triumph of Reason

If most people were asked to name a Greek philosopher, the chances are that they would come up with the name of one of the Athenian philosophers—Plato, Socrates, or Aristotle.

46

From the point of view of the development of scientific thought, however, these men (with the possible exception of Aristotle) are not as important as they might be in other areas of philosophy. In the case of Plato, in fact, the net contribution to science is actually negative! He introduced several reasonable (but wrong) ideas which had to be overcome before modern science could be developed.

With Plato, we have the first full statement of the philosophical position that most of us tend to associate with Greek thought—the idea that the correct way to learn about the world is through reason rather than observation. In the *Phaedo,* he puts it this way:

> If we are ever to know anything absolutely, we must be free from the body and behold the actual realities with the eye of the soul. . . While we live, we shall be nearest to knowledge when we avoid intercourse and communion with the body.

In other words, true knowledge can be gained only through the use of reason, by using the "eye of the soul." Knowledge gained by experience or observation which uses the "eye of the body" is suspect, and perhaps even misleading. This way of looking at the world is stated even more clearly in the *Republic,* where he likens the human condition to a man sitting in a cave watching shadows on the wall. He might think that what he was seeing was real, but actually the true world was not available to his senses. It could be perceived only through reason.

As a philosophical system, this way of thinking may have some merit, but when it is applied to science, it is disastrous. Plato gives this advice to astronomers in the *Republic:* "In astronomy, as in geometry, we should employ problems, and let the heavens alone if we would approach the subject in the right way."

The idea that an astronomer isn't supposed to look at the stars is a little hard to digest today, but given the Platonic outlook, it is easy to see how such advice could be given. After all, looking at the stars would only tell us about the shadows of reality, and the true reality could only be perceived with the mind.

How did astronomy proceed in the Platonic scheme of things? First, it was felt that since the heavenly bodies were far from earth, they had to be "perfect"—immutable and unchanging. They had to move in a "perfect" figure. What figure does reason tell us is most "perfect?" The sphere or circle, of course. From arguments such as this, Platonic philosophers concluded that the heavenly bodies must be embedded somehow in a huge celestial sphere which rotated around the earth. Circular motion of this type, since it was "perfect," could go on forever without any outside interference and without slowing down.

We know that on the earth a wheel that is set in motion will eventually stop. Why didn't the wheel of the heavens stop as well? The Platonist would have answered by pointing out that a wheel on the earth is not "perfect," since it is here and immediately available to our senses. Hence, it will not behave in the same way as the heavenly sphere, which reason tells us will turn forever without stopping.

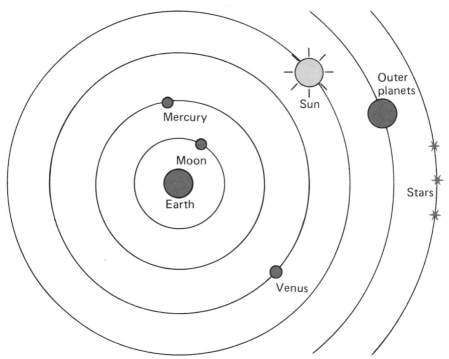

Figure 3.5. Spheres of Eudoxus.

The only problem with this way of looking at things was the fact that some heavenly bodies—the planets—do not seem to stay in place, but move around with respect to the stars. The Greek word for planet means "wanderer," or "vagabond," but a better flavor for their original meaning might be obtained by translating it "hippie." The planets didn't seem to fit neatly into the Platonic ideal.

The first attempt at solving this problem was made by a student of Plato named Eudoxus. He devised a scheme in which all of the heavenly bodies rotated about the earth, but on *different* spheres, so that some would appear to move with respect to the others. The spheres of Eudoxus are shown in Fig. 3.5.

While this scheme did indeed explain some things about the planets, it failed to explain others. For example, while it provided a simple explanation of why the planets appeared to move against the background of the stars, it did not explain the fact that some planets appear brighter at certain times of the year than at others. (In our modern way of looking at the solar system, this happens because at some time during the year, the earth is closer to a given planet than at others.) This would not happen in the universe designed by Eudoxus, because there each planet is always the same distance from the earth. Furthermore, the fact of variable brightness was *known* to astronomers long before Eudoxus proposed his model.

This way of dealing with the world—ignoring what we see and relying solely on what we think—is so far removed from the Ionian way of doing things that it is hard to believe that the two are representatives of the same culture. One important change in Greek civilization between the time of Thales and the time of Plato was the introduction of slavery. In Ionia, a man was expected to be involved in the practical affairs of the world, for this was the lifeblood of his city. In Athens, on the other hand, such mundane things were left to slaves, while a "gentleman" was expected to devote himself to "higher" pursuits. It is hard to believe that this fact did not influence Athenian thought and play some role in the evolution of the Platonic world view.

From the point of view of the development of science, then, the legacy of Plato is mixed. He was very much a man of the philosophical tradition, but with an important difference. While a Babylonian or Egyptian priest relied on sacred

49

writings for his answers to questions about the world, Plato asserted the right of each human being to use his own mind to find his own answers. With his work, the method of seeking the truth shifted from the authority of revealed writings to the mind of man himself. This was an important step in human development, and one that was essential for the development of science.

On the negative side, the refusal to trust observations of the world led to the introduction of some false ideas that served as stumbling blocks in the development of science for 1,500 years. It was not until the 17th century that the Platonic idea of the "naturalness and perfection" of circular motion was finally laid to rest, and many brilliant men spent their entire careers trying to reconcile the observed facts of astronomy with the dictum that the orbits of the planets had to be circular. The Platonic prejudice, which led men to believe that the heavens were "perfect" while things on earth were not, while it did allow "gentlemen" to study astronomy provided that they went about it in the right way, discouraged the study of the behavior of objects on the earth that might have provided enlightenment for astronomers.

Plato founded a school in Athens, which was called the Academy. His most important pupil—a man who spent 20 years studying there—was Aristotle. In his later years, Aristotle was the tutor of Alexander the Great, so his lifetime spans the end of the great age of Athens. He was a man of prodigious intellect, and his lectures cover the entire spectrum of human intellectual inquiry. He had a profound influence on the development of science because his lectures formed the basis for the curriculum of Greek, Roman, and medieval European higher education. His *Physics,* for example, was not only still in use in the 17th century, but it was the basis for attacks on the work of Galileo! Few science textbooks can expect this long a lifespan.

Although he started out as a student of Plato, Aristotle was too great an intellect to stay in anyone's shadow very long. One of his favorite areas of study was biology, and he spent a lot of time on the Greek coast collecting specimens of aquatic life. A man with this kind of outlook is not going to take the Platonic strictures against observation very seriously. Observation is a central feature of the biological sciences.

But if he did break with pure Platonism on the question of observation, he still retained many of the Platonic ideas about the place of reason in the sciences. He taught, for example, that heavy objects would fall faster than light ones. This conclusion was not reached on the basis of experiment or observation, but on the basis of the fact that it seemed "reasonable" that this should be the case.

Actually, the whole body of Aristotelian thought on the motion of material objects is very interesting, both from the historical point of view (it exerted a tremendous influence on the development of science) and from the point of view of developmental psychology discussed in the first chapter. Aristotle's rejection of experimental evidence is not as capricious as it might appear. It was based on a distinction that is very important in the biological sciences—the distinction between natural and disturbed states of nature.

For example, if you want to know something about the behavior of a particular type of animal, you must observe it in nature without disturbing it. Up until the last few decades, scientists had totally incorrect ideas about the social behavior of primates because they had observed them only in zoos, not in their natural homes. Thus, in dealing with biological systems, it is important that the interference with nature be kept to a minimum.

Aristotle simply extended this argument to physical systems as well. He felt that there were two kinds of motion—natural motion, which an object would undergo if it were left to itself, and violent motion, which was imposed on the object by some outside agency. He argued that every object had an innate tendency (sort of a life force) which impelled it to seek its "natural" place in the universe. For Aristotle, this "natural" place could be nowhere else but the very center of the universe, which was located at the center of the earth. The larger the body, the more life force, and hence the faster the fall.

The application of this sort of reasoning to concrete physical examples was carried out by many different men over the centuries. It resulted in a body of knowledge that goes by the name "Aristotelian physics." We shall look at two examples to get some feeling for the type of arguments that were given to explain physical phenomena.

When a rock is thrown, or a projectile is shot from a sling, we know that it first goes up and then comes down.

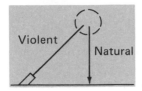

Figure 3.6.

Figure 3.7.

To an Aristotelian, this process was divided into two distinct steps. First, a violent motion is imposed on the projectile by an outside agency. Then, at some point, the "imposed" force becomes weak enough so that the natural tendency of the projectile to seek its rightful place in the center of the earth takes over, and the fall begins. Thus, the path of a projectile should be similar to the one pictured in Fig. 3.6. Does this agree with your own observations of these things?

Another ancient philosophical problem for which the Aristotelians had an answer (although not the right one) was the problem of how a projectile could keep on moving after it had left its starting point and nothing was pushing on it. They felt that such motion occurred because the air that had been in front of the projectile flowed around in back of it, and pushed it forward, as illustrated in Fig. 3.7. It is interesting to compare this idea to the reply elicited by Piaget from an early rational stage child to the question: "What makes the clouds move?" The answer was "It's the air which they (the clouds) make. . .it chases the clouds."

In terms of the developmental stages of Piaget, this sort of reasoning has many characteristics of the late transitional and early rational phase of development. The elevation of hunches and guesses to the status of unquestioned assumptions is characteristic of all stages of development, but it is perhaps most obvious with the Athenian philosophers and scientists whose thinking so strongly influenced all that was to follow.

With the coming of Alexander the Great, the political and military dominance of Greece ended. The centers of learning moved away from Athens to the new centers of empire. However, the influence of Plato and Aristotle was very strong in the Hellenistic period, and, in a sense, much of the work of later Greek scientists was devoted to working out ideas that were first developed by these men and their schools.

SUGGESTED READING

Most of the texts cited in Chapter 2 contain sections describing the Greek sciences. In addition, the following texts should prove useful:

Farrington, Benjamin. *Greek Science.* Baltimore: Penguin Books, 1961. A thoroughly readable account of the Greek sciences as

Sunrise at Stonehenge. Many scientists believe this monument a tool to aid in the construction of a calendar.

"Earthrise" as seen from the surface of the moon.

Zeno

Great Library at Alexandria.

they developed from the Ionians onward. There is an excellent discussion of the founding of the Athenian Schools.

Hutchins, Robert M. (Ed.) *Great Books of the Western World.* Chicago: Encyclopedia Britannica, 1952. This multivolume collection is one of the best sources for the original works of many of the men we shall be studying. In particular, students wishing to read the actual words of the men discussed in this chapter should consult Vol. 7 (Plato) and Vols. 8 and 9 (Aristotle).

Mason, Stephen F. *A History of the Sciences.* New York: Collier Books, 1962. A concise (but complete) history of sciences with sections on the Greeks.

Wheelwright, Phillip (Ed.) *The PreSocratics.* New York: The Oddessy Press, 1966. One of the best available sources on the pre-Socratic Greek thinkers. Each philosopher is dealt with in three ways: first by his direct quotations, then by quotations of what other Greeks said about him, and finally by a short essay which gives biographical background. The use of primary sources makes this a very valuable text.

QUESTIONS AND DISCUSSION IDEAS

1. The idea that the planets must move in circles because circles are "perfect" could be called a Greek "unquestioned assumption." What are some of *your* unquestioned assumptions?

2. A group of men are asked the question "Why did it rain last night?" Give an answer typical of each of the following:
    a. an Egyptian scribe
    b. an Ionian engineer
    c. a Babylonian priest
    d. an Eleatic philosopher

3. How would Eudoxus have answered the objection to his system that we raised (that it could not account for the fact that the planets appear at different brightnesses at different times of the year)?

4. Thales is supposed to have derived the "law of reflection." This law says that when, for example, a pool ball bounces off a cushion at the edge of the table, the angle at which it leaves the edge will be the same as the angle at which it arrived (see figure). To get a little more insight into Thales' mind, let's work the following problem (you may want to use the Pythagorean theorem to work out some of the lengths). We have a 20-foot pool table, and a ball at point B 10 feet up one edge. We want the ball to

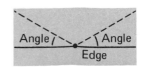

Angle   Angle

Edge

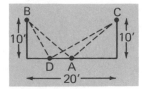

arrive at point C, a distance 10 feet up the opposite side.

 a. If the ball hits the edge at point A, the two angles will be equal. How far does the ball travel along the path BAC?

 b. Consider the point D 5 feet from the edge of the table. How far does the ball travel on the line BDC?

 c. Can you find any location for the point D where the distance the ball travels is less than the distance along BAC?

Thales' idea was that the ball would get from B to C by traveling the shortest possible distance. This is an example of something that is now called a "variational principle" in physics.

5. Think of an experiment that you might do to test Aristotle's idea that heavier objects fall faster than light ones. Do the experiment and report your results.

6. How do you think Aristotle would have reacted to Problem 5?

7. We have neglected a large number of important Greek philosophers in our discussion. Discuss the importance of the following men in the development of Greek thought:

 a. Archimedes

 b. Strato

 c. Anaximenes

8. Could Zeno cross a room?

9. The following statement is read to a group of men: "The force that holds the moon in its orbit is the same as the force that makes an apple fall to the ground." Give a comment that might be typical of each of the following:

 a. Plato

 b. Thales

 c. a Babylonian priest

10. Which of the schools of philosophers studied would have made the following statements?

 a. Earthquakes are caused by the drying up of the soil.

 b. Everything in the world changes.

 c. Nothing in the world changes.

 d. Smallpox is a disease like any other disease.

CHAPTER IV

THE FLOWERING AND DEATH OF GREEK SCIENCE

"If the Lord God Almighty had consulted me before embarking on the Creation, I would have recommended something simpler."

Alfonse the Wise of Castile,
upon being introduced to the Ptolemaic system

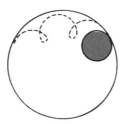

A. Alexandria: The Last Act

At his death, Alexander had conquered most of the known world. His generals divided this huge empire into three parts—that ruled by the Antigonids (comprising Greece, Italy, and Macedonia), that ruled by the Selucids (comprising modern Turkey and Iran), and that ruled by the Ptolemys (comprising Egypt and North Africa). The day of the small city-state had passed, and the era of large political units with large financial resources was at hand. This meant that a ruler who was inclined to support projects in the arts or sciences had the revenues of an empire, as opposed to a single city, to draw upon. In Alexandria, this led to a flowering of Greek science and learning.

Alexander decided that he wanted to build a new city to be the center of his new empire. In the Nile delta, the enormous wealth of the conquered territories was funneled into the city of Alexandria. After the death of Alexander, the Ptolemys (they were the ruling family in the Egyptian part of the empire, and shouldn't be confused with Claudius Ptolemy, the astronomer, who will be discussed later) continued the grand plan, and adopted Alexandria as their capital.

The general plan of the city was as shown in Fig. 4.1. It lay along a lagoon, with two main streets crossing at the center. At this intersection was the Royal Enclosure, in which were located the two buildings of interest to us—the Museum and the Library.

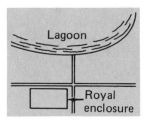

Figure 4.1. Map of Alexandria.

These institutions were founded by the Ptolemys as centers of learning. Their function was to bring glory and renown to the new state. You could think of them as a pre-Christian moon shot. Whatever the rulers' purpose in setting them up, the Library and the Museum played an important role in the development of Hellenistic science.

The Library, as the name suggests, served both as a place where manuscripts were collected from all over the world and as a model for other libraries which sprang up in other parts of the empire (the most notable of these being at Pergamon). The Museum, or Temple of the Muses, was a place where scholars gathered to study and advance the cause of learning. It was not like a modern university, since there were no students trained there. It was probably more like a large government laboratory or research institute (think of

56

Oak Ridge or Brookhaven Lab) which was devoted to research alone.

Clearly, the Ptolemys had provided an atmosphere in Alexandria that was very conducive to the advancement of learning. The science of the Alexandrian period was not the indigenous Egyptian science, but, like most of the cultural advances throughout the Macedonian (and later the Roman) Empire, it was Greek in character, based on Greek thought and ideas, and written in Greek. Greek was considered to be the language of educated men, much as French was the language of the aristocracy in pre-revolutionary Russia. Men were advised to learn Greek so that they would not bore their mistresses, and the education of a Roman gentleman was not considered complete until he had spent some time studying in Greece. Even the founding of the Museum was carried out by an Athenian named Strato, who was, at the time, the head of the Lyceum.

There are two Alexandrian scientists who, from the point of view of their influence on later work, stand out from the others. They are Euclid, who wrote the text *Elements of Geometry,* and Claudius Ptolemy, the astronomer. It is one measure of the enduring worth of the Alexandrian scientific institutions that these two men, who we normally consider to be practically contemporaries, actually were as separated in time as were Columbus and Napoleon!

In the third century B.C., Euclid began the work of collecting the available knowledge of geometry and putting it together in one cohesive development. He discovered that all of the geometrical theorems that were known at that time could be derived as consequences of 11 basic postulates. You probably are familiar with these postulates from high school geometry. They are:

1. Things which are equal to the same thing are equal to one another.
2. If equals be added to equals, the wholes are equal.
3. If equals be taken from equals, the remainders are equal.
4. If equals be added to unequals, the whole are unequal.
5. If equals be taken from unequals, the remainders are unequal.
6. Things which are doubles of the same thing, are equal to one another.
7. Things which are halves of the same thing, are equal to one another.

8. Magnitudes which coincide with one another—that is, which exactly fill the same space—are equal to one another.

9. The whole is greater than its part.

10. All right angles are equal to one another.

11. Two straight lines which intersect one another cannot be both parallel to the same straight line.

From these 11 postulates, everything else follows. For example, once we have these postulates, we don't have to assume that the angles of a triangle add up to 180°, we can *prove* it. The same goes for the theorem of Pythagoras or any other geometrical result.

The axioms themselves, however, are not proved. They are assumed to be true. To the Greek, they were dictated by "reason," just as the spherical motion of the stars was dictated by reason. Indeed, they seemed so reasonable to mathematicians that they were not seriously questioned until the 19th century. And the idea that the way to present scientific or mathematical results was to start from a few axioms and then show all results as conclusions from these axioms is still very much in style today.

We know comparatively little about Claudius Ptolemy, other than that he was taking astronomical observations from 127 to 151 A.D. Like Euclid, he collected the available body of astronomical knowledge and put it together into a single self-consistent (if complicated) picture of the solar system. The book that he wrote is now called the *Almagest* (from the Arab prefix "Al" and the Greek "megiste," meaning great). Later scholars referred to it as "The Great Astronomer," but Ptolemy himself called it "The Mathematical Composition."

It is sometimes surprising to the modern reader to find out just how much knowledge of the world the Alexandrians had. For example, in about 200 B.C., Eratosthenes estimated the radius of the earth by measuring the shadows cast by the sun in various Egyptian cities, and came up with a value that was only a few percent different from modern measurements. In Book I of the *Almagest,* Ptolemy titles a section "That the Earth, taken as a whole, is sensibly spherical," and argues that "if it were flat, then the stars would rise and set for all people together and at the same time... But none of these things appear to happen." The Alexandrians knew that the earth was round, and had developed good arguments to back this knowledge up!

The central feature of the *Almagest* was Ptolemy's description of the solar system. Like virtually all Greek scientists, Ptolemy accepted without question that (1) the earth stood at the center of the universe, and (2) the motion of the heavenly bodies had to be spherical in nature. However, it soon became obvious to him as he compiled astronomical data from the time of the Babylonians onwards (and added his own observations) that simple realizations of these ideas (such as the spheres of Eudoxus) would simply not be able to explain what was observed. Consequently, Ptolemy took it as his task to "save the appearance" by producing a system that contained both of the features dictated by reason and yet, at the same time, predicted observations like the ones that were actually recorded.

The construct that allowed him to do this was the epicycle. The details of how it worked will be left to the following technical digression, but in essence Ptolemy assumed that the spheres of Eudoxus did not themselves contain the planets, but served as the guides for other spheres which did. Thus, a planetary orbit would look like the one shown in Fig. 4.2, with one sphere rolling on another.

As discussed in the following section, such a system did, indeed, "save the appearances." But to do so, it was necessary to make the solar system very complicated. It turned out that a single epicycle like that shown in Fig. 4.2 would not suffice, and the second circle would itself be taken as a center for yet another circle, and so forth. The final picture is a universe of wheels within wheels within wheels, each wheel spinning around at a different rate, and each planet having its own collection of wheels. In the final analysis, Ptolemy had over 70 such wheels in the solar system!

So the appearances were saved, and a picture of the solar system had been developed that would stand as scientific "truth" for 15 centuries. But what a complicated, ugly system it was!

Planet or epicycle

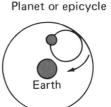

Figure 4.2. An Epicycle.

B. Technical Digression: Retrograde Motion, Epicycles, and All That

In this section we will discuss one aspect of the apparent movement of the planets—retrograde motion—and the method devised by Ptolemy to save the Platonic system of astronomy with its reliance on the concept of movement in

Figure 4.3.

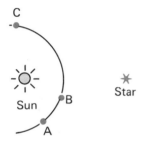

Figure 4.4.

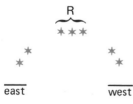

Figure 4.5.

spheres. We will split the discussion into three parts: (1) what is observed, (2) what kind of system Ptolemy devised to explain the observation within the system of Platonic thought, and (3) the modern explanation of the effects.

Suppose that we were to make the following set of observations: each night at a specified time (say midnight), we go out and record the position of a particular star in the sky. Then if we plotted these positions against the horizon lines, we would get a set of points something like those in Fig. 4.3. The reason for this result is obvious. In Fig. 4.4, we see a picture of the solar system and the star that is being observed. When the earth is at point A, the star first appears in the evening sky on the eastern horizon. (Why?) As the earth moves around its orbit to point B, the star will appear higher in the eastern sky. This trend will continue until point C, when the star will cease to be visible in the night sky at midnight. The points in our observations corresponding to these three positions of the earth are labeled in Fig. 4.3.

What we would see each star do over the six months that it would be visible at evening, then, would be to come up in the east and move toward the west. (Remember that this motion does not refer to the way it moves on a given night, but to the way its position at a given time changes from one night to the next.) Furthermore, since the earth's motion around the sun is approximately uniform, we would expect the apparent motion of the star to be uniform as well, so that it would be higher in the sky by equal intervals of angle for equal intervals of time. For example, if it climbed by $30°$ in one month, it would climb by another $30°$ during the next month, and so on.

This sort of motion would be very easy to explain in the Greek picture of things. The stars are embedded in the celestial sphere, which turns around the earth. It must turn a little bit less than one complete turn each day to explain both the motion of stars across the sky on a given night and the shift in position from one night to the next.

The motion of a planet, though similar in some respects to that of stars, is sufficiently different to have made life very difficult for astronomers for a long time. If we performed the same set of observations that led to Fig. 4.3 on a planet such as Mars rather than on a star, we would find a result something like that in Fig. 4.5. On the first day, the planet would appear in the east and on successive days it would appear higher and higher, moving to the east faster than the star. Some time around the middle of the period of observation (labeled R), however, there would be an obvious "bunching" of the observed locations, corresponding to the fact that during this period, the movement of the planet from one night to the next appears to slow down.

Thus, while the stars appear to move in steady increments across the sky, the motion of the planets is irregular. The period of slowing down (labeled R in Fig. 4.5) is called retrograde motion. This name comes from another way of representing the phenomenon. Suppose that instead of asking how far the planet moves each night with respect to the horizons, we ask how far it moves with respect to one of the steadily moving stars.

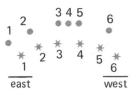

Figure 4.6.

In Fig. 4.6 we show the successive positions of both a planet and the stars observed at the same time each night. During the first few months, the planet will appear to rise faster than the stars, so that *relative to the stars* it appears to move to the west. In other words, although both planet and stars are progressing to the west during this phase, the planet is moving farther west than the star.

During the period that we have defined as retrograde motion, however, the apparent movement of the planet toward the west slows down, while that of the stars does not. During this period, the stars "catch up" with the planet, and then pass it. Relative to the stars, then, the planet appears to move eastward during this phase, even though it is still moving westward relative to the horizons. Finally, in the final phase, the planet again moves to the west relative to the stars.

This means that during the period of retrograde motion, the path of the planet plotted against the background of fixed stars would look something like the path shown in Fig. 4.7. The loop is what was so puzzling to Greek astronomers, because it is hard to see how such a thing could come from the motion of spheres.

A couple of points should be made about retrograde motion before we move on to Ptolemy's explanation of it. First, if the orbits of the earth and Mars were exactly in the same plane, we wouldn't see a loop, but a line, with the parts

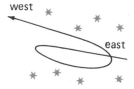

Figure 4.7.

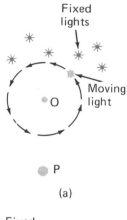

(a)

(b)

Figure 4.8.

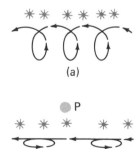

(a)

(b)

Figure 4.9.

of the line corresponding to R traveled more than once. Because of the slight tilting of the orbits, we actually see a small loop.

Second, the apparent backward motion of the planet is seen *only* when we plot its position against the background of fixed stars. The planet itself does *not* reverse its direction in the sky. That is, if Mars appears 55° up in the east one night, it will *not* appear up 50° some time later. The apparent backward motion is purely a result of the fact that the stars move across the sky at a steady rate. The slowing down of the movement of the planet, then, leads to an apparent backward motion because it is being compared to a steady, ongoing motion.

Obviously, if we consider the motion of the spheres of Eudoxus (in which the planets went around the earth in circular orbits), no such irregularities in the motion could occur. Just as obviously, the motions *did* occur. How could the two central features of Greek astronomy—geocentrism and motion in spheres—be reconciled with this stubborn fact?

It was the genius of Ptolemy that he was able, by modifying the system of Eudoxus, to come up with an explanation. To understand his explanation, let us consider how motion in a circle would look to an observer standing outside of the circle. In Fig. 4.8a we see a light moving steadily in a circle, whose center is at O, and an observer at point P watching the light. Suppose also that there is a wall in back of the circle on which some fixed lights are located. The motion of the moving light against the fixed lights will appear as in Fig. 4.8b—a back-and-forth type of motion. (The motion would all appear on the same line, but in the figure we have displaced the lines slightly so that they are visible.) The light will appear to move to the left when it is on the upper half of the circle, and to the right on the lower half.

Now consider what would happen if the center of the circle at O were to move constantly to the left. Then the light would trace out a series of loops (called cycloids) as shown in Fig. 4.9a. Against the fixed background, the observer at P would then see the type of motion shown in Fig. 4.9b. During the time when the light was making the loop in the cycloid, it would appear to move backwards against the fixed background!

It was this realization that gave Ptolemy the key ingredient of his system, which he called the epicycle. The idea here was that both geocentrism and circular motion could be reconciled with the fact of retrograde motion if it was assumed that the planet did not itself ride on the sphere of Eudoxus, but rather rode on another sphere which rolled on the first sphere (see Fig. 4.10). Both the larger and smaller sphere would be turning, and it would then be possible to adjust the rate of turning of the two spheres so that at some point in the motion, the planet would appear to move backward against the background of the fixed stars, just as the light appeared to move backward in Fig. 4.9b. This was the Ptolemaic explanation of retrograde motion, and if it leads to a very complicated picture of the universe, at least Ptolemy cannot be accused of ignoring the things he saw in the sky when he put together his theory.

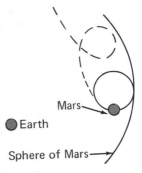

Figure 4.10.

Fence Posts

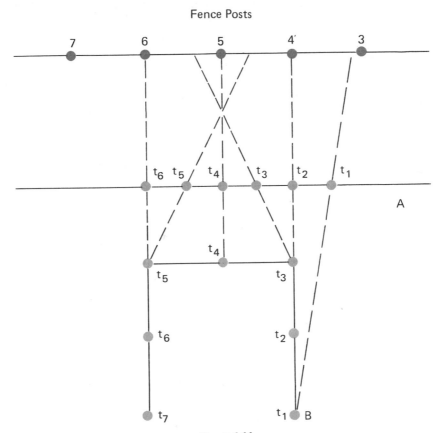

Figure 4.11.

63

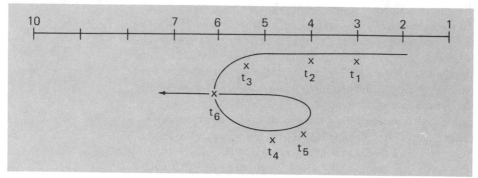

Figure 4.12.

For the sake of completeness, let us now discuss the understanding of retrograde motion that goes with our modern idea of the solar system, in which planets are going around the sun. Some understanding of how retrograde motion might arise in such a system can be gained by considering an analogous situation which is conceptually simpler.

Suppose that we had two cars on an open field (see Fig. 4.11). Suppose further that along one edge of the field are a series of fenceposts labeled 1 through 10. Suppose the first car, labeled A, is moving along to the left at a steady rate, and the second car (labeled B) moves twice as fast around three corners of a square located farther into the field. Suppose that at time t_1 the cars are at the points labeled t_1, at time t_2 at the points labeled t_2, and so forth. Since car B is moving faster, it will travel farther in each time interval.

Now suppose that at the end of each time interval, the driver of car B sights along a line to car A, and records on a chart the position of car A. At t_1, he will see that car A is between fenceposts 3 and 4, at t_2 just opposite fencepost 4, etc. The important event occurs when car B turns the corner and begins to overtake car A. Although car A continues to move at its usual rate, at t_4 it will appear to the driver of car A to be opposite fencepost 5, whereas at the previous measurement, it was in front of the same fencepost. Thus, to the driver of car B, it appears that *relative to the fenceposts* car A has started to move backward, even though no changes of velocity were involved for either car. The rest of the apparent motion is easily marked out, and is summarized in Fig. 4.12.

64

Now if we look at Fig. 4.13, we see that exactly the same situation occurs when the earth passes Mars in its orbit—an event that happens during each year when Mars is visible. The reader will immediately see that if we consider an observer on earth to be analogous to car A, and the fixed stars to be analogous to the fenceposts, the results of Fig. 4.12 can be taken over directly to the solar system, and the retrograde motion of the planet is seen to be due to the fact that both the observer on earth and the observed planet are in motion.

This means, then, that the extreme complexity of the Ptolemaic system of the universe was primarily a result of the fact that Ptolemy insisted on a geocentric universe, with an unmoving earth at its center. Once this requirement was dropped (over 1,000 years later) by Copernicus, retrograde motion ceased to be a problem in astronomy. It is simply another example of how starting with a wrong idea can lead to all sorts of problems in dealing with the world.

C. The Legacy of the Greeks

Before the Greek experience, man's knowledge of the world around him was fragmentary, and men who were actively engaged in working with the material world—men of the technological tradition—had little to do with men who dealt with philosophical questions. It was felt that arriving at the truth was largely a matter of seeking divine revelation, either through religious ceremonies or through

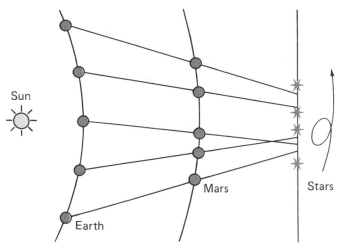

Sun

Mars

Stars

Earth

Figure 4.13.

consulting the sacred texts. In such a world, there was little premium put on finding new and innovative ways to deal with problems.

With the Greeks, the dead hand of tradition was overthrown, and a new way of looking at the world developed. In spite of the important differences between them, both the Ionians and the Athenians insisted that human beings could come to their own answers to questions without recourse to religious authority. For the first time in history, men were free to think!

Of course, the *ways* that these two different schools of philosophers chose to come to their answers were different. To the Ionians, starting as they did from the technological tradition, it was obvious that the way to learn about the world was to observe it, and learn its principles by experimentation. To the Athenians, starting from the philosophical tradition, the correct way to approach a question was to apply human reason and, if necessary, to ignore observations. But both schools would have agreed that it was man who had the ability to answer these questions, and both would have rejected the idea that it was necessary to consult some higher authority to test an idea.

By the time of the Alexandrians, this principle was so ingrained that it is seldom stated or debated. The Alexandrians devoted themselves to working out and codifying the great areas of knowledge opened up by their prodecessors. Thus, Ptolemy did not question the basic assumptions of earlier astronomy, but rather bent all of his considerable talents to making these assumptions fit the observed data. Euclid showed how all of Greek geometry could be contained in a few simple postulates, and worked out virtually all of the important consequences of these postulates. Later Alexandrian mathematicians carried this work further, but never questioned the postulates themselves.

Aside from plane geometry, very little of Greek science has stood the test of time. This is not surprising, since they were the first men to try to explain the world in a rational way. The main concept that we have inherited from them has to do with the *way* in which they went about finding this explanation. The idea that men are free to seek out their own answers to questions has survived long after epicycles have been relegated to the realm of historical curiosities.

One very interesting question about the Greeks has to do with Alexandria. In spite of its first-rate scientific establishment, and in spite of the fact that many Hellenistic scientists like Archimedes also spent a great deal of time devising practical uses for scientific principles, science was never put to work in Alexandria the way it is in modern America. For example, we know that the Alexandrians knew about the steam engine, but it remained a toy for the nobility and was never used to lighten the work of the common people. Similarly, clever hydrodynamic devices were invented to run public fountains, but the principles were not widely applied in irrigation systems. The question of why a people who developed science to the point where it would remain for a thousand years got so little benefit from it is a puzzling one, and one which the reader may want to ponder.

D. Interlude: The Middle Ages

The institutions of Alexandria continued to function for hundreds of years after the work of Ptolemy, but new forces were at work in the world that would destroy the marvelous edifice of Greek culture. The years 200-400 A.D. are generally considered to encompass two historical events—the fall of Rome and the rise of Christianity. Both of these events played a role in the destruction of Greek science.

The fall of Rome destroyed the political stability on which scientific establishments such as that at Alexandria depended. Christianity, in addition, was profoundly hostile to all forms of "pagan" learning, and many Greek libraries and schools were burned by Christian mobs during this period. Perhaps the best example of this was the fate of the last great Alexandrian scientist—a woman named Hypatia, who was famous in her time as a mathematician and astronomer. She was murdered by a Christian mob in 415 after a protracted conflict with Cyril, the archbishop of Alexandria. Little is known about her, but in 1753 John Toland published a novel (complete with portrait on the frontispiece) entitled *History of a most beautiful, most virtuous, most learned and every way accomplished lady; who was torn to pieces by the clergy of Alexandria to gratify the pride, emulation, and cruelty of their archbishop, commonly but undeservedly styled St. Cyril.*

Hypatia of Alexandria is worth noting not so much for her scientific achievements, or even for her martyrdom, but because she is one of the first women scientists of whom we have any record. Only Alexandria, of all of the cultures we have so far studied, allowed women the kind of freedom that would result in their making contributions to learning and science. Conditions such as these were not to come again until our own time.

Little progress in science was made either during the Roman Empire or the "Dark Ages" (usually taken to be roughly from 400–1000 A.D.) following the fall of Rome. Part of the reason, illustrated by the fate of Hypatia, was the fact that science, and learning in general, was associated with the pagan Greeks, and hence was considered inferior by Christians. More important than this, however, was the advent of a new way of looking at the world which made science not so much evil as irrelevant. To a medieval churchman, the goal of gaining knowledge about the world around him was just not very important. What *was* important was the salvation of souls. Insofar as science did not interfere with this higher goal, it might be tolerated.

In 397, Ambrose of Milan (the teacher of St. Augustine) wrote: "To discuss the nature and the position of the earth does not help us in the hope of life to come." Later, St. Bonaventure (the head of the Franciscan order) wrote in a similar vein: "The tree of science cheats many of the tree of life, or exposes them to the severest pains of purgatory."

This attitude toward science—that it is harmful to the acquisition of faith and salvation—persisted throughout the Middle Ages. It is not unknown in our own time, as current problems with teaching evolution in the schools illustrate.

So Greek learning was largely lost and ignored in Europe. Fortunately for us, however, much of it was preserved by the Arabs who conquered most of the Middle East during the first wave of the expansion of Islam. Thus, when Europeans began to have extensive contact with the Arabs during the Crusades, they were able to rediscover much of what had been lost. For example, in 1085 the Spanish city of Toledo was retaken from the Moors, and a center for the translation of Arabic texts was set up. In 1175, Gerard of Cremona translated the *Almagest.* Thus, after a silence of hundreds of years, the voices of the Greek thinkers and scientists were heard in Europe.

The rediscovery of the Greeks is sometimes referred to as the "Twelfth Century Renaissance." It gave rise to a kind of scholastic science based on the writings of Aristotle and other Greeks. It was an arid sort of thing, concerned mainly with the study of ancient texts and proofs of various statements by appeal to authorities. The famous debate about how many angels could fit on the head of a pin was perhaps the most excessive example of this.

The new orthodoxy of Aristotle combined with the teachings of the church fathers set the tone for the background against which the development of science was to take place. The actual events leading to it bear a striking similarity to the development of the first glimmerings of pragmatic thought in Ionia. Like Ionia in 600 B.C., Italy was a commercial and industrial center during the late Middle Ages. We are all familiar with the flowering of the arts that we call the "Renaissance" that took place there at the end of the 13th and beginning of the 14th centuries. It was a period in which new ideas abounded and new thoughts were possible. The discovery of Plato and other lost classical writers convinced people that exclusive dependence on the writings of Aristotle was not necessarily a good thing. It was a period of discovery—new worlds, new forms of art, new ways of seeing things. Although it produced very little in the way of science itself, the Renaissance opened doors that allowed science to develop.

Like Ionia, this period in Europe was one dominated by technological changes. We tend to think of the Middle Ages as a static period, but actually it was marked by a continuous, slow buildup of technological competence in Europe. The paper mill (1189), gunpowder (1249), the mechanical clock (1232), and the printing press (1436) were just a few of the inventions that revolutionized life and made modern industrial civilization possible. And, as was the case in Ionia, practical men of affairs—men of the technical tradition— began initiating scientific inquiry. We shall mention two examples to illustrate this point.

The compass was used as an instrument of navigation, but very little was known about how it worked. In 1581, Robert Norman, an English sailor, published a paper called "The Newe Attraction" in which he discussed experiments he had performed on magnets. He weighed magnetized iron to

69

see whether magnetism had weight (and found that it didn't). He floated magnets on corks in water to see if they were pulled toward the north (they weren't). In short, be began asking questions about magnetism, and tried to discover things about it by observation rather than by appealing to the writing of Aristotle.

A similar incident is recorded in Italy in 1537. The chief engineer of the Duke of Milan, a man named Tartaglia, was told to find out what had to be done to get the maximum range from a cannon. We have already discussed the Aristotelian theory of projectiles, which Tartaglia found totally inadequate for his project. In spite of the accumulated weight of authority, cannonballs just *didn't* fall straight down, but traveled in some sort of arc. To fulfill his commission, then, Tartaglia went out to the artillery range and had the chief bombadier shoot off cannons to *see* how to get maximum range.

Actually, Tartaglia guessed that the best way to work things was to shoot the cannonball up at $45°$. He arrived at this conclusion by noting that if the cannon were aimed at $0°$ or $90°$ (straight up), the cannonball wouldn't get anywhere. He guessed that halfway between there would be an optimum, and actually went out to see if it was true. Against the background of scholastic science, this was a revolutionary way to proceed, and it is significant that once again it was made by a technician rather than a philosopher.

So this is what was happening in Europe around 1500. After centuries of stagnation, the scientific idea was breaking out again. During the next two centuries, most of the problems that had plagued ancient scientists would be solved, and man was to gain a new vision of the universe and his place in it. Five men played particularly significant roles in this development, and we list them here:

Nicolaus Copernicus	1473–1543
Tycho Brahe	1546–1601
Galileo Galilei	1564–1642
Johann Kepler	1571–1630
Isaac Newton	1642–1727

SUGGESTED READING

The texts cited in Chapter 3 all contain sections on the Alexandrian sciences, and Mason's book has an excellent section on medieval science. In addition, the following should be very useful:

Mason, Stephen F. *A History of the Science*, New York: Collier Books, 1962. Although the main part of this book is concerned with modern science, it has an excellent review of science in the Middle Ages.

Sarton, George. *A History of Science.* Cambridge: Harvard University Press, 1959. An excellent source of material on Alexandrian and Hellenistic science, but it contains much more information than the average browser may want.

Sarton, George. *Ancient Science and Modern Civilization.* Lincoln, Neb.: University of Nebraska Press, 1954. A series of lectures summarizing many of the ideas in the more complete text cited above.

Sarton, George. *Introduction to the History of Science.* Baltimore: Williams & Wilkins, 1927. An exhaustive list of every known scientist through the 14th century. One of the few reliable sources on Hypatia of Alexandria.

Hutchins, Robert M. (Ed.) *Great Books of the Western World.* In Vol. 11 of the *Great Books* series you will find the works of Euclid, and in Vol. 16, the complete *Almagest.* The latter is especially worthwhile (at least the first few sections), and is not difficult reading at all. It gives a good impression of just how advanced the Alexandrian scientists were.

QUESTIONS AND DISCUSSION IDEAS

1. How many planets did the Greek astronomers know about? Why didn't they know of more?

2. Why is it that some constellations are visible in the evening only at certain seasons of the year? For example, why is Orion in the evening sky only in the winter?

3. Given your answer to problem 2 above, why is the North star always visible at night in the northern hemisphere?

4. What do you think the reaction of Greek scientists would have been if Ptolemy had *not* been able to reconcile the observed motion of the stars and planets with a geocentric universe and spherical motion in the heavens?

5. Prove the following:
   a. Except for a brief period when it is on the other side of the sun, every star is visible from earth at some time during each night.
   b. Each star is visible at a given time each night for only six months each year.

6. Make a drawing of the positions of the two cars in Fig. 4.11 as they would be seen against the background of the fenceposts by an observer standing at P. Hence

71

discuss retrograde motion as it would appear to an observer at the sun.

7. Is it ever possible to see the planet Venus overhead at midnight? If so, explain how; if not, explain why not.

8. Why can an eclipse of the moon occur only near full moon? What is the phase of the moon during a solar eclipse?

9. The geometry of Euclid seems to us to be pretty reasonable, and all of the postulates seem eminently sensible. Let us examine postulate 11, however, and see if our reason is giving us correct impressions. An alternate statement of postulate 11 is that the sum of the angles in a triangle must be 180°. Consider a triangle on the surface of the earth. Prove that such a triangle can have more than 180° in angle. (HINT: Take a triangle with the equator as one side.)

   The realization that Euclid's geometry applied only to the plane, and there were surfaces where it did not apply was realized by the 19th-century Russian mathematician Nicolai Ivanovitch Lobatschevsky, and led to the field of mathematics now called non-Euclidean geometry.

10. Write a short paragraph summarizing the scientific contributions of the following Hellenistic scientists:
    a. Archimedes
    b. Hero of Alexandria
    c. Aristarchus of Samos

11. List some other technological advances made between the fall of Rome and the Renaissance.

CHAPTER V

THE COPERNICAN REVOLUTION

He stopped the sun and set the earth in motion

Description of Copernicus

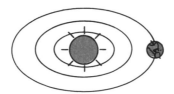

A. Copernicus and the Heliocentric Universe

It is hard to imagine anyone less likely than Nicolaus Copernicus to revolutionize the scientific world. He was born in 1473 in the Polish city of Torun into a well-off local family. His uncle, Lucas Watselrode, was the bishop of the area, which is located on the Vistula River and was at that time a bone of contention between Germany and Poland. From 1496 to 1506 he studied in Italy, preparing himself for a career in the church. It is believed that he received an M.D. during this period, and while in Italy he was exposed to the new ideas that were sweeping Europe.

He returned to Poland to take up a position as canon of the cathedral in the town of Frauenburg, located where a branch of the Vistula enters the Baltic. As canon, he was responsible for the affairs of the church estates, which were considerable. We know that he achieved some local fame as a politician, for he was invited to head a group charged with reforming the currency, and may even have been the discoverer of Gresham's Law (which states that good money drives out bad). He held a free clinic in the cathedral, and at one time was called upon to lead troops against invading German knights. He was, in short, a man of many parts, and it is difficult, looking at what he did in his everyday life, to see how a picture of Copernicus as a lonely, star-gazing recluse could have grown up. But grow up it did, as anyone who has read books about him can testify.

It was expected in those days that every educated man would interest himself in the new classical learning. Having spent so much time in Italy, Copernicus had returned to his native land resolved to do so himself. After completing an ultimately forgettable translation of the letters of an obscure Byzantine writer into Latin, he turned to astronomy as an avocation.

Despite the time consumed by his duties at the cathedral, he studied the writing of Ptolemy. He also set up a small observatory in a remote tower on the cathedral wall. Most of his work, however, did not depend on his own observations, which, in any case, were few in number. What he did was to take the same data on the positions of the stars and planets that Ptolemy had started from, and show that there was an alternative way of interpreting them which was as consistent with the facts as was the universe of the Ptolemaic tradition.

74

There were two main ingredients in the prevailing ideas about the universe at that time: (1) the earth was at the center of the universe (geocentrism); and (2) all heavenly motion was spherical or, at worst, worked on the principle of epicycles. The "Copernican Revolution" did not overthrow both of these ideas at once to produce, in one step, our modern view of the solar system. Science just doesn't proceed that way. Copernicus was very much a transitional man, with one foot in the Middle Ages and one foot in the future. It was quite enough for such a man to challenge the geocentric idea.

And this is precisely what Copernicus did. Over the years, he worked out a system showing that all of the data that had been gathered by astronomers could be explained by assuming that the sun (not the earth) was at the center of the universe, and that the apparent motion of the stars was due to the motion of the earth itself.

How could Copernicus, in remote Frauenburg, have come up with such an idea? Scholars have speculated that during his visits to Italy he might have run across the ideas of Aristarchus of Samos, a Hellenistic astronomer who had suggested (but never worked out) a heliocentric universe. For us, where he got the idea is much less important that what he did with it.

In the Copernican system, the planets moved about the sun in the order in which we now think they do—Mercury closest to the sun and Saturn (the outermost planet known at that time) farthest from it. In this system (as in our present one), the apparent motion of the stars was not due to the rotation of a celestial sphere, but to the rotation of the earth itself. The motion of the planets, and retrograde motion in particular, were then easily understood in terms similar to our modern ideas.

The main difference between the Copernican system and our own was that in the Copernican system the planets still moved on epicycles! In fact, when Copernicus was finished explaining all of the detailed data on the motion of the planets, his system had almost as many wheels within wheels as did Ptolemy's. But the hold that the Platonic idea of circular motion had on the minds of men was much too strong, and Copernicus, as original as he was, was able to break only one of the two ancient "truths" about the place of the earth and the order of the solar system.

Although Copernicus now had a system in which things such as retrograde motion were easily explainable, he realized that there were many objections to his system that would be raised by the Aristotelian scientists at the universities. These could be put into two classes: astronomical objections, which dealt with problems relating to observed phenomena in the heavens, and physical objections, which dealt with problems about the motion of bodies on the earth.

The primary astronomical objection dealt with the lack of parallax in the stars. If you hold your finger in front of you and look at it first through one eye and then the other, you will see that it appears to move against the background of the wall. This apparent motion is called parallax. To understand it, look at the diagram in Fig. 5.1.

Figure 5.1.

Object A is observed from two points, B and C. When the object is being observed from B, it appears to be in front of point B' on the wall in back of it, but when it is observed from C, it appears to be in front of point C'. If we were moving from point B to point C, the object would appear to move from point B' to point C'. The Aristotelians argued that if the earth were really going around the sun, then the nearer stars should appear to move with respect to the background of father stars. In this argument, points B and C would be the positions of the earth in summer and winter, respectively, and object A would be a near star. They argued that since such an apparent motion was not seen, the earth could not be moving.

We know now that the stars are too far away for parallax to have been seen with the instruments available to astronomers at that time. This argument, however, was not accepted by the Aristotelians, and the objection was simply ignored by Copernicans. This illustrates an important point. In any new theory, there are bound to be problems and objections that cannot be immediately answered. The only way that progress can be made is to ignore these problems and hope that some time in the future the answers will be found. This is just as true today as it was then.

The physics objections had to do with the consequences of the earth's rotation. If the earth were really turning, why didn't everything fly off, as bits of clay fly off a potter's wheel? Why wasn't there a constant wind as the earth turned underneath the air?

The problems associated with circular motion had plagued scientists for centuries, and they would not really be solved until the time of Newton. All that Copernicus had to go on was Aristotelian science, and he attempted to answer these objections in Aristotelian terms. He argued, for example, that we didn't fly off the earth because circular motion is "natural" and hence cannot be "disruptive."

But Copernicus is not remembered because he answered all of the objections to his ideas. He didn't. He is remembered because he introduced a new idea—heliocentrism—to the scientific world, and he worked out the system to the point where it could compete, in terms of rigor and detail, with the prevailing Ptolemaic system. He thought the unthinkable and had the intellectual courage to carry through with an idea that would have been considered ridiculous by any reputable astronomer of his time.

Before leaving Copernicus, we should mention one famous controversy that still surrounds him—the problem of the Preface. Copernicus had worked out his system in isolation, with very little contact with the outside world. Late in his life, a young Lutheran mathematician named Rheticus came to work with him. Under the urging of Rheticus, he agreed to publish his work. A summary was published first, and then the main draft of *On the Revolutions of the Spheres* was put together. Rheticus took the manuscript to Nürnberg to supervise the printing, but was forced to leave after becoming involved in a homosexual scandal. This left the printer, Andreas Osiander, in charge of the text. He thought it was too revolutionary and, fearing official sanctions, wanted to include a preface stating that the book should not be taken seriously but was simply a mathematical exercise. The preface which he wrote, said, among other things, that "these Hypotheses need not be true or even probable: if they provide a calculus consistent with the observations, that alone is sufficient."

To later generations, this seemed like a "cop-out" on the part of Copernicus. However, we know that he never saw the Preface. In one of those dramatic moments that occur occasionally, the finished book was rushed into his room just a few hours before his death, when he was too sick to read it. Johann Kepler claimed to have seen some of the correspondence between Copernicus and Osiander, and claimed that

Copernicus had rejected the idea of the Preface. But the correspondence is now lost, and so the question is debated by historians even today. Whether we ever find the answer, however, is not as important for science as the fact that after Copernicus there were two detailed systems of the universe competing with each other, and it was necessary to choose between them.

The choice was made very difficult because within the limits of accuracy in observation available at the time of Copernicus, both his system and the system of Ptolemy did equally well at describing the motions of the heavens. When confronted with this sort of a situation—two theories explaining the facts—scientists have traditionally used one of two sets of criteria for choosing. They can choose on the basis of aesthetics (one theory is more beautiful or more elegant or more reasonable than the other) or they can improve the observations to the point where distinctions can be made on the grounds of new data. Since the Ptolemaic and Copernican systems differed primarily in the initial assumptions and were equally complicated, the first type of choice could not be made. This meant that the ultimate distinction would have to be made on the basis of observation, and hence would depend critically on the error inherent in those observations. The concept of error has not been encountered yet—indeed, this is the first case in history where it was important. Therefore, we will discuss it first before going on to describe the resolution of the controversy started by the publication of the work of Copernicus.

B. Technical Digression: The Idea of Experimental Error

With the development of the Copernican system, scientists for the first time in history were confronted with two equally valid theories of the universe, each proceeding from a radically different assumption about the nature of the solar system, and each doing an equally good job of explaining the observed motion of the planets. There was literally no way of choosing between the two theories except by an appeal to experiment and observation. In the next section we will discuss what actually happened in this case, but the advent of this new element into the scientific picture is so important for modern science, and it is such a departure from the Greek way of doing things, that it deserves a special discussion.

Let me begin by making a categorical statement. There is no such thing as a scientific fact that is totally free from uncertainty. *Every* idea that we have about the world is only an approximation, and is limited by the precision to which we can carry out experiments to test it. Thus, the "fact" that the earth was at the center of the universe was known by the Greeks only to the extent that it did not conflict with the measurement of Ptolemy and others. Similarly, our present knowledge about the orbits of the planets is only good to the limits of our present experimental ability which, although it is much better than Ptolemy's, is still not perfect.

In order to understand what is meant by the term "limits of experimental ability" and to see what effect it has on our ability to test ideas about the world, let us consider an experiment that was actually carried out by a group of students at the University of Virginia. The purpose of the experiment was to test the Aristotelian notion that heavy objects fall faster than light ones. The way we went about doing this was to measure the time it took an object to slide down an inclined plane, measure the time it took a heavier object to do the same, and then compare these times to each other. The inclined plane was used (rather than letting the objects fall) because it takes longer for an object to "fall" down an inclined plane, so we had a better chance of measuring the time. As a matter of historical interest, this is precisely the way Galileo tested this idea almost 400 years ago.

The experimental procedure was quite simple. We had a stopwatch, which we would start when the object was released at the top of the plane and stop when it got to the bottom. On the first try with the first weight, we found that the watch read 4.5 seconds. Being careful souls, we wanted to check this result, so we repeated the operation. This time, however, our reflexes weren't so quick when the weight was released, or perhaps we gave the weight a slight, indiscernible shove. For these, or any number of other reasons, we found that after the experiment had been repeated, we had a different time—4.3 seconds.

Which of these is right? To find out, we had to do the experiment again. But, as before, we couldn't get everything to be *exactly* the way it was the first time, and still a third time resulted—4.6 seconds. At this stage, we realized there was nothing for it but to repeat the experiment many times, and see what happened. When we did so, we got the following results for the time it took a single weight to slide down the plane:

4.0 sec. — 1 time
4.1 sec. — 2 times
4.2 sec. — 2 times
4.3 sec. — 5 times
4.4 sec. — 5 times
4.5 sec. — 9 times
4.6 sec. — 5 times
4.7 sec. — 1 time
4.8 sec. — 1 time
4.9 sec. — 1 time

These results are shown graphically in Fig. 5.2, in which the number of times each reading is obtained is plotted for the time range of 4.0 to 5.0 seconds. A picture like this is called a histogram.

The first thing that comes to our attention when we look at the results of these repeated trials of the same measurement is that they are different from each other. Why should two people measuring the same thing with identical clocks come up with different numbers?

Many reasons for this can be advanced. In the first place, different people will have different reaction times and

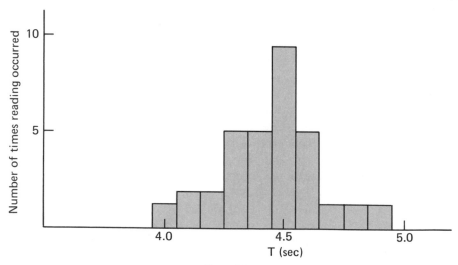

Figure 5.2.

this will affect the starting and stopping of the stopwatch. Even a single person taking repeated readings will react more quickly some times than others, so that different results will be obtained for different trials. As mentioned above, one trial may involve a slightly different method of releasing the weight. Dust may settle on the table between trials, increasing the friction and thereby giving different times. Finally, the stopwatch itself reads only to the nearest tenth of a second (i.e., has its face marked off in tenths of a second). Most of the readings will not have the hand pointing exactly at one of the numbers on the clock face, but will have the hand pointing between two numbers. Thus, some judgment is necessary to translate the position of the clock hand into a number for our histogram.

The net result of all of these sources of uncertainty in the measurement is that we can only describe the experiments within certain limits. We knew, for example, that it did not take 3 seconds or 6 seconds for the weight to fall. However, it is clear that there is a spread in the reading that we obtained. If we had to guess at what the correct time of fall was, we would probably choose the average of all of the readings which, in the case of Fig. 5.2, is 4.5 sec. The spread in the readings, as shown in the figure, is about 0.2 sec. This spread, inside of which most of the readings fall, is called the *experimental error* associated with the measurement. The fact that there is this much error associated with the experimental finding is usually incorporated into the result of the experiment by writing the final result as

$$t = 4.5 \pm 0.2 \text{ sec.}$$

This way of writing tells us two things: it tells us that the average time was 4.5 seconds, and it tells us that the experimental error (sometimes called the standard deviation of the measurement) was 0.2 second. The technical meaning of the result is that if we were to repeat the entire experiment, there is a 68% probability that the new average would be between 4.4 and 4.7 seconds, and a 95% probability that it would be between 4.1 and 4.8 seconds. For our purposes, however, it is enough to think of the error as telling us roughly the range of values that the average would have were we to repeat the entire experiment and generate a new set of readings in Fig. 5.2.

81

In any given experiment, this error can be reduced by redesigning the apparatus used. For example, if we had timed the fall of the weight with a photocell (like the one that activates automatic doors), the element of human reaction time would be eliminated from the experiment, and much better readings would result. We might be able to reduce the error to .0002 second instead of 0.2 second. Nevertheless, there is always some error in any experiment, and all we can do is to make it smaller. The source of this irreducible error may be due to human limitations, like the error due to reflex time we discussed above, or it may be due to the measuring instrument itself, but it is always present in any measurement, and must be accounted for when we interpret the results of the experiment.

Now that we understand that every measurement has an error associated with it, we can talk about how we can use measurements to distinguish between scientific theories. In our sample experiment, we were interested in seeing whether a weight which was heavier than the one we used actually fell faster, as Aristotle predicted. In fact, we used a weight which was twice as heavy as the one used to generate the histogram in Fig. 5.2. Therefore, if we took Aristotle seriously, we would expect it to take only half the time to slide down the plane.

However, we realize now that we have to be more careful. Saying that the weight "takes half the time" implies that we will be measuring the time of fall, and every measurement involves an error. Thus, rather than finding a precise time of fall for the heavier weight, we would expect to find some spread of times if the measurements were repeated for this weight, but with an average about half that shown in Fig. 5.2.

On the other hand, if, as Galileo thought, the time of fall were independent of the weight of the object, we would expect to find a spread of reading with an average of about 4.5 seconds. In both cases, however, we would expect the experimental error to be about 0.2 second, since the sources of this error do not depend on the weight of the object going down the plane. Thus, the two theories of falling objects would predict a set of results similar to that shown in Fig. 5.3.

One important point has to be made about Fig. 5.3. The predicted values of the time of fall for the two theories are farther apart than the experimental error associated with the

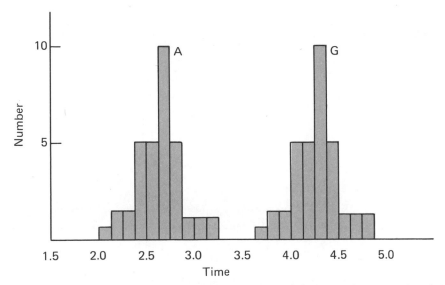

Figure 5.3. Predictions of the Aristotelian (A) and Galilean (G) theories for the results of the experiment to be performed with the heavier weight.

measurement of the time. In other words, the two histograms do not overlap. Whenever we get a situation like this, where the theories make predictions that are substantially different compared to the experimental error, it becomes possible to see which theory is correct by performing an experiment.

When we did the same measurements for the heavier weight as we had done earlier for the light one, we found a set of results shown as the dotted line in Fig. 5.4. We found that the time of fall for the heavier object was

$$t = 4.3 \pm 0.2 \text{ sec.}$$

Thus, our experiment proves conclusively that the Aristotelian notion of falling bodies is wrong. Does it also prove that the Galilean notion is right?

This is a somewhat more difficult question to answer. The average results for the two weights are not exactly the same. However, each experiment involved an error. In particular, we could state the effect of the error in this way: the heavier body fell in a time that was between 4.1 and 4.5 seconds, while the lighter one fell in a time that was between 4.3 and 4.7 seconds. Thus, we would say that the times of fall are equal *to within experimental error.*

83

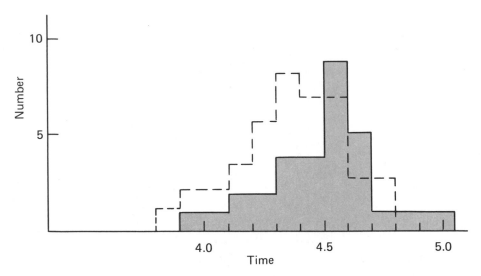

Figure 5.4. The dotted line shows the times we obtained for the heavier weight to slide down the inclined plane. The histogram for the lighter weight is shown as a solid line.

This way of stating the conclusion is very important, because it reminds us that when we measure two numbers in order to find out if they are equal or not, the result will depend on how accurately we can make the measurement. Thus, a difference of 0.2 second in this experiment is not considered significant (that is, if the results of the two experiments come out this close together, we say that the numbers are equal), but it might be extremely significant if the experimental error was reduced to 0.002 second (in which case we *could* say that the results were different within the experimental error). In short,

> *Scientific facts are valid only up to the limits of the experiments that are performed to establish those facts.*

Now that we understand the role that experimental error plays in the process by which one scientific theory is accepted and another is rejected, we can return to the story of how our present ideas about the solar system came into being.

84

Gest. v. Ant. Karcher. Manh. 1796.

NIC. COPERNICUS.

geb. d. 19 Feb. 1473. gest. d. 24 Mai 1543.

Nicolaus Copernicus (1473- 1543).

Galileo Galilei (1564-1642).

C. Tycho Brahe and Johann Kepler: The Odd Couple

There are some people who seem to attract strange happenings the way a magnet attracts iron. From the time of birth into a noble Danish family in 1546, Tycho Brahe seems to have been such a man. His childless uncle, Jorgen, an admiral in the Danish navy, had made an agreement with Tycho's father that he would be allowed to raise one of Tycho's father's sons. Tycho himself was a member of a set of twins, but, unfortunately for all concerned, his brother was stillborn. Feeling himself cheated, the admiral kidnapped young Tycho Brahe and raised him.

As a member of a noble family, Tycho was sent to the university to study philosophy. His real interests, however, were in the sciences. So great was his interest in mathematics that at the age of 20 he fought a duel with a fellow student over the question of which one was a better mathematician. It seems a strange way to settle such an argument, and the duel certainly did Tycho no good. In fact, the tip of his nose was sliced off, and he was forced to make a silver substitute which he had to keep glued to his face.

At the end of his first year at the University of Copenhagen, he observed an event that was to alter his life. That year there was an eclipse of the sun visible from northern Europe. He was deeply impressed by the event—not so much the obliteration of the sun itself, but by the fact that men could *predict* events in the heavens. He began buying books on astronomy, much to the consternation of his family. Astronomy (and learning in general) were considered slightly disreputable for men of noble birth, who were expected to spend their time hunting and attending the court. When he was sent to Leipzig for further study, he was accompanied by a tutor under explicit instructions to prevent him from studying science. As might be expected from his family background, he was not about to be prevented from doing what he wanted to do.

After completing his studies, he returned to Denmark and lived quietly until 1572, when another celestial event occurred which shook the foundations of classical astronomy. On Nov. 11, 1572, a new star appeared in the constellation of Cassiopeia!

According to the Greek ideas then prevalent in Europe, such an event was impossible. Change and corruption were possible on the earth, but the heavens were supposed to be the realm of the pure, the true, and the unchanging. And yet here was an event that looked very much like a change in the heavenly order. Astronomers set about feverishly to try to measure any movement the new star might have. After all, if it moved then it was probably an atmospheric phenomenon, and the Aristotelian establishment was safe.

The techniques used by the established astronomers to decide this question were unbelievably crude by modern standards. For example, one man lined up a string so that it went through the nova (as the new star was called) and two other stars, and then watched to see if the star moved off the line of the string. Because all such measurements depended on sighting with the eye, and because the motion of the nova could be very small, different astronomers published different conclusions about it.

But while the problem was being attacked with these peashooter techniques elsewhere, Tycho Brahe was setting up the heavy artillery at the family estate in Denmark. He had just built a sextant (an instrument for measuring the angle of elevation of a star) that was 5½ feet long, and on which he had done studies to allow him to correct for some of the errors in the instrument. With this new advance in instrumentation, he was able to make measurements accurate enough to prove that the nova was in the region of the stars and not in the region between the earth and the moon.

Why did the new instrument make this possible? Why should increasing the size of an astronomical instrument increase its accuracy? To answer this question, let us consider how we would go about measuring the angle at which a star appears to be above the horizon. The usual way of doing this until the invention of the telescope was to sight along a rod in the direction of the star (see Fig. 5.5). This is something like sighting along a rifle barrel. Once we have the rod lined up with the star, then the angle of elevation, A, can be determined by measuring the two lengths x and y (or, equivalently, measuring x or y and the length of the rod). Thus, measuring the position of a star in the sky eventually comes down to measuring a length in the instrument. In particular, if we wish to know if the star is moving with respect to the other stars, we have to measure its position on several different occasions and see if it is different.

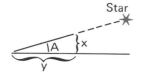

Figure 5.5.

But we know that in order to come to a definite con-clusion in such an experiment, it is essential that the error involved in the measurement be small enough so that the error histograms representing the position readings on suc-cessive days do not overlap. Thus, in order to determine whether the Greek ideas about the structure of the universe are correct, we are reduced to talking about the errors in-volved in measuring lengths. What would Plato have thought about that!

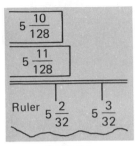

Figure 5.6.

In measuring a length, there is always an error involved. If we are using a ruler the markings on the scale of the ruler will not be infinitely divisible, but will have a smallest grad-ation. It might, for example, be 1/32 of an inch, or perhaps 1 mm. With such a ruler, it would be impossible to say whether a particular length was, for example, 5 10/128 of an inch or 5 11/128 of an inch. The easiest way to see this is to imagine two sticks whose lengths differ by 1/128 of an inch laid side by side, and the ruler laid down next to them (see Fig. 5.6). The ends of both sticks would fall between the markings on the ruler.

Thus, in our example of the ruler whose smallest grada-tion is 1/32 of an inch, distinguishing between lengths much less than that involves guesswork. Even if there were no other sources of error in the experiment, we would expect an error of something like 1/32" (or perhaps half of this if we are good at estimating distances by eye). Certainly, it would be unrealistic to expect an error of .00001" in such a measurement.

Once we have an idea of how big the actual error must be in measuring a length, we can calculate the per-centage error involved as well. For example, if the length is 10" and the error is 1/32", then the percentage error is

$$\frac{\frac{1}{32''}}{10''} = \frac{1}{320''} \cong 0.3\%.$$

If this particular length were involved in measuring the angle A, then it is obvious that we couldn't measure the angle to better than 0.3% either.

Consider what happens, however, when the length we

87

are measuring, instead of being 10" is 100". Then the percentage error is

$$\frac{\frac{1}{32''}}{100''} = \frac{1}{3200} \cong 0.03\%,$$

or ten times smaller! This means that the percentage error in a length measurement (and hence the error involved in measuring the angle of elevation of a star) can be reduced by making the measuring instrument bigger. This explains why Tycho succeeded where others had failed in establishing the nova as a star beyond any reasonable doubt. When he published his results in a book titled *De Stella Nova* in 1573, there was no way for other astronomers to refute them.

As a matter of historical interest, the remnants of the nova that Tycho measured constitute what we call today the Crab Nebula. The first pulsar was recently discovered in this Nebula, so it has not yet finished giving us new informations about the heavens.

The observations on the nova established Tycho's reputation as an astronomer. Frederick II of Denmark decided that having such a man as a member of his court would add to his renown, so he granted a large sum of money and the island of Hven, near Copenhagen, for the construction of an observatory. This observatory was the first modern example of a government funded research institute. In addition to astronomical instruments of a huge scale (including an instrument for measuring angles that was 14 feet long), it contained everything that was needed for astronomical research. It even had a printing press and a paper mill for the publication of new results!

For the next 20 years, Tycho held court at the palace of Uranienborg, which he built on the island, and carried out astronomical measurements of an accuracy never before achieved in the annals of astronomy. King Frederick died in 1588, however, and Tycho's abrasive character soon caused trouble with the new king Christian IV. Consequently, in 1597, he packed his instruments and went into a self-imposed exile in Europe, eventually settling at the court of the Emperor Rudolf in Prague. There he hired a young German mathematician, (also an exile) named Johann Kepler to help him analyze and understand the data that he had so laboriously accumulated. This collaboration would complete the revolution in astronomy that had been started by Copernicus.

Johann Kepler was born in 1571 in Weil-der-Stadt, a town in southwestern Germany. He was a Protestant, and in those times of religious upheaval, this religious identity often made it impossible to hold jobs in areas ruled by Catholic princes. In 1599, Kepler was forced to leave a post in Graz, Austria. Tycho Brahe was established at that time in Bohemia, and offered the young mathematician a post analyzing his astronomical observations. Today, we would call this sort of job a post-doctoral fellowship.

Actually, Kepler had been thinking about astronomical problems for some time before leaving Austria. He had a mystical frame of mind and was obsessed by the idea that there was some sort of overall scheme to the universe. He spent much of his life searching for this scheme, which he believed would reveal the ultimate beauty of nature. In all, he produced three models of the universe, but he is remembered only for the one that resulted from his work on Tycho Brahe's data. To get some idea of how his mind worked, however, we shall discuss his other two ideas first.

His first system of the universe was developed while he was in Austria. It involved the fact that there are things which geometricians call "regular solids." These are defined to be solid figures whose sides are composed of polygons, with the edges of the polygons of equal length. For example, a cube would be a regular solid because its sides are squares, and the edges of a square are all of equal length. It had been proven by Euclid that there were precisely five regular polygons that could be constructed (see Fig. 5.7).

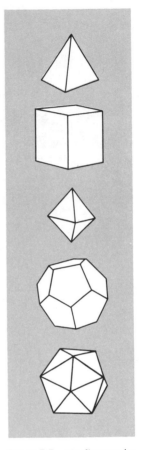

Figure 5.7 The five regular solids.

Kepler knew this fact, and he associated it with the fact that there were six known planets (with five spaces between them). He argued that each space between the planetary orbits must be just large enough to allow us to fit in one of the regular solids. In this way, the universe would be a geometrical work of art, and its order would be derivable from geometrical principles. The similarity to the work of Eudoxus and his followers is obvious, and, indeed, at this stage Kepler felt that the planetary orbits had to be spheres. He did, however, incorporate the heliocentric theory of Copernicus into his ideas, and would remain a staunch supporter of these ideas throughout his life.

Johann Kepler (1571-1630).

Actually, not much came from this first attempt to describe the universe. He quickly discovered that the data on planetary orbits could not be fit into his neat scheme, and he dropped the project. The only practical outcomes seem to have been an attempt on his part to convince Frederick, Duke of Wurtenberg, to have a model constructed of hollow metal spheres and solids, each holding a different kind of liquor, and the publication of his first book, *The Mystery of the Cosmos.* This episode did, however, focus his attention on the need for precise data on the planetary positions and on the departures of the planets from circular orbits. This, in turn, led him first to correspond with, and then to work for, Tycho Brahe.

Later in life, he would return to these mystical ideas about the universe and attempt to construct a system in which each planetary orbit represented a musical theme, and the solar system was constructed in such a way as to produce pleasing harmonies from the "Music of the Spheres." This system didn't work out too well, either.

But in between these two flights of fancy came a period when he was in intimate contact with observational data more accurate than had ever before been available. The data seemed to have the effect of channeling his imagination away from mystical ideas like those discussed above and toward the more mundane task of finding out exactly what the planetary orbits were. Even though his motivation in this study may have been mystical, his results were not. In fact, they formed the cornerstone of the new order of the universe which was devised later by Newton.

To understand the problem facing him, let us consider what the orbits of the planets are. In fact, they are almost circular. We know now that they are elliptical, but only slightly so. This means that in order to deduce the fact that they are not circular, extremely precise measurements of their positions are needed. Up until the time of Tycho, the best measurements of the positions of the planets had an error of about 8" of arc. This corresponds to something like 1/10 of the diameter of the moon as seen from the earth.

As we now know, the size of this error is very important in the process of choosing between two theories. As it turned out, the predictions for planetary positions made by the Copernican theory on the one hand and the Ptolemaic

90

theory on the other differed by *less* than 8″ of arc. Thus, from our discussion of experimental error, we know that given the old data, it would be impossible to distinguish between the two theories.

With the new data from Tycho Brahe, however, Kepler was able to prove conclusively not only that the solar system was heliocentric, but that the concepts of epicycles (and hence of circular motion in the heavens) was not correct.

In 1609, in his *New Astronomy,* he published the results of these findings in the form of two "laws" of planetary motion. These laws were:

1. Planets move in elliptical orbits with one focus of the ellipse at the sun.
2. The planets move in such a way that a line drawn to the planet will sweep out equal areas in equal times.

Ten years later, a third law was published. This stated:

3. The square of the period of a planet is proportional to the cube of its distance from the sun.

The first of these laws is simply a statement of fact about the deductions he made from Brahe's data. To understand it, we note that an ellipse is defined as a curve constructed such that the sum of the distances from two points (called foci) to the curve is the same for every point on the curve. One way of constructing an ellipse, for example, is to tack a length of string down at two points (see Fig. 5.8) and then catch a pencil in the string as shown. Moving the pencil around the points will then produce an ellipse, and the two tacks will be the two foci. Kepler's first law simply states that this geometrical figure happens to be the one that planets follow in their motion around the sun.

This is a truly revolutionary statement coming, as it did, after thousands of years of belief that only circular motion was allowed in the heavens. There is, after all, nothing "perfect" about an ellipse. Furthermore, Kepler had no idea *why* nature had chosen elliptical courses for the planets. All he could do was state that observations showed that this was the case.

We can get some idea of the difficulty that Kepler faced in establishing the first law by noting how easy it

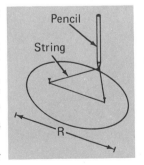

Figure 5.8.

91

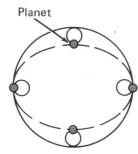

Figure 5.9.

would be to get a curve that is almost elliptical with an epicycle. In Fig. 5.9, we show how an epicycle can be arranged on a larger sphere to give a curve that has a flattened-out appearance. It was this sort of thing that Kepler was able to rule out, not just the simple motion on a single sphere.

The second law is a statement about how the planets move around their elliptical courses. What it says is that planets move slowly when they are far away from the sun, but speed up as they swing in close to it, and then slow down again as they move away. If the point labeled A in Fig. 5.10 is the position of the planet on one day, and the point labeled A' is the position 24 hours later, then the shaded area (called A_1 between the two lines is the "area swept out" by the line to the planet in one day. When the planet is closer to the sun (at point B), it is moving faster, so that it will get all the way to B' in a single day. As before, the area (now called A_2) that it sweeps out in a day is shaded in. Kepler's law simply states that A_1 and A_2 are equal.

The last law is a statement about how far away from the sun the various planets lie, and how long it takes them to get around their orbits. It says that if it takes a time T for a planet to go around the sun, and R is half the distance shown in Fig. 5.8, then the mathematical ratio T^2/R^3 is the same for all of the planets. For example, for the earth, T is one year and R is about 193 million miles. Working out the ratio for other planets is left to the problems, but it is clear that this is simply a statement that the farther away from the sun a planet is, the longer it will take it to get around.

So with the completion of the work of Kepler, the overthrow of the two basic tenets of the Ptolemaic system was complete. The heliocentric hypothesis of Copernicus had been combined with the observations of Brahe to give three laws of planetary motion. These laws were first and foremost based on *observation,* whereas the Ptolemaic system had been primarily designed to accommodate a system based on reason. But even though the new ideas had been stated, much work was left to be done. There was still no answer to the objections to a spinning earth that had been raised against the work of Copernicus. There was likewise no understanding of why the planets moved as Kepler stated that they did. Kepler himself had a vague notion of something he called

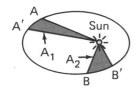

Figure 5.10.

"virtue," which kept the earth moving around the sun and prevented it from flying off into space, but this was really not a satisfactory solution.

It was necessary, therefore, that more understanding be achieved before men could say with certainty that they understood the laws that governed the universe in which they lived.

D. Galileo the Astronomer

Galileo Galilei was born in 1564. He has often been called the "first experimental scientist," but most people asked about him would probably think first of the famous heresy trial in which he was forced to recant his statements that the earth moved around the sun. From the point of view of the development of science, he played two distinctly different roles at different times in his life. First, he introduced the use of the telescope into astronomy (although he did not invent it himself) and popularized the Copernican system with his writings. Second, he began the systematic study of the motion of material bodies—the science that we now call mechanics. His major work on this topic was actually published after the trial.

Galileo started out in life in the cultured environment of the Italian minor nobility. His family was not wealthy, however, and he was sent to the university to study medicine. While there, however, he discovered the law of pendulum motion. Legend has it that he discovered this law—the statement that the time it takes a pendulum to swing back and forth does not depend on the size of the swing, but only on the length of the pendulum—from watching objects hanging from ropes in a cathedral that was under construction. In what way he made the discovery, however, it launched him on an academic career that led him to a professorship in mathematics at Pisa and, eventually, to the post of chief mathematician to the Medici court in Florence.

In 1609, a series of events occurred which diverted Galileo's attention from the mechanics of the pendulum to astronomy. After reading about a new Dutch invention called the telescope, he built one for himself and turned it toward the heavens. This was the first time anyone had scanned the skies with anything except the naked eye. He saw some things that put the standard Aristotelian view of the universe

93

into grave doubt. Among the things he saw were the moons of Jupiter, mountains on the moon, the phases of Venus, and sunspots.

Why should these things, which we regard as everyday and ordinary, have had such an impact when Galileo saw them? The answer, of course, lies in the Platonic idea of the perfection of the heavens. Mountains and other natural parts of the landscape were regarded as the products of the corrupt processes of change which, while they might exist on earth, certainly could not exist in the heavens. Thus, the appearance of mountains on the moon caused consternation among the scientists and philosophers of the time.

Similarly, the phases of Venus, indicating as they did that regular changes occurred in the planetary spheres, cast further doubt on the idea of a perfect, unchanging heaven. But the real blow to the Aristotelians came with the discovery of the moons of Jupiter. The geocentric picture of the world arose from the idea that everything moved around the earth and, indeed, this is precisely the way things appear to the observer on the earth. However, this argument began to be suspect when it was discovered that other heavenly bodies had satellites, and could therefore serve as the centers of their own little universes. You will recall that the entire Aristotelian theory of falling bodies, for example, depended on the assumption that every body tried to move toward the center of the universe, which they took to be the center of the earth. But here were some bodies—the moons of Jupiter—which seemed quite content to travel in orbits around a planet.

It is difficult for us to realize the impact that this sort of discovery had on scientists in the 17th century. There is a famous anecdote concerning a dinner party at which Galileo produced his telescope and invited the guests to see the moons of Jupiter for themselves. Two famous philosophers of the time, Cremonini and Libri of the University of Padua, actually refused to look, since observing such a phenomenon could only confuse their minds as to the truth of Aristotle. In a famous (and typical) remark on the death of Libri, Galileo said that the philosopher ". . .did not choose to see my celestial trifles while he was on earth. Perhaps he will do so now that he has gone to heaven."

This, then, was the background against which the famous heresy trial took place. Evidence for the heliocentric ideas of Copernicus had slowly built up, and it was well known among scientists that the geocentric Aristotelian picture of the universe was in deep trouble. These ideas had not, however, become common knowledge among educated people because of the highly technical nature of the arguments and because the debate was carried out largely in Latin. With Galileo, all of this changed. He was a gifted writer, and had the knack of being able to explain scientific ideas to people who did not possess a scientific training. Furthermore, he wrote in Italian, rather than Latin, so that his works became accessible to a large readership in Italy. He was a skilled debater and possessed the kind of sarcastic wit that was not satisfied with simply showing an opponent to be wrong, but insisted on demolishing him as well. While this may have won debates for him, it also made him many enemies among the more conservative clergy.

In 1616, Galileo came to some sort of understanding with the Church about his public expositions of the Copernican system. Historians still debate whether he actually agreed to stop his public writings in this field, but this agreement was the grounds for his trial in 1632. The immediate cause of the trial was the publication, in 1630, of his *Dialogue Concerning the Two World Systems.* This was a brilliant exposition of the arguments in favor of the Copernican system, cast in the form of a series of conversations between three men—Salvatio (the man who had the answers), Sagredo (the "straight man"—an earnest seeker after truth), and Simplicio (the Aristotelian). The fact that many of the Aristotelian arguments advanced by the Pope were put into the mouth of Simplicio didn't help Galileo's case with the clergy.

The rest of the story is common knowledge. Galileo, old and sick by this time, was brought to trial by the Church and forced to recant the heretical doctrine that the earth moved. He spent the rest of his life under virtual house arrest, pursuing work that we shall discuss in the next chapter.

Why was the Church so opposed to the heliocentric ideas espoused by Galileo? Many writers and historians have tried to answer that question. Some have blamed the character of Galileo, others the ignorance of the Church officials. I think, however, that the cause of the Church's hostility to the new ideas can be found in the medieval attitudes toward science discussed earlier. The Churchmen were not so much interested in the truth of falsehood of the Copernican idea, as they were in the question of whether allowing these ideas to be promulgated would interfere with the infinitely more important work of salvation. In his novelized life of Galileo called *The Star Gazer*, de Harsanyi puts these words into the mouth of Cardinal Bellarmin, the chief opponent of Galileo:

> "The whole cosmos relates to humanity—to man's struggle to save his soul. Nothing can be of more significance than the salvation of a single individual . . . Can I permit the thought to enter his mind that the earth is only a tiny satellite and the sun the real center of the cosmos? . . . This first doubt will be followed by a hundred others. The whole structure would crumble."

In other words, if the most important goal in life is salvation, the question of whether or not a particular scientific idea is true or not is not as important as the effect it would have on the work of salvation. While this attitude may seem strange to us, a little reflection should convince the reader that there are many instances of this same attitude toward scientific investigations in modern America.

Most modern theologians, of course, would reject the notion that the worth of the human soul is somehow tied to the idea that the earth is at the center of the universe. Religious thinkers are quite capable of adapting their ideas to new scientific discoveries and have done so many times since the trial of Galileo. And while the trial may not have been all that important as a scientific event, it has attained a symbolic importance in our minds as an example of how not to go about solving differences between science and religion.

E. Summary

As far as our ideas about the solar system go, the period we have just discussed was truly revolutionary. At its beginning, the Ptolemaic system reigned without question, but by the end, a picture of the earth's place in the scheme of things had emerged that was very much like our own. To a large extent, this change was a result of the work of the four men we have talked about. Each one contributed a piece of the picture; when they were finished, the view that men had of their world was permanently changed.

96

Copernicus was the first to question the Ptolemaic system. Today the only thing that remains of his elaborate scheme of a system of epicycles centered around the sun is the heliocentric idea itself. His contribution to the process was to make heliocentrism intellectually and mathematically respectable, without actually going all the way to the final solution. He was unable to free his mind of the epicycle fixation that had been accepted as scientific fact for a thousand years. In the final analysis, his single great thought—that the earth was not the center of the universe—remains his monument.

Tycho Brahe contributed the first advance in experimental techniques that had occurred since the Babylonians. For the first time, men had data more accurate than Ptolemy's to work with and to test the Ptolemaic ideas against. The use of these new techniques to prove that changes did indeed occur in the heavenly spheres further weakened the classical Greek picture of the cosmos, and made it possible for more scientists to question accepted theories.

The epicycle idea was finally overthrown by Johann Kepler who, by careful analysis of Tycho Brahe's data, was able to prove that the orbits of the planets were not circles, or even circles-within-circles, but were an entirely different geometrical form—the ellipse. He enunciated three laws of planetary motion based on the observational data, and these three laws are still believed to be true today. In fact, one of the great triumphs of Isaac Newton was to show that these three laws of planetary motion could be derived from the law of universal gravitation.

Kepler's other schemes—the harmony of the spheres and the geometrical universe—have not fared so well, and remain today as curiosities in the history of science.

Galileo introduced the telescope and made a number of discoveries with it. From the point of view of the changing picture of the universe, however, Galileo's main contribution was not in the area of new knowledge but in making the reading public aware of the tremendous changes that were taking place in astronomy. After his writings were published, the ideas that up until that time had been the domain of a few scientists became general knowledge, and the comfortable medieval world was no more.

97

SUGGESTED READING

In addition to the general texts listed in the Introduction, all of which deal with this period, the following should be interesting:

Galilei, Galileo. *Dialogue Concerning the Two Chief World Systems.* Translated by Stillman Drake. Berkeley: University of California Press, 1967. An extremely readable translation of the dialogue, with a foreword by Albert Einstein.

Hall, A.R. *The Scientific Revolution 1500–1800.* Boston: Beacon Press, 1954.

Hutchins, Robert M. (Ed.) *Great Books of the Western World.* Volume 16 of the *Great Books* series contains not only the *Almagest,* but translations of *On the Revolutions of the Heavenly Spheres* by Copernicus (including the Preface) and Kepler's *Music of the Spheres* (*The Harmonies of the World*).

Koestler, Arthur. *The Sleepwalkers.* New York: Grosset & Dunlap, 1959. A semi-novelistic approach to the period which suffers from Koestler's tendency to skip over facts that do not fit in with his preconceptions. For example, the political life of Copernicus is totally ignored in order to paint him as the "Timid Canon." The discussion of Kepler's work on the orbits of Mars is very good, though.

QUESTIONS AND DISCUSSION IDEAS

1. List as many ways as you can of reducing the errors in the experiment we discussed in Section B. Can you think of any way of eliminating them entirely? Why or why not?

2. Why did we say that the experiment discussed above ruled out the Aristotelian notion of falling bodies?

3. From what you know of Greek astronomy, make an educated guess as to why Aristotle might have argued that comets were in the atmosphere and not in the heavens. Check your guess against Aristotle's *Meteorology.*

4. Discuss an example from modern life in which belief and science are pitted against each other. Are there analogies to the trial of Galileo?

5. Why did the existence of sunspots contradict the Aristotelian picture of the universe?

6. Work through the phases of the planet Venus in analogy to the phases of the moon which we did earlier.

7. Which planets besides Venus exhibit phases?

8. In the discussion of Kepler's third law, we wrote down the value of T^2/R^3 for the earth. Find out what T and R are for some other planets, and verify Kepler's law for yourself.

9. One of the main objections to Copernicus was the idea that, if the earth were turning, we would be able to "feel" it. In fact, we can. Look up the following two topics and discuss how they bear out the Copernican hypothesis:
   a. Coriolis Force
   b. Foucault Pendulum

10. Make up a response that Tycho Brahe might have made to Plato and Eudoxus had he lived in their times.

CHAPTER
VI

THE SCIENTIFIC METHOD

"The whole Burden of Philosophy seems to consist of this: from the phenomena of motion to investigate the forces of nature, and then from these forces to demonstrate other phenomena."

Isaac Newton, Principia Mathematica

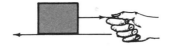

A. Galileo the Physicist

In many ways, Galileo played the same role in physics that Tycho Brahe played in astronomy. The spirit of the times was moving away from philosophical arguments and discussion as a way of arriving at scientific truth and toward the idea that the proper way to investigate scientific questions was by precise, quantitative experiments. Galileo was by nature a highly skilled technician. We know, for example, that he made a number of inventions, such as the proportional compass (an instrument used in drafting) and improved the thermometer. Consequently, he was exceptionally well qualified to follow this new direction in science.

Leaning Tower of Pisa.

As we mentioned above, his interest in mechanics began at a very early age with his studies of the pendulum. This was, of course, a very practical problem at the time, since the pendulum figured in many devices (such as clocks). This, in turn, led him into the general study of mechanics. The famous legend about dropping the weights off the Leaning Tower of Pisa belongs to this period in his life (although it is doubtful that he ever actually performed the experiment). Later, after the trial, he returned to mechanics and, using the dialogue format, collected his work on mechanics into a book entitled *Dialogue Concerning the Two New Sciences.* In this book, physics as an experimental science began to emerge.

The "hot" topic in physics in those days concerned the motion of projectiles. This work had important military applications. It was not the first time, nor would it be the last, that military needs influenced the problems that scientists worked on. The basic problem was the idea that motion could be categorized into "natural" and "violent," and that projectile motion somehow involved a switch from the latter (immediately after the cannonball left the cannon) to the former (when the cannonball began to fall). As long as the argument was couched in these terms, it was impossible to predict, for example, how far a cannonball would go for a given muzzle elevation. What Galileo did was to ignore the categories set up by Aristotle and try to discover exactly what the motion of a projectile was, independent of whether it was "natural" or "violent." This step was analogous to the rejection of epicycles by Kepler.

First, he started trying to understand the way in which a body falls under the influence of gravity. This involves a phenomenon called acceleration, which is simply a change in the velocity of the moving object. The study of this type of motion had always been confusing to scientists, because it is difficult to think about a velocity that is not always constant. For example, if we drive in a car at 20 miles per hour, we know that at the end of an hour we have gone 20 miles. Similarly, if we are traveling at a constant velocity and go 20 miles in one hour, we know that we have been traveling at 20 mph. But what if we start from rest and at the end of an hour are going 60 mph? How far will we have traveled, and how fast will we be going at any point along the way? The answers to questions such as these are not so obvious, and it was Galileo who first provided the answers.

When a body falls, we know that it moves faster and faster the farther along it has gone. To see just how much faster, Galileo set up an experiment similar to the one pictured Fig. 6.1. There was a long inclined surface, and large, precisely machined balls were rolled down it. Because the tilt of the surface could be adjusted, the balls could be made to roll very slowly, so that the time it took to travel given distances down the plane could be measured with a clock. Obviously, in the limit that the angle of the plane (denoted by A in the figure) is 90°, we would be dealing with a freely falling body—that is, an object falling under the influence of gravity alone.

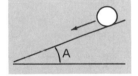

Figure 6.1.

When he performed these experiments, Galileo discovered a number of regularities. He found that if the ball rolled one foot in one second, then it would have traveled four feet at the end of two seconds, nine feet at the end of three, 16 feet at the end of four, and so forth. He also discovered a rather interesting fact about the speed at which an object is moving at the end of each second. In the same example used above, in which the object had covered one foot in the first second, he found that at the end of that second it would be moving at a speed of two feet per second. At the end of the second second, it would be moving at the speed of four feet per second, and at the end of three seconds it would be moving at the speed of six feet per second. (See Problem 12 at the end of this chapter for more on this subject).

These findings can be expressed simply in two mathematical expressions. The first one, which relates distance traveled to time elapsed, can be written

$$D = KT^2,$$

where D is the distance, T is the time, and K is a constant discussed below. In the same way, the result for velocities can be written

$$V = 2KT,$$

where V is the velocity at the end of time T.

The constant K, which appears in these equations, is not mysterious; it can be determined quite simply from measurement. For example, in the discussion above we talked about an object that traveled one foot in the first second, four feet in the first two seconds, and so on. We can determine the constant K for this motion by using this information. For example, when T = 1, we have D = 1. Substituting these into Galileo's equation, we find

$$1 \text{ ft} = K \cdot 1 \text{ (sec)}^2$$

so that

$$.K = 1 \text{ ft/sec}^2.$$

Similarly, if we used the fact that at T = 2, D = 4, we would have

$$4 \text{ ft} = K (2 \text{ sec})^2 ;$$

then we would get

$$K = 1 \text{ ft/sec}^2,$$

which is what we expect.

The constant K, then, is a number that can be determined from a single time-distance measurement. Once this number is known, the rest of the motion can be predicted (see Problem 13).

The case for K = 1 corresponds to a particular acceleration, or, in the case of the inclined plane, to one particular

104

setting of the angle A. If we changed the angle, we would get another acceleration, and hence a different value of K. If we increased the angle to 90°, as we have discussed, we would then have the case of a falling object. In this case, for historical reasons, the equation is written

$$D = KT^2 = \frac{1}{2}gT^2.$$

The number "g", which is just twice the constant K for the case where A is 90°, is called the "acceleration due to gravity," and is very important in the theory of projectiles. It has a value

$$g = 32 \text{ ft/(sec)}^2 = 980 \text{ cm/(sec)}^2$$

so that, for a falling body,

$$D = 16 T^2 \text{ ft.}$$

From Galileo's result on velocities, this means that the speed of a falling object will be increasing according to the formula

$$V = gT.$$

To convince yourself that you understand the meaning of these two equations, see if you can understand the following table, which gives the distance an object has fallen and the speed it has achieved as a function of time.

Time since release	Distance traveled	Speed Achieved
1 sec	16 ft	32 ft/sec
2 sec	64 ft	64 ft/sec
3 sec	144 ft	96 ft/sec
4 sec	256 ft	128 ft/sec

The fact that the velocity is increasing uniformly with time means that the acceleration, which is defined as the change in velocity per unit time, is a constant. To see this, let us calculate the acceleration during the T + 1st second using the above formulae

The velocity at the end of T seconds is

$$V = gT$$

while the velocity at the end of T + 1 seconds is

$$V = g(T + 1) = gT + g. \, 1.$$

Therefore, the change in velocity during the T + 1st second is

$$g \, (T + 1) - g(T) = g.$$

Since the change in velocity divided by the time is the acceleration, we have

$$a = g/1 = g.$$

In other words, the velocity changes by 32 ft/sec in each second of fall. We would say, then, that the acceleration (which is just the rate of change of velocity) was 32 ft/sec/sec or, as it is more usually written, 32 ft/(sec)^2. Because the acceleration of a freely falling body does not change, even though the velocity does, the problem that Galileo solved is usually called the problem of "uniform acceleration."

Two important points about this result have to be noticed as far as Aristotelian physics is concerned. First, neither of these laws depends on how heavy the object is. Two bodies released and allowed to fall will fall at the same rate regardless of their weight. This experimental finding is a direct contradiction of the very "reasonable" Greek idea that a heavy body should fall faster than a light one because it has more "force." The second point is that although we are dealing with freely falling objects, it was never necessary to introduce the concepts of "natural" or "violent" motion.

But what does this have to do with cannonballs? As it turned out, it had a great deal to do with cannonballs. This is because cannonballs are, after all, objects that are moving under the influence of gravity. It is clear that if we shot a cannonball straight up, it would come to a stop at some height. On its way down, then, it would have to obey the same law of falling bodies that we discussed above. The real question concerns how we describe the motion of a cannonball shot at an angle—that is, one that is moving under the influence of gravity *and* is moving in the horizontal direction as well.

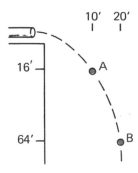

Figure 6.2.

Working in the same way as he had for falling bodies, by close attention to experimental evidence, Galileo discovered how to take account of such motions. He formulated the result in what we now call the "law of compound motion," which states that

> The motion of a body in the vertical direction is completely independent of the motion of that body in the horizontal direction.

To understand this law, let us consider the motion of a cannonball shot off the edge of a cliff (see Fig. 6.2). Suppose that the muzzle velocity of the ball were 10 feet per second, and that the cannon were originally aimed in a perfectly horizontal direction. If the ball didn't fall, we would have no difficulty describing the motion. The ball would travel 10 feet in the first second, another 10 feet in the next second, and so on. In fact, if we neglected things like friction and wind resistance, it would go on at that rate forever.

On the other hand, if the cannonball were not moving horizontally, but were just dropped, we know that it would have traveled 16 feet in the first second, 64 feet by the end of two seconds, and so forth. We can combine these two different motions according to the law stated above. The law says that the horizontal and vertical motions are independent of each other. Thus, at the end of one second, the cannonball will have moved 10 feet in the horizontal direction and 16 feet down. This is the point labeled A in the figure. At the end of the second second, the cannonball will have traveled 20 feet in the horizontal direction and 64 feet down, and will be at the point labeled B. In this way we can construct the path the cannonball will follow, as shown.

Is this motion natural or violent? The question is meaningless now. The old categories simply do not apply. In fact, Galileo showed that the path that the cannonball would follow is a well-known geometrical shape called a parabola. From the answer to the problem of the cannonball shot from a cliff given above, we can construct the solution to the problem of a cannonball shot upward at an angle. In the vertical direction the cannonball will simply go up and fall down, while in the horizontal direction it will move

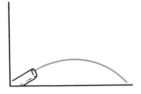

Figure 6.3.

with a constant velocity, resulting in a trajectory similar to the one in Fig. 6.3.

One consequence of the law of compound motion (which many people find troubling) is that the cannonball shot off the cliff will strike the ground at precisely the same time as a cannonball that is dropped! This is true regardless of how fast the cannonball is shot out horizontally, since the time to fall depends only on the vertical motion and this is the same for any cannonball. Many people will not accept this result even when it is demonstrated experimentally, but it can be made plausible by remembering that even though the cannonball that is shot out horizontally must travel farther than the one that is dropped, it is traveling faster as well. Thus, it should not be impossible that they might hit the ground at the same time.

From the point of view of physics, then, the legacy of Galileo lies in his introduction of a new way of attacking problems. His method was to carry out experiments and to summarize the results of these experiments in mathematical form, rather than to set up philosophical categories and argue about them. As we shall see, this technique is one important component of the scientific method.

B. The World as It Appeared before Newton

The advance of science had been steady since the time of Copernicus. No longer did serious scholars consider that the Ptolemaic view of the world, in which the planets moved in epicycles around the stationary earth, could be made to explain the observed facts of planetary motion. The hard, detailed, scientific work of Kepler, who analyzed the data taken by Tycho Brahe, had made it impossible to explain the new observations of the planetary positions by the old ideas. The invention of the telescope and its use by Galileo in astronomy also made it much less likely that the earth could occupy a special place in the solar system. The fact that other planets (such as Jupiter) had moons, and were therefore themselves the centers of motion of other heavenly objects, made it very difficult to argue on logical grounds that the earth had to be special and unique in the universe.

At the same time, the ancient idea that the motions of the planets and the motions of the stars must be circular had come under special attack from the work of Kepler. Again, detailed observation and hard mathematical work had shown that the orbits of the planets, although *approximately* circular, were in fact elliptical in shape. This meant that the idea that we could know about the world exclusively through the use of reason, which had been so strong in the minds of the Greeks, had to be discarded. After all, there is no "reasonable" way of showing that the orbit of a planet must be elliptical. The ellipse is not a very special figure in the mind of man.

But Kepler had gone further than just showing what the orbits of the planets were. He had summarized the knowledge that he had gained from his analysis of the data of Tycho Brahe in the famous three laws of planetary motion. These laws summarized, in succinct mathematical form, all of the known facts about the motion of planets. Although Kepler's motivation in deriving these laws may have been more mystical than scientific, the laws themselves have stood the test of time.

There is one important point about Kepler's laws that we want to keep in mind. They are not deductions from more general principles. They are generalizations from collections of data. To understand what is meant by that distinction, ask yourself how you would go about proving the first law, the one stating that the motion of a planet is elliptical with the sun at one focus of the ellipse. Clearly, the way that Kepler proved this law, and the way that we would prove the law if we had to do it at this stage of our understanding of the world, would be to look at each planet in turn, find out what its orbit was, and then compare them. We would find, as Kepler did, that each planet moved in an ellipse, and that each ellipse had one focus at the position of the sun. This is similar in principle to the discovery by the Egyptian scribes that it was possible to write down an approximate formula for the area of a circle. It is very different in principle from the derivation of the area of a circle from the postulates of Euclid. In the first case, we have the discovery of a truth by trial and error, with no general principles involved. In the second case, we have the deduction of a truth from general postulates. One of the great contributions of Isaac Newton was to move Kepler's laws from generalizations of data to deductions from more general laws. Kepler's contribution to astronomy and to the science of the solar system was to tell us, in succinct mathematical terms, the facts about the motion of the planets.

While this work was going on in astronomy, another piece of important work was being done in a seemingly unrelated area—the area of physics. To the Greeks, there had been no connection between the science of astronomy, which dealt with the motion of heavenly bodies, and the science of physics, which dealt with the motion of objects on the earth. The idea that there might be some connection between the path that a rock takes when you throw it and the path that the moon follows in going around the earth was patently ridiculous to the mind of the Greek. Neither Galileo nor Kepler, the two giants of pre-Newtonian physics, ever questioned this basic assumption seriously. Even though Galileo was both a physicist and an astronomer, in the modern sense of the words, he never devoted a great deal of thought to the proposition that the two might be connected in some fundamental way. To him and his colleagues, the ancient Aristotelian division between events on earth and events in the heavens remained as important as it had been to the Greeks.

But, just as Kepler had discarded the approach of "pure reason" in his astronomical work, Galileo discarded the approach of reason in his physics. When he studied the laws of falling bodies, the law of the pendulum, and other physical phenomena, he was less concerned with what the results should be than with what the results actually were. As we have seen, he did experiments to determine whether the Aristotelian idea that the speed of fall of a projectile was related to its weight was valid. He sorted out the rather complicated problem of compound motion and proved, for the first time, that a projectile moved in a particular geometrical path—a parabola. As with the work of Kepler, these laws were purely experimental laws. They were inferences or generalizations of experimental findings. They were not in any sense of the word deductions from some more general principle. And, as with the laws of Kepler, they were a necessary piece in the development of the ideas of science. After all, it is very difficult to think of the motion of material objects if you persist in using the Aristotelian categories of natural and violent motion.

So by the middle of the 17th century we had a situation in both physics and astronomy in which a number of general laws were known. These laws were derived from observation—from analysis of data—and were statements

about the nature of natural phenomena. They were statements about "what is." Thus, the broad new outlines of a
new way of looking at the world were becoming clear. The
first steps in what we shall come to know as the scientific
method had been taken by these early pioneers. The fact
that they did not finish the work that they began does not
reflect on their abilities in any way, but is simply an indication of how difficult it is for the human mind to conceive
of new ways of thinking.

C. Isaac Newton: The Man Who Put It All Together

Galileo died in 1642. Isaac Newton was born in 1642.
This coincidence is symbolic of the continuity in thought
between the two men. There was nothing in Newton's early
life to indicate the achievements that were to follow. He
was born to moderately well-to-do parents in England, and
read natural philosophy (what we would call today science
and mathematics) at Cambridge University. In the year
1665, however, a very important series of events took place.
This was the year in which the Black Plague was sweeping
through England. There was no known cure for the Plague
at that time, and the only way to avoid contagion was head-
long flight from the centers of population. Like many other
people of his class, Newton left the large city (in his case,
Cambridge) and returned to his family manor at Woolsthorpe. He had already completed his studies at Cambridge
at this time, and had a position there comparable to a junior
faculty position at an American university today. During
the years 1665–1666, Newton lived alone at Woolsthorpe.
He had very little contact with the outside world or with
the other scientists in his country. It is hard to conceive,
however, of a two-year period in which more progress was
made in science. That it was made by one man working
alone with virtually no help from others makes it all the
more remarkable.

Let us listen to the words of Newton himself as he
describes this period in his life:

I found the Method of fluxions by degrees in
the years 1665 and 1666. In the beginning of the
year 1665 I found the method of approximating

Series and the Rule for reducing any degree of any Binomial into such a series. The same year in November I had the direct method of fluxions, and the next year in January had the Theory of Colours, and in May following I had entrance into ye inverse method of fluxions. And the same year I began to think of gravity extending to ye Orb of the Moon, and having found out how to estimate the force with which a globe revolving within a sphere presses the surface of the sphere, from Kepler's Rule of the periodical times of the Planets being in a sesquilaterate proportion of their distances from the centers of their Orbs I deducted that the forces which keep the planets in their Orbs must be reciprocally as the squares of their distances from the centers about which they revolve; and thereby compared the force requisite to keep the Moon in her Orb with the force of gravity at the surface of the earth, and found them to answer pretty nearly. All this was in the two plague years 1665 and 1666, for in those days I was in the prime of my age for invention, and minded Mathematics and Philosophy more than at any time since.

What do all of these new discoveries mean? The "method of fluxions" and "inverse method of fluxions" are known today as the differential and integral calculus. This was a major step forward in mathematics. The calculus is the basic tool used by modern mathematicians to describe physical phenomena. It is as important to modern physics as the science of geometry would be to the Egyptian carrying out land surveying in his time.

As an interesting aside, at the same time that Newton was developing the calculus in England, a philosopher-scientist named Gottfried Wilhelm Leibniz was making the same discovery independently in Germany. This simultaneous discovery of an important mathematical technique caused a great deal of difficulty throughout the next few centuries. The reason, of course, was that continental scientists, and particularly Germans, were very reluctant to credit Newton with having developed the calculus. At the same time, English scientists were very reluctant to credit Leibniz.

The result was a long-standing feud which probably did a great deal to retard the advance of science during the next few centuries. As a matter of fact, it turns out that the techniques developed by Leibniz were somewhat easier to use, and that what we know as calculus today does not use the notations and techniques of Newton but those of Leibniz.

During this same period, according to Newton's account, he also discovered the universal law of gravitation. This law is nothing less than the long-sought-after connection between physics and astronomy. With the law of universal gravitation, it becomes obvious that the force which causes a piece of chalk to drop when we let go of it is the same force as that which holds the moon in its orbit around the earth. After 1,500 years of separation these two sciences were united for the first time by the work of Isaac Newton.

The discovery of the theory of colors (the idea that what we normally consider white light is actually a combination of all the colors in the spectrum from red to blue) was another of his achievements during these years, as well as a couple of mathematical theorems, which he mentions in passing. Any one of these achievements taken singly would be considered ample reward for a lifetime of work by most scientists. The fact that Newton accomplished them all within a short space of time, working alone, is evidence enough of the genius that this man possessed.

But this was not the only contribution that Isaac Newton was to make to science. In fact, his greatest work still lay ahead of him in 1666. He returned to Cambridge, and, as his work became more widely known, achieved the fame that he so richly deserved. He was, however, one of those men who, once they start on a problem, seem to be unable to leave it. For the next 20 years he continued his work in physics, and especially his work in the science of mechanics. Finally, in 1687, he summarized and codified his work in a book entitled *Principia Mathematica.*

This work bears a striking, logical resemblance to Euclid's *Elements of Geometry.* The science of mechanics, which concerns itself with the motion of material objects, is presented in such a way that the solution to any physical problem is seen to follow as a logical deduction from a few basic principles. These principles are (1) the three laws of motion and (2) the universal law of gravitation. The three laws of motion, to use Newton's exact words, are stated as follows:

I. Every body continues in its state of rest, or of uniform motion in a right line, unless it is compelled to change that state by forces impressed upon it.

II. The change of motion is proportional to the motive force impressed; and is made in the direction of the right line in which that force is impressed.

III. To every action there is always opposed an equal reaction: or, the mutual actions of two bodies upon each other are always equal, and directed to contrary parts.

The law of universal gravitation can be stated in modern terminology as follows: if two bodies of mass m_1 and m_2 are located at distance d apart, then there will be a force between these two bodies which is given by the equation

$$F = G \frac{m_1 m_2}{d^2},$$

where G is a universal constant.

What does all of this mean? Let us begin by looking at the laws of motion one at a time. For all of its innocuous wording, the first law contains a clean break with 2,000 years of traditional Greek thought. For what Newton has done in the first law is to redefine what is meant by "natural" motion. To the Greek mind, and to all the scientists before Newton, there was something "natural" about motion in a circle. It was felt that if a body were left to itself, it would move in a circle. What the first law of motion says, however, is that if a body is left to itself, it will move in a straight line!

What Newton has done with the first law is to get rid of the obsession with circular motion that had been such a hindrance to physics and astronomy for so long. He noticed that objects on the earth, when left to themselves, most emphatically did *not* move in circles. This fact had, of course, been known for millennia. Everybody knew that if you tied a rock to a string and twirled it around your head and then let go, it would certainly not continue to move in a circle.

The rock would move in a straight line. What had been missing was the realization that this simple fact, familiar to everyone through everyday experience, was a key to the riddle of the motion of the planets.

The second law was also based on everyday experience. We all know that if you push something that isn't tied down, it will move. If you keep pushing, it will begin moving faster and faster. In order words, if we exert a force on an object, it will be accelerated. We also know that if an object is heavy, it requires a much larger force to make it move as fast as an object that is light. These two facts, which are obvious to most of us from experience, are combined mathematically in the statement of the second law.

The only question about the statement concerns what exactly is meant by the term "change of motion." There was a considerable debate about how to characterize motion. Newton used the intuitive arguments given above to define a "quantity of motion" called momentum which is defined as mass times velocity, or

$$p = mv.$$

When a force acts on a body, its effect is to change this "quantity of motion," and this change in momentum is what is meant by "change of motion." For most of our considerations, we can consider the mass of an object to be fixed and unchanging, so that a change in momentum would correspond to a change in velocity (or acceleration). For this case, Newton's second law can be written

$$F = ma,$$

where F stands for the force involved, m is the mass of the object, and a is the acceleration, or rate of change of velocity, caused by the force.

There is another, philosophically interesting way to look at the first two laws of motion. The first law can be interpreted as a statement that tells us *when* a force is acting. The second law then becomes a statement that relates the speed which the body reaches under the action of a particular force to the mass of the body and to the size of the force which is acting. In other words, the second law can be taken as a definition of what we mean by a force.

115

The more astute reader will have realized by this time that there is a certain circularity in what we have discussed so far. That is, we have defined force in terms of acceleration and mass. Acceleration is something that is easy to measure. For example, if the body starts at rest and after a certain time, T, has achieved the velocity V under the influence of the force, then the acceleration is simply given by the equation

$$a = \frac{V}{T}.$$

This means that we can measure acceleration simply with a ruler and a stopwatch. But, how do we measure mass? Have we done anything more than define one unknown concept (force) in terms of another unknown concept (mass)?

We all have a simple intuitive idea of what mass is—it has something to do with the resistance of a body to motion. A very massive body is one that requires a large force to get it moving while a less massive body will get to the same speed with a smaller force exerted. But, like most intuitively obvious concepts, we run into dilemmas such as those discussed above when we try to tie them down more tightly.

The answer to this puzzle can be obtained only by looking at the third law of motion. This law, as we have stated it, does not seem very profound. It simply states that when a body, A, exerts a force on a body, B, then the body B will exert an equal and opposite force on the body A. For example, when you stand on a floor, your weight is exerting a certain force on the floor. At the same time, the floor is exerting an equal and opposite force on you. In fact, you can feel this force in the soles of your feet. If you think about the second law for a moment, you will realize that such a situation has to exist when you are standing on a floor. After all, you do not fall through the floor. This means that you have no acceleration in the vertical direction. From the second law, this must mean that there is no net force acting on you, because if there were a net force acting on you, you would move. Since we know that gravity is always exerting a force downward—a force equal to your weight—the fact that you do not move means that there must be another object exerting an equal and opposite force. This "other object,"

14th Century Cologne.

IMPRESSIO LIBRORVM.

Poteſt vt vna vox capi aure plurima: Linunt ita vna ſcripta mille p. inas.

Medieval printing press.

A contemporary drawing of Tycho Brahe (1546-1601) in his laboratory. Note the size of the instruments.

of course, is simply the floor. It exerts a force that is equal in magnitude to your weight, but is directed upward (that is, in a direction opposite to the force of gravity). Thus, the two forces that are acting on you cancel each other exactly, and there is no net acceleration. You don't move.

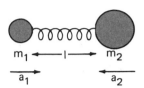

Figure 6.4.

But what does this have to do with the question of defining mass? The solution to this problem can be seen by considering an arrangement such as that shown in Fig. 6.4. Two objects of different size, which we have labeled m_1 and m_2, are connected by a spring. The spring is stretched to some length, l. When we let go of the masses, we know that they will both be accelerated toward each other, since the spring will contract and pull the masses with it. However, the spring itself will not accelerate. From the second law, this means that no net force is being exerted on the spring, just as no net force is being exerted on you when you stand on a floor. This means that the force exerted on the spring by mass m_1 must be equal and opposite to the force exerted on the spring by mass m_2. From the third law, it then follows that the force exerted *by* the spring on mass m_1 is equal and opposite to the force exerted *by* the spring on mass m_2.

We are now in a position to define mass. In the above example, we have a situation where two objects of different size find themselves acted upon by equal and opposite forces. Under the influence of these forces, the objects are accelerated. With a ruler and stopwatch, we could measure the acceleration, as discussed above. Let us call the acceleration of body m_1, "a_1," and the acceleration of body m_2, "a_2." Then, from the second law, we must have

$$m_1 a_1 = m_2 a_2.$$

This equality follows from the fact that both bodies have the same force exerted on them.

We see, then, that we can define the mass of object m_2 in terms of its acceleration, the acceleration of body m_1, and the mass m_1. We could, in principle, take any other object, m_3, and go through the same procedure, defining its mass in terms of some measurable numbers and the still undefined mass, m_1. This means that using the three laws of motion, we can define any mass anywhere in terms of one single standard mass, which we have called m_1. Suppose, now, that

117

we simply define mass m_1 to be "one unit of mass." An example of such a definition would be to call mass m_1 "one kilogram" or "one pound mass." Once this had been done, the apparent circularity in the definition of mass can be eliminated from Newton's laws.

As an interesting aside, the reader should be aware of the fact that a block of metal called the standard kilogram is kept in the vaults of the National Bureau of Standards in Washington. Of course, the guardians of these standard weights would not allow them to be used in an experiment of the type we are describing here, although they could in principle be so used.

The "definition" of mass, seen in this way, is simply a way of relating the masses of objects to the mass of a single object we have chosen as a standard. Thus, a two-kilogram mass is simply that mass that would be accelerated half as fast as the standard kilogram by a given force. We have not defined "mass" in any absolute sense. This technique of setting up a standard for comparison and then ranking everything else in terms of that standard is quite common in the sciences. We shall encounter it again when we study the definition of electrical charge.

Before leaving this subject, we should distinguish between mass and weight, two concepts that are often confused. As we are using the term, mass refers to the resistance of a body to being accelerated. Weight, on the other hand, is simply the force exerted on an object by gravity.

Weight and mass are related to each other, of course. If an object of mass m is located somewhere where gravity can act on it, it will be accelerated (see second law). At the surface of the earth, we know from the work of Galileo that this acceleration is $g = 32$ ft/sec^2. For a spaceship in orbit, it would be less than this, and for a spaceship far from any large masses, it might be zero. Nevertheless, whatever the acceleration due to gravity is, the weight and mass will be related by the equation

$$W = m a,$$

where a is that acceleration. In particular, on the surface of the earth, this equation will read

$$W = m g,$$

where g is defined above. If the object were in space or on a different planet, where the value of g is different from what it is on the surface of the earth, its weight would change, but its mass would not. It would always be equally difficult to set the body in motion, no matter what its weight. Thus, if Mount Everest and a ping-pong ball were in space, they would both have zero weight, but Mount Everest would still have a large mass, and thus would be difficult to move, while the ping-pong ball would still have a small mass and would be easy to move.

The final important ingredient in Newtonian mechanics is the law of universal gravitation. Let us now consider exactly what this law means. From the words of Newton that we saw earlier, we know that he arrived at the law of gravitation by considering the orbits of the planets. However, it is a little bit easier to understand the law if we just consider two objects in isolation. Let us consider the situation shown in Fig. 6.5. If we have two bodies, which we shall call m_1 and m_2, a distance d apart, then the law of gravitation tells us that these two bodies will be attracted toward each other. It tells us that this attraction will be such that if the bodies were free to move, they would move toward each other. It also tells us that the size of the force that is exerted on each body by the other is given by

$$F = G \frac{m_1 m_2}{d^2},$$

where G is a number known as the constant of gravitation.

To understand what this formula means, let us consider a few examples. If we doubled mass m_1 in Fig. 6.5, we would double the force between the two bodies. A similar statement holds true for mass m_2. If we doubled both masses at the same time, however, the force would be increased fourfold. All of these statements follow from the fact that the force depends on the product of the two masses involved.

The dependence on the distance separating the two objects is a rather common one in physics, and is called an "inverse square law." The fact that the force depends on $1/d^2$ means that if the distance between the two objects is doubled, the force is cut by four. If the distance is cut in half,

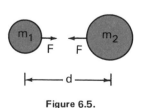

Figure 6.5.

119

the force is increased fourfold. Obviously, the gravitational force gets larger as the two objects near each other.

The constant G plays much the same role in the equation defining the gravitational force as the constant g did in Section A of this chapter. It is a number that cannot be deduced, but must be measured. In units where distances are measured in meters, masses in kilograms, and force in Newtons (see Appendix A for these units), the value of G is

$$G = 6.67 \times 10^{-11} \text{ N m}^2/\text{Kg}^2.$$

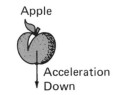

Apple

Acceleration Down

Acceleration "Up"

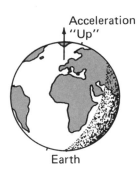

Earth

There are some interesting consequences of the law of gravitation. For example, from the second law we know that the fact that this force exists means that if we allow two objects to be near each other, they will be accelerated toward each other. If object m_1 were the earth, and object m_2 were the famous apple, then there would be a force exerted on the apple by the earth, causing it to fall. Not as obvious, but equally true, is the statement that there will be a force exerted by the apple on the earth, causing the earth to move toward the apple. Of course, the mass of the apple is relatively small, so that a relatively large acceleration will be attained by the time the apple reaches the head of Isaac Newton. The earth, on the other hand, has a very large mass, and its acceleration is not even noticeable.

If the two objects in question are not of greatly unequal size, however, the situation is rather different. For example, consider the case where mass m_1 is the earth, and mass m_2 is the moon. In this case, if the earth and the moon were to be stationary, they would fall toward each other at a rather high acceleration. Only the fact that the earth and the moon are revolving around each other prevents this from occurring. The discussion of orbits and circular motion in general is contained in the next chapter.

We shall conclude this discussion of Newton's mechanics by considering the constant that we have labeled G. By now, the method by which a constant such as this can be measured should be obvious to the reader. The general technique is to take two masses of known size, place them a known distance apart, and then measure the acceleration on each mass as they move toward each other. From the acceleration and the mass, we can then use the second law to determine the force between the two objects. From this force, the size of masses m_1 and m_2, and a measurement of the distance d, we can then determine the gravitational constant from Newton's Law.

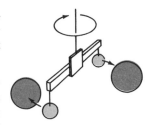

Figure 6.6.

The actual method by which this constant was measured, however, was not nearly so straightforward. It was done in 1798 by Lord Cavendish in England. He used an apparatus similar to that pictured in Fig. 6.6. Two small masses, m, were suspended on a dumbbell arrangement held up by a cord. Two large masses were brought near the two small masses, as shown in the figure. The large masses were held down, so that they could not move. Each of the large masses then exerted a force on the small mass, causing each small mass to move toward the nearest large mass. This resulted in a twisting motion of the dumbbell, as each of the two small masses in the above figure would tend to move in a clockwise direction. A twisting force of this type is called a "torque."

If you have ever tried to twist a stick that is attached to a rope, you will realize that as the rope is twisted, eventually it begins to resist being twisted further. In the language of Newton's mechanics, we would say that when a torque is exerted on a rope or cord to cause it to twist in one direction, eventually the rope or cord will exert a force of its own, tending to resist the applied torque, and to prevent itself from being twisted any further. Thus, the amount of twist in the dumbbell apparatus would depend on two things—the gravitational attraction between the small and large objects and the amount of resistance in the cord. In this way, we can determine the gravitational force between two objects if we know the characteristics of the cord. This, in turn, determines G.

What we have discussed in this section, then, is the essence of Newtonian mechanics. Whereas before Newton, we had only a group of uncorrelated experimental facts, we now have three general principles and one general law of nature. From these general principles, the entire spectrum of classical Newtonian mechanics was derived by Newton and his followers. Just how far-reaching and important these advances were in the development of science will become obvious as we begin to trace what happened after Newton had published the *Principia Mathematica.*

D. The Scientific Method

Probably the most important consequence of the development of Newtonian mechanics was the realization that after centuries of trial and error, mankind had at last hit upon a way of dealing with questions that had to do with the external world. This method of asking and answering questions about the world is what we call today "the scientific method." We can examine the development of mechanics as a paradigm case of the scientific method.

The first step in the development of mechanics was not taken by Newton, but by the men who preceded him. It consisted of the slow, careful gathering of observational data about the world. For example, it would have been impossible for Newton to enunciate the three laws of motion had he not known of the work of Galileo, and had he not been aware that the old Aristotelian ideas of compound motion and motion in circles were incorrect. The first stage of this gathering of data is generally a very haphazard and accidental one. Experiments are done, observations are made, and results are recorded without any apparent overall connection between the three. The work of Tycho Brahe in astronomy would be an example of such a process. Brahe observed the positions of the planets and stars with heretofore unequaled accuracy. However, when his work was completed, it consisted simply of a long series of statements about observations of the solar system. There was no overall cohesion—no unifying idea. At the same time, the work in mechanics of the men who preceded Galileo took this same form. As we have seen, a great deal of empirical work had been done on the motion of projectiles by artillery men in the time immediately preceding the work of Galileo. This work, too, was in the form of a long series of uncorrelated facts that were known about nature.

The second phase of the data-gathering stage in the scientific method is the condensation and generalization of the known facts about the world into precise mathematical statements. For example, the three laws of Kepler summarized, in precise mathematical form, the observations of Brahe. In the same way, the work of Galileo on falling bodies and on compound motion summarized in a mathematical way the experiments and observations that had been made on material objects over a long period of time. However, it is important to realize that even though Kepler's laws and the laws of compound motion are stated in mathematical form, they are *not* mathematical deductions. They are simply statements using the language of mathematics. To understand this distinction, you can ask the question "From what general principle does the law of compound motion follow?" Clearly, it does not follow from any general principle at all. It is simply a restatement of the experimental facts that were observed by Brahe and his predecessors.

The statement by Newton of the universal law of gravitation is another example of a general statement about the universe that is not a mathematical deduction but it simply a summary of experimental facts. Although Newton himself did not perform experiments to verify the law of gravitation, he used the observed laws of planetary motion to deduce the law of gravitation. He did not derive the law of gravitation from other mathematical principles.

At the completion of the first stage of the scientific method, then, we have statements in mathematical form that summarize the laws of nature as they have been observed. As an exercise, the reader should convince himself that the three laws of motion that were stated by Isaac Newton, and that were discussed in the previous section, are further examples of this sort of development.

In terms of the categories that we have been discussing throughout this course, this first stage of the scientific method can be thought of as the contribution to science of the technical tradition. That is, it is a contribution that is based on experiment and observation, and not based on the use of pure reason. Thales of Miletus would probably have felt very much at home with Galileo. It is doubtful that Plato or Eudoxus would have.

Once the first stage of the scientific method has been completed and the general laws of nature are both known and stated in mathematical form, work can begin on the second stage. This second stage consists of taking the general laws that have been derived from experiment and using the known laws of mathematics to deduce consequences. For example, once the three laws of motion were known, it was possible to find answers to the old objections that had been raised to the Copernican system concerning the question of why objects did not fly off of the earth. Other examples of such deductions will probably spring to mind, and a number will be discussed in the next section. The point as far as we are concerned here, however, is that these conclusions that are drawn are bonafide mathematical deductions. They follow from the laws of motion just as surely as a geometrical theorem follows from the axioms of Euclid.

This last statement should make it obvious to the reader that, whereas the first stage of the scientific method can be thought of as the ultimate legacy of the technical tradition, the second stage can be thought of as the legacy of the philosophical tradition. Once the laws of nature have been derived from observation, the consequences can be derived from the pure use of mathematical reasoning. In other words, once we have discovered the laws of nature by *experience,* we deduce all of the consequences by *reason.* Thus, what we now call the scientific method is a blend of both the philosophical and technical traditions. Each has its place, and each is important. Neither alone would lead to science as we know it. The old Greek ideal of understanding all of nature through pure reason is replaced by a system in which we use our powers of observation to find the basic laws of nature, and then use our powers of reason to draw all possible conclusions and deductions from these laws.

In Isaac Newton we have the final coming together of these two important strains of human endeavor. A new and important technique had been developed to help man in his dealings with his environment. So, for better or for worse, mankind had taken an enormous step forward. The world would never be the same again.

E. Consequences of the Newtonian Revolution

1. *Astronomy*

Once the laws of motion had been stated, the law of universal gravitation discovered, and the remaining dead wood of the Ptolemaic system thrown aside, real progress could be made in astronomy. You will recall the distinction that was made between physics and astronomy up until the time of Newton. If you look at the three laws of motion, however, you will see that nowhere do they say that these laws to not apply to stars or planets or any other object. In fact, the laws of motion are supposed to apply equally to everything. This means that the apple falling or the moon in its orbit are both special cases of one general law, the law of universal gravitation.

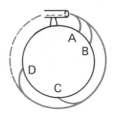

Figure 6.7.

One rather amusing development which illustrates this point concerns the deduction by Newton that planets must move in elliptical orbits. The Oxford astronomer, Edmond Halley (the discoverer of Halley's comet), was visiting Newton in Cambridge. He began to question Newton about the consequences of the law of universal gravitation for the solar system. In the words of Newton's cousin, John Conduitt, here's what happened after he asked Newton what the shape of a planetory orbit had to be:

> Newton immediately answered, an ellipse. Struck with joy and amazement, Halley asked him how he knew it? Why, replied he, I have calculated it; and being asked for the calculation, he could not find it, but promised to send it to him [Halley].

In other words, the essential scientific question of a millennium—the question of how the planets move—was such a trivial consequence of Newton's laws that he had derived the result and simply laid it aside and forgotten it!

The way in which Newton arrived at this unification of the laws of physics was actually rather interesting. He reasoned as follows: Suppose we had a situation such as that shown in Fig. 6.7 with a cannon located on top of a high mountain. Now if we shot off a cannonball, we all know what would happen. The cannonball would move out a certain distance from the mountain and then would fall, perhaps at the point labeled A in the diagram. Now suppose that we shot the cannonball with a much greater muzzle velocity. In this case, it would travel farther before it returned to earth, perhaps reaching the point labeled B. In both cases, the tendency of the particle to move in a straight line would be modified by the existence of the force of gravity. The force of gravity, in other words, would pull the particle down until it finally struck the earth.

Now we can imagine taking this example to an extreme. If we made the muzzle velocity much higher, perhaps we could get a cannonball halfway around the world—finally hitting the earth at point C. In principle, there is no reason why we could not get it to go more than halfway around the world—perhaps striking at point D, as shown in the figure.

Revolutionary war cannon.

Finally, it would seem reasonable that if we put in just enough powder to get just the right muzzle velocity, a cannonball would go all the way around the earth and would come back to the position of the cannon itself. Now ask yourself what would happen to the cannonball after it had made one revolution. Well, if there was nothing to slow it down when it returned, it would have the same velocity as it had when it left the cannon in the first place. In such a situation the cannonball would continue to go around and around the earth. This, in essence, was Newton's explanation of the motion of a satellite like the moon around the earth, or a planet around the sun. The essential point in this explanation is that it is the gravitational attraction of the earth (or in the case of planetary motion, the sun) that holds the satellite in. But there is no new physics involved in this explanation. The gravity that holds in the moon is precisely the same gravity that causes an object to fall on the surface of the earth.

Although the actual details of the proof that planets, under the influence of a gravitational force of the sun, will move in elliptical orbits with the sun at one focus involve the use of mathematics which is beyond the scope of this text, it is as much a mathematical consequence of the laws of motion as the theorem of Pythagoras concerning the sides of a right triangle is of the axioms of Euclid. In much the same way, every undergraduate physics major in the 20th century has been asked, at some time in his career, to prove that Kepler's other laws of planetary motion can also be derived as mathematical consequences of Newton's three laws of motion. Thus, once the groundwork of the laws of nature is understood, experimental observations that had previously been unexplainable become simple consequences of the general principles. In exactly the same way, the laws of motion of falling bodies and projectiles that were derived by Galileo are also mathematical consequences of Newton's laws.

So it isn't just the case that Newton brought together the philosophical and technical tradition when he developed the scientific method. He also brought together the sister sciences of physics and astronomy after 1500 years of separation. This is a very beautiful thing, because it tells us that the laws of nature as we know them on the earth are the same laws that hold everywhere else in the physical universe.

This means that if physicists on the surface of Mars or near Alpha Centauri study the same phenomenon, they will come to the same conclusions about the world in which they live as we do about our world. The artificial distinction between the heavens, which were perfect, and the earth, which was not, could no longer be maintained in the face of this tremendous intellectual breakthrough. The poet Samuel Rogers was so taken with this idea that he penned the following lines, which state the excitement of this idea much better than I can:

> That very law which molds a tear
> and bids it trickle from its source
> That law preserves the earth a sphere
> and guides the planets in their course.

2. A new precision in thinking about the world

One very beneficial consequence of Newtonian physics was that it was no longer necessary to speak in vague terms about what was happening in nature. For example, one could discuss the attraction between the earth and the sun in terms of the force of gravity, which could be measured and quantified in the law of universal gravitation, rather than in terms of a vague concept like "virtue." It also meant that when new phenomena were encountered in nature, it was not necessary to return to vague terms. For example, it had long been known that when a compass was laid down on the earth it would turn and point toward the north. Instead of talking about a "magnetic virtue" or some other obscure term, however, a Newtonian physicist could look at this entirely new phenomenon and could reason in Newtonian terms. For example, he could say "since the compass moves, there must be a force acting on it. This new force must have properties which I can measure. There is no new idea here, simply a new kind of force."

Similarly, people had known for a long time that if an object was very large and heavy it was difficult to get it moving, but once it was moving, it was difficult to get it to stop. Many scientists had talked vaguely of a "quantity of motion" which was somehow conserved. By this they meant no more than what we have already said—i.e., that once something is

127

moving, it tends to remain moving, and once something has stopped, it tends to resist being put into motion.

But what was this "quantity of motion?" The French philosopher-scientist René Descartes argued that the "quantity of motion" was simply the velocity of the object. This notion is clearly false, as the reader can verify. Imagine a large bowling ball moving toward a ping-pong ball which is at rest. After the two collide, we would expect the ping-pong ball to be moving quickly, and the bowling ball to be relatively unaffected. Thus, the total velocities after the collisions will be different from those before. It is an interesting comment on Descartes that he did not consider this a valid objection to his idea. This attitude is a throwback to an earlier era in the development of science. In Newtonian physics, the "quantity of motion" is momentum.

If we state Newton's second law in the form

Force = change of momentum per unit time,

as we discussed earlier, then some interesting consequences follow. For example, if no forces act on a system, the momentum cannot change. Thus, for an isolated system, velocities cannot change unless the masses change in just such a way as to keep the momentum fixed. This sort of reasoning (which will be discussed in greater detail in the next chapter) leads to what are called "conservation laws." In this example, the momentum stays the same ("is conserved") if no forces act.

The concept of momentum is something that has a great deal of physical meaning to all of us, even though we may not be aware of it all the time. For example, we have the feeling that something like a rifle bullet, which is very light but which moves very rapidly, is capable of exerting a great force. In other words, our everyday experience tells us that it's more than just velocity or just mass that enters into the ability to exert forces. In fact, it must be some combination of these two. Just which combination is specified by Newton's laws of motion.

In this same way, other concepts for which we have an instinctive feeling can be given very precise mathematical definitions in the Newtonian picture. For example, if we lift a heavy weight, we feel that we have done work. In Newtonian physics, work is defined precisely as the force that is exerted times the distance over which the force moves or, symbolically,

$$W = F \times d.$$

In lifting a weight, it is necessary to counteract the force of gravity, and exert just a little bit more force in order to get the object to move up. This means that in lifting a weight I have exerted a force roughly equal to the weight of the object over the distance equal to the height to which I have lifted it. The units or work, then, will be in force times distance. For example, in the English system of units, a unit of work is the foot-pound, and is the work required to lift one pound one foot.

We come now to a term that is very much in vogue these days, the term "energy." In Newtonian physics, energy is simply the ability to do work. For example, if I lifted a thousand kilograms one meter off the ground, and attached it to a pulley, I could let the weight fall and use the pulley to drive an electric generator.

Energy of the type discussed in this example is given a special name. If a weight is lifted off the ground, it has the potential to do work. However, until I actually set up some kind of apparatus to get the work from it, that remains a potential only. Such energy, which is stored and not used, is called "potential energy." It is possible to store energy by lifting weights to a height, as in our above example. This is sometimes called "gravitational potential energy." Another example of gravitational potential energy would be water that is backed up behind a dam, and then allowed to fall and run electric generators.

Another kind of energy that can be stored is chemical energy, which we shall discuss later. Gasoline and other petroleum fuels are examples of stored chemical energy. Of course, in the case of gasoline, the energy is released, not by allowing it to fall, but by allowing it to burn.

Once the weight in the original example is allowed to fall, it will return to ground level. Obviously, it has no more potential energy left. Yet it is clear that it has energy. For example, if we allowed the object to fall into a tub of water, it would splash the water around and heat it, and then we could use the heated water to run an engine of some kind. This means that when an object falls, it does not lose its ᴜ energy. It simply converts the energy from one form to another. We speak of the energy of motion as "kinetic energy." In the problems at the end of this chapter, you will have an opportunity to discuss different kinds of potential and kinetic energy, and to discuss how they can be converted one into the other.

The concepts of momentum and energy which we have introduced here will be discussed in more detail in the next chapter.

SUGGESTED READING

Hall, A.R. *The Scientific Revolution 1500–1800.* Boston: Beacon Press, 1954.

Hall, A.R. *From Galileo to Newton.* New York: Harper & Row, 1963. Traces the development of the laws of motion. It discusses other sciences besides physics, and is a good reference for Newtonian philosophy.

Hutchins, Robert M. (Ed.) *Great Books of the Western World.* Volume 28 of the *Great Books* series contains a translation of Galileo's *Dialogues Concerning the Two New Sciences.* The *Principia Mathematica* of Newton is in Vol. 34 (*Mathematical Principles of Natural Philosophy*).

Mailer, Norman. *Of a Fire on the Moon.* New York: Signet, 1971. The second part of this book (which describes the Apollo moon landing) contains an extremely good discussion of Newton's laws.

Mason, Stephen F. *A History of the Sciences.* New York: Collier Books, 1962. Contains a good description of this period, and a nice account of the founding of the Royal Society and Newton's role therein.

QUESTIONS AND DISCUSSION IDEAS

1. Galileo's book was called *Dialogue Concerning the Two New Sciences.* The study of the laws of motion was one of the sciences. What was the other?

2. A boy stands on the edge of a cliff. In order to find out how high it is, he drops a rock off it. The rock hits the bottom in four seconds. How high is the cliff?

3. Another boy on the same cliff decides to make the measurement in a different way. He throws the rock out as far as he can. How long will it take his rock to hit the ground? How high will he think the cliff is?

4. If I push a cart along a road and then let it go, I observe two things. First, it moves along in the direction

I was pushing it, and, second, it eventually stops moving. Does the fact that it stops violate Newton's first law of motion? If not, why not?

5. One of the problems of Aristotelian physics was to explain why a projectile that is thrown continues to move after it has left your hand. Give a Newtonian explanation of this phenomenon.

6. What do you think the following men would have made of the law of universal gravitation?

    a. a Babylonian priest
    b. Thales
    c. Aristotle
    d. Copernicus
    e. Kepler

7. Two bodies of mass m_1 and m_2 are held together by a spring that is stretched to a length D. When the masses are released, it is found after a very short time that the body of mass m_1 is moving with velocity v_1 to the right, and the body of mass m_2 is moving with velocity v_2 to the left, and that

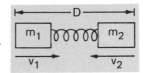

$$v_1 = \frac{m_2}{m_1} v_2.$$

    a. If I double m_1, what happens to v_1 (all other things being kept constant)?
    b. What if I double m_2 instead?
    c. The above equation can be written

$$m_1 v_1 = m_2 v_2.$$

What conservation law does this represent?

8. We carry out the following experiment: A man on a moving railroad car drops a ball bearing as he passes an intersection. At the same time, a man on the ground drops a ball bearing that weighs twice as much. Neglecting air resistance and other complications (the experiment is complicated enough already!), what is the time difference between the times that the two ball bearings will hit the ground?

9. Trace the transformations of energy in the following sequence: A fire is lit, which heats some water. The steam rises, and condenses in a pot near the roof. When

the pot is full, it is emptied into a pan, which squirts the water over the blades of a paddlewheel, which makes a toy boat move forward.

10. What are the main features distinguishing the scientific method from other methods of knowing about the world that we have studied? In particular, compare it to the Babylonian, Ionian, and Athenian methods.

11. Resolve the following paradox:

Consider a horse pulling a cart. We know that the horse is exerting a force forward on the cart. Call this force F. By the third law, the cart must be exerting a force equal in magnitude to F, in the opposite direction. This means that the total force on the system must be exactly zero and, by the first law, this must mean that the system cannot move. Therefore Newton's laws lead us back to Zeno.

HINT: Think about what happens if you try to pull something when you are standing on ice.

12. We cheated a little in the discussion of Galileo's measurements on falling bodies, because he did not actually measure the velocity of an object at some time during its motion, but measured the *average* velocity over a time interval. To understand this distinction, let us take the case of motion for K = 1 discussed in the text.

a. How far will the object travel in the time interval between two and three seconds?

b. If the object were not accelerating, but were moving at a constant velocity, how fast would it have to be moving to cover the same ground in one second as it did in 12a? This is known as the average velocity.

c. According to Galileo, how fast will the accelerating object be moving at 2.5 seconds (midway into the time interval)?

The fact that the average velocity over a time interval is equal to the velocity halfway through the interval is known as the Merton Rule, after the medieval scholars of Merton College, Oxford, who first discovered it.

13. For the example of K = 1 discussed in the text,

a. How far will the object have traveled in 10 sec?

b. How fast will it be going at the end of 10 sec?

c. How long will it be before the object is moving at more than 30 ft/sec?

CHAPTER VII

THE NEWTONIAN WORLD

"In any event, he said, no one was really accepted in Chicago until he'd rubbed someone out. It was time for Aristotle to get his."

Robert Pirsig, Zen and the Art of Motorcycle Maintenance

A. Introduction

The introduction of the scientific method was a totally new way of dealing with the world. It had been a long time developing, but once it was available it made possible a fundamental change in man's perception of his world. Whereas there had been virtually no progress on many scientific questions from the time of Aristotle, the work of Newton opened new doors and showed the way out of century-old dilemmas.

We have already discussed one of these resolutions—the solution of the problem of natural versus violent motion. We saw that the ."solution" to this problem found by Galileo didn't really "solve" the problem at all. We *still* can't say when a projectile motion stops being "violent" and starts being "natural." What we *can* say is that the categories "natural" and "violent" have nothing to do with the motion of real particles in the real world. The attempt to impose these categories on projectile motion resulted in an impossible problem. Instead of asking "How does a projectile move?" scientists have been trying to classify projectile motion and were making very little progress in doing so. What the scientific revolution did was to dispose of the old categories and to ask the correct questions.

After all, we really aren't so much interested in developing philosophical categories for phenomena in the world as we are in trying to develop an understanding of the basic laws that govern them. The tremendous success of the scientific method in the time since Newton is a testimony to how far the human mind can go when it is freed from false categories.

Newtonian science also gave rise to a very different way of looking at the world than what had preceded it. The realization that the entire solar system could be understood in terms of a few simple laws led to a burst of confidence that extended far beyond the boundaries of science. The view of the world that eventually evolved was one in which the universe was thought of as being something like a clock—everything was ticking along according to natural laws. In a sense, God had "wound up" the universe, and then let it run.

Like many of the scientists who followed him, Newton was a deeply religious man. Many scientists of this era believed that their studies were actually directed toward learning about God through the study of the natural laws that He had set up. Thus, the schism between religion and science that we saw with Galileo had been at least partially resolved by the time of Newton's work.

The picture of an orderly universe in which everything was unfolding according to a divine plan was characteristic of the art and literature of the period following Newton, as well as of its science. The music of Bach and Mozart said much the same things about the world as Newton had, and emphasized once more the fact that the kind of science people develop is closely related to the rest of their culture.

In this chapter we shall look at some of the new ideas that Newton introduced and try to get some feeling for what Newtonian mechanics was all about.

B. The Wicked Dance in Circles

It is difficult for us today to comprehend the strong hold that the idea of the circle and sphere had on Greek and medieval scientists. As we saw when we discussed the cosmology of Plato, it was felt that since the circle was the most "perfect" figure, a body free to move (i.e., on which no restraints were placed) would naturally move in a circular path. This, of course, was the basic reason for the tremendous difficulty experienced in freeing astronomy from the grip of epicycles.

The idea of the circle, as we have seen, had the place of an "unquestioned truth" in early science. It would not have occurred to an Alexandrian astronomer to question it, any more than it would occur to a modern scientist to ask whether it was really possible to describe the physical world in terms of a rational thought process. (This latter idea is one of the unquestioned truths of the 20th century. It is as "obvious" to us as the idea of circular motion was to the Greek mind.)

The essential feature of Newton's laws of motion was the realization that in the real world, unrestrained motion more closely approximated a straight line than a circle. By the second law, this means that a particle must be accelerated to go in circular motion. In this section, we will discuss how such a state of affairs could come about.

The real question comes down to this: In what way will a body move when there are no forces acting on it?

The Greeks would have said a circle, and Newton would have said a straight line. In order to choose between these two answers, we could do a simple experiment.

Consider a ball on the end of a string. Suppose we twirl the ball around our heads so that it goes in a circle. After it has started up, a steady circular motion will result. Isn't this the motion in a circle that the Greeks talked about?

Not quite. Think about what happens if we let go of the string. When there is nothing touching the ball. Does it continue to move in a circle? Clearly, it will not. In fact, it will fly off in a straight line tangent to the circle at the point where it was let go. Thus, on the basis of one simple experiment, we can convince ourselves easily that circular motion is in no way the "natural motion" of an unrestrained object.

The idea that this motion is accelerated, however, is a little harder to grasp (remember—according to the first law it *must* be accelerated if it is not moving in a straight line). Suppose we did the experiment in a slightly different way. Suppose instead of holding the string in our hand, we attached it to a scale. Then, when we twirled the ball around our head, the scale would read a certain number of pounds, just as if a weight were hanging from it. This would happen because the spring inside the scale would be stretched by the whirling ball. Even without the scale we would know that there was a force on the string because we could feel it pulling on our hand.

So far, so good. While the ball is moving in a circle, there is a force acting on it. But, from the second law, this means that the ball *must* be accelerating. This is the essential new ingredient that allowed Newton to unravel the problem of circular motion.

A good way to think about the circular motion of a ball is illustrated in Fig. 7.1. If we were to let go of the string at point A, the ball would move in a straight line to point B in some small time T. But if we want to prevent it from going in that straight line (i.e., if we want to keep on in the circle), we'll have to pull it back to the circle at point C. From C, it would proceed to D unless we pulled it back to the circle at E, and so forth. It is easy to see that as the time T gets very short, the zig-zag path described above becomes a true circle, and the force toward the center has to be exerted continuously to keep the body in circular motion.

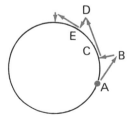

Figure 7.1.

136

We see, then, that in order to have circular motion, it is necessary that something exert a continuous force on the moving object in order to overcome its tendency to move in a straight line. In the example we have chosen, this "something" is the string and, ultimately, the hand that is holding it. Other examples of agencies that can exert this force will be dealt with in the problems. The force exerted by an outside agency to maintain the circular motion is sometimes called the "centripetal force."

We have all had experience with another aspect of circular motion. If we are sitting in a car that is going around a curve, we know that we will feel as though we have been thrown to the outside of the car. We even use the term "centrifugal force" to describe this feeling. But what is actually happening?

Before the car comes to the curve, we are moving in a straight line. From Newton's first law, we know that we will continue to move in that line until forced to do something else. As the car starts to turn, we continue to move in a straight line until some part of the car—the seat or the door—exerts a force on us, pulling us around the curve with the car. What we feel when this happens is not what has just been described, however. We do not feel as if the car were turning and pushing on us as we move in a straight line. What we experience is a sensation of being pushed against the car. It is almost as if a force were being exerted on us, throwing us to the outside.

What we have, then, is a situation in which the person sitting in the car feels that a force is acting on him, but a person standing on the ground sees no such force but only the tendency of the person in the car to move in a straight line. In modern language, we would say that whether or not a force is perceived to be acting depends on where the observer is standing. If he is in the car, he sees a force, but if he is on the ground he does not. This kind of distinction is one of the important ingredients of the theory of relativity, which we shall study later. For the moment, however, we simply note that there is a certain ambiguity in the notion of a "centrifugal force," an ambiguity that has caused some physicists to call it a "fictitious" force as opposed to "real" forces like the centripetal force. This distinction simply emphasizes the dependence of the force on the observer.

137

Having made a careful distinction between "real" forces and "fictitious" forces, and argued (convincingly, I hope) that centrifugal force is "fictitious" and not "real," we can now do what most physicists do and drop the distinction between real and fictitious forces, and between centripetal and centrifugal forces. These terms are used only when one is being exact to the point of tediousness; usually the term "centrifugal force" is used in a much more sloppy way than we have been using it here. For example, it would not be at all unusual for a physicist to discuss the problem of the ball and the string, which we dealt with earlier, by saying that the force exerted by the string is exactly sufficient to cancel the centrifugal force on the circling ball. What he would mean by such a statement, of course, is that the centripetal force exerted by the string overcomes the tendency of the ball to move in a straight line, and that the fictitious force perceived by someone sitting on the ball would be equal in magnitude (but opposite in direction) to the centripetal force exerted on the ball by the string. The physicist would be saving a lot of words at the cost of a slightly imprecise use of terms.

One common thing preventing people from accepting the idea that circular motion is accelerated comes from the observation that the speed of the circling object (like the ball) doesn't change. Since acceleration is a change in velocity, how can we say that the ball is accelerated?

The answer to this question lies in the difference between velocity and speed. Velocity, as it is understood in physics, implies both a rate of travel *and* a direction. In technical terms, it is a vector. Speed, on the other hand, is just a rate of movement. Thus, to be correct, we could report the *speed* of a car as 50 mph, but its *velocity* would have to be 50 mph in a given direction.

In this way of looking at things, the velocity of the ball in the circle is changing all the time because its direction is changing, even though its speed is constant. Thus, in Fig. 7.2, if the velocity at point A on the circle is as shown, and the velocity at point C is as shown, there is a net change in velocity between these two points which is represented by the different direction of the two arrows, even though the length of the arrows is the same.

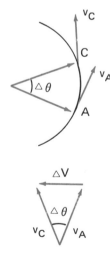

Figure 7.2.

DANGER :::::::: MATHEMATICS AHEAD

For those who want to follow it, we present here a simple derivation of the formula for the centripetal force according to Newton. The less mathematically inclined reader can skip ahead.

Suppose we imagine sliding point C down to point A. Then it is easy to see that the difference between v_A, the velocity at A, and v_C, the velocity at C, is just what we have labeled ΔV in Fig. 7.2. A little geometry will also convince you that the angle between v_A, and v_C, is just $\Delta\theta$, the angle between A and C measured at the center of the circle. Now, from Newton's second law,

$$F = m\frac{\Delta v}{\Delta t},$$

where Δt is the time required to cause the velocity change. But from geometry,

$$\Delta v = v\sin\theta \approx v\,\Delta\theta$$

so that

$$F = mv\frac{\Delta\theta}{\Delta t}.$$

What we have to do, then, is find $(\Delta\theta/\Delta t)$, the rate at which the angle at the center of the circle is changing. To do this, we note that if the ball is moving with a speed v, then the time it takes to go around the circle once, which we shall call T, is just

$$T = \frac{2\pi r}{v}$$

because the distance around the circle is just $2\pi r$. Now in general, when the ball has traveled a length of arc AC, it will have traversed an angle

$$\Delta\theta = \frac{AC}{r}$$

(this is the definition of an angle in the so-called radian measure). In the example in which the ball goes through one revolution, the length of arc traveled is just

$$\overline{AC} = 2\pi r$$

so that the total angle traversed (in radian measure) is

$$\Delta\theta = \frac{2\pi r}{r} = 2\pi.$$

To find $\Delta\theta/\Delta t$, then, we simply have to divide the angle traveled by the time it took to travel it, so that

$$\frac{\Delta\theta}{\Delta t} = \frac{2\pi}{\left(\frac{2\pi r}{V}\right)} = \frac{V}{r},$$

which gives

$$F = \frac{mV^2}{r}.$$

END OF MATHEMATICS

This is the formula that Newton finally derived, and that once and for all ended the massive confusion in physics about circular motion. Once we have this concept of centrifugal and centripetal forces firmly in mind, we can instantly see the solution to Copernicus's difficulty with the turning of the earth, and can also see why it is that the earth can turn without having everything fly off it.

It really comes down to a question of two forces acting on a body on the surface of the spinning earth:

1. Centrifugal force, which tends to make the body fly off in a straight line while the earth turns underneath it, and which has a magnitude

$$F_C = \frac{mV^2}{r},$$

where r is the radius of the earth.

2. The force of gravity, which acts toward the center of the earth, and tends to keep the body fixed to the earth. If the earth has mass M_e, this force is

$$F_g = \frac{G M_e m}{r^2}.$$

A quick calculation will show that the gravitational force is much greater than the centrifugal one, so that the reason the earth can spin and not throw things off is that it doesn't spin fast enough!

To give another illustration of this effect, which has more relevance to Copernicus, lets talk about the objection that was raised to his idea—the objection that clay flew off a potter's wheel when it turned, so everything would fly off the earth if *it* turned.

What are the forces acting on a piece of clay sitting on a spinning wheel? They are (1) centrifugal force, as above, and (2) force of cohesion, which makes the clay stick to the wheel. When the wheel is not turning, the centrifugal force vanishes, and the force of cohesion makes the clay stick to the wheel. (We wouldn't expect the clay to fly off the wheel when it is stationary, after all.)

If we now turn the wheel very slowly, the centrifugal force, while not zero, is still very small, and the force of cohesion is still strong enough to overcome the centrifugal force. In this case, the clay will still remain on the wheel. As we speed the wheel up, the centrifugal force gets bigger and bigger, until, at some critical speed, it becomes equal to the force of cohesion. At this point, there is a change in behavior. The forces holding the clay to the wheel are now less than the forces tending to throw them off, and the clay will leave the wheel.

The answer to Copernicus's opponents, then, is simply to point out that clay doesn't *always* fly off the potter's wheel—it depends on how fast the wheel was turned. The situation would have to be the same with the earth. Since everything *doesn't* fly off, the speed of the earth's rotation must be below the critical speed discussed above. This, we now realize, is precisely the result we can calculate using Newtonian mechanics.

C. Energy, Momentum, and Conservation Laws

In the previous chapter, we saw how the so-called "conservation laws" could arise from Newtonian mechanics. In the case of momentum, we saw that in the absence of an external force, the momentum must be unchanged (conserved) in time, so that it has the same value any later time than it

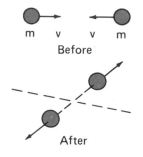

m   v   v   m

Before

After

Figure 7.3.

has right now. Let us explore a few of the consequences of this law.

Consider as an example the situation shown in Fig. 7.3, in which two billiard balls of equal mass are rolling toward each other with equal velocity. The momentum of the billiard ball on the left is mv, and is directed toward the left (which is the direction of its velocity). The billiard ball on the right has momentum mv, directed to the right. The total momentum of the entire system is then

$$mv \text{ (to the right)} + mv \text{ (to the left)} =$$

$$+mv - mv = 0.$$

In other words, the momentum of one ball exactly cancels the momentum of the other.

When the two billiard balls collide, they will exert forces on each other. Hence, by the second law, the momentum of each ball will change. However, nothing is exerting a force on the entire system. This means that the total momentum of both billiard balls must remain what it was before the collision—zero. Consequently, if we saw one billiard ball moving to the upper right as in Fig. 7.3, we would know, even without looking, that the other had to be moving to the lower left, as shown. This is the only way it could be moving and still exactly cancel the momentum of the ball we observe. Thus, the law of conservation of momentum in this case allows us to state what will happen to both billiard balls after the collision, even though we only observe one of them.

Momentum is not the only quantity that is conserved in Newtonian physics. In the previous chapter, we introduced the concept of energy, or the ability to do work. There is also a conservation law associated with energy. To illustrate this law, consider a roller coaster on a track similar to the one in Fig. 7.4. At point A, at the top of the track, the roller coaster is stationary. This means that it has no kinetic energy. However, something had to lift it up to this height, so that a force was acting over a distance. From the discussion in the previous chapter, this means that the roller coaster has potential energy when it is stationary at point A.

To find out how much potential energy, we simply have to ask how large a force would have to act over how

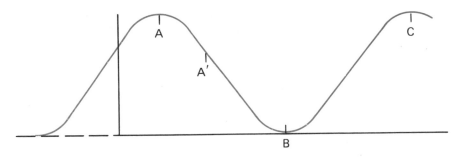

Figure 7.4.

much of a distance to get the roller coaster to A. If the roller coaster started at ground level, and point A is at a height h, then it is obvious that the work needed to lift the roller coaster up to A is simply

$$W = Mgh,$$

where M is the mass of the roller coaster. This is simply the force Mg needed to counteract the pull of gravity times the distance h. If there are no sources of loss in the system, the roller coaster would be capable of doing this much work if it falls, and hence the potential energy of the roller coaster at A is also Mgh.

When the roller coaster starts down the hill, however, the situation changes. At point B, it will be back at ground level. This means that relative to its original starting point (which was also ground level) the roller coaster no longer has any potential energy, since it would take no work to get it from the starting point to the height of B. Does this mean that the roller coaster has lost the energy it posses at A? Of course not, because we know that at B, even though the potential energy is zero, the roller coaster is moving. All that has happened between A and B is that the potential energy has been converted into kinetic energy. During the descent (for example, at A'), the initial potential energy has been only partially converted, and the total energy is split up—some is kinetic and some potential. However, the total energy itself does not change. All that happens is that the energy changes from one type to another. If there were no friction in the wheels, the roller coaster would go from point B back to point C, which is at the same height as A. At C, the kinetic energy that existed at B would be converted back to its original form of potential energy.

This sort of analysis makes us realize that there must be a conservation law involving the total energy of a system, even though there is no such law for potential energy alone or for kinetic energy alone. The law of conservation of energy is usually stated in the form:

> Energy cannot be created or destroyed in any physical process.

We will discuss this conservation law later in connection with the laws of thermodynamics (where friction and heat are taken into account) and the laws of relativity (where the concept of mass as energy is introduced). The important point we can gain from our discussion here, however, is the concept of a fixed amount of energy being shifted back and forth between different categories in such a way that the total amount remains constant even while the amount in each category is constantly changing. The best analogy to fix this idea in mind is to think of the categories of energy as a series of sinks, one labeled "potential energy," one "kinetic energy," and so on. If we had one gallon of water, we could divide it up among these sinks, shifting it around at will. However, no matter how the division was done, there would always be a gallon of water in the sinks. In this analogy, the water plays the role of the total energy. As we shall see later, scientific advances in this century have not changed the law as stated above, but merely added to the number of "sinks" into which energy can be classified.

One more point should be made before we go on. We have discussed potential energy only in terms of lifting objects up against the force of gravity. However, it should be obvious that there is a potential energy associated with *any* system where it is possible to do work by having force act over a distance and then recover this work. In the next chapter, for example, we shall discuss the existence of an electrical force. With this force, moving an electrically charged object around requires that a counteracting force act over a distance, just as with gravity a counteracting force is required to lift an object. Consequently, we could define an electrical potential energy in complete analogy to the gravitational potential energy we have discussed here.

There is one more energy-related concept that we should introduce at this time. We would all recognize that there is a difference between lifting a weight quickly and lifting the same weight slowly. For example, a human being could easily lift a 10 pound weight one foot in one second, while it would take a fairly sophisticated machine to do the same amount of work in one-thousandth of a second. The difference between these two situations is not the amount of work that is done, since that is simply 10 foot-pounds in either case. The difference is the *rate* at which the work is done. The rate at which work is done is called power, and is defined by

$$\text{power} = \frac{\text{work done}}{\text{time it takes to do the work}}.$$

For example, when we spoke of lifting 10 pounds one foot in one second, the power expended during that second would be

$$P = \frac{10 \text{ foot-pounds}}{1 \text{ sec}} = 10 \text{ foot-pounds/sec.}$$

Actually, the unit of foot-pounds/second is not familiar to most of us as a unit of power. When steam engines were first being introduced into coal mines in England, they were used to replace horses. Engineers rated their engines according to how many horses they could replace, and reckoned (with an adequate safety margin) that a horse could do 550 foot-pounds of work each second. This became the standard unit of power in the English system, and has the more familiar name of "horsepower."

Units of work and power in the metric system are discussed in Appendix A.

D. Technical Digression: The Idea of a Limit,
   or, How Do We Cross a Room, Anyway

In addition to allowing him to deduce Kepler's laws from his own more general three laws of motion, the introduction of the calculus by Newton marked a major step forward in our ability to deal with physical systems mathematically. In this section we shall present some of the main ideas of the calculus and try to give the reader some notion of how it works in a physical system. As an added bonus, we shall see that one of the important concepts in the calculus—the idea of the limit—also is the key to resolving Zeno's paradox.

To understand the idea of a limit, consider a man who is driving from Charlottesville to Washington, a distance of about 100 miles. Suppose he makes the drive in two hours. We would all agree that he had been driving at an average rate of 50 miles per hour.

Let us examine for a moment the process by which we came to this conclusion. When the car left Charlottesville, we started a stopwatch. We had previously measured the distance to be traveled, and when the car reached its destination, we stopped the clock. Thus, we had measured both the distance traveled and the time elapsed. Then we took the distance the car had traveled, in this case 100 miles, and divided by the time it took to travel that distance, in this case two hours. Using the formula we all learned in grade school,

$$\text{velocity} = \text{distance/time} = 100 \text{ miles/2 hours} = 50 \text{ mph.}$$

This is the average velocity of the car. In terms of a graph (see Fig. 7.5), we know that at the beginning the distance traveled is zero. After two hours has elapsed, 100 miles have

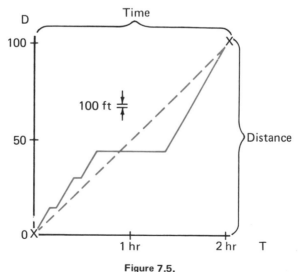

Figure 7.5.

been traveled. These two statements mean that we know the position at two different times—at zero and at two hours. This knowledge is represented by the two X's on the graph.

If the car traveled steadily at the average velocity, after one hour, it would have gone 50 miles, after an hour and a half, 75, and so on. If we plotted its position at each time for this case, it would be the straight line shown in Fig. 7.5. The average velocity is D/T, which is also sometimes called the "slope" of the line.

It is highly unlikely however, that the car could travel at a steady 50 mph. It would have to stop for traffic lights, and make up for lost time by going a little faster elsewhere. The knowledge that the car traveled the distance at a certain average velocity tells us nothing at all about how fast it was traveling at any particular time.

One way to zero in on the velocity at a particular time would be to take a much shorter interval over which to measure the average velocity. In our example of the car, suppose that instead of 100 miles, we measured a distance of 100 feet. Obviously, in the shorter distance, it would be much harder for the car to make radical changes in its velocity. The driver couldn't for example, stop for lunch in the interval and still maintain a 50 mph average velocity over the 100 feet.

In Fig. 7.6 this 100-foot interval is shown. Now suppose we went through the same process for this interval that

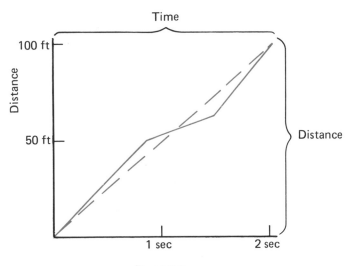

Figure 7.6.

we did for the 100-mile interval. That is, suppose we start a clock when the car enters the interval, and stop it when the car leaves. Once again, we could calculate an average velocity for the car in the 100-foot interval, and once again it would not be possible for us to say exactly how fast the car was going at any specific time. However, it is obvious that there is a lot less room for maneuvering in 100 feet than there is in 100 miles, so that the average velocity over the 100-foot interval will be a lot closer to the velocity at any point than was the case for the 100-mile interval. This point is made in Fig. 7.6.

Now suppose that we still weren't satisfied with the way things were going, and we wanted to know the velocity more precisely than we could by using the 100-foot interval. What would we do? It is obvious how we would have to proceed. Instead of an interval of 100 feet, we would pick a much smaller one—one inch, for example. We would go through the same procedure outlined above, and we would have determined the velocity of the car as it traversed that inch much more closely than we had been able to do previously. It is obvious that the average velocity over the one-inch interval would be very close to the actual velocity at any point in the interval. However, it is also obvious that there is still some small leeway—there is more than one way a car can travel one inch in a set period of time, even as there was more than one way it could travel 100 miles. It's just that in the case of the one-inch interval, the differences in velocities at any point along the way would have to be fairly small.

Now we can, in our minds, continue this process further. It is clear that each time we shorten the interval, we find an average velocity over the interval which is closer and closer to the actual velocity at any point in the interval. We can imagine taking shorter and shorter intervals, measuring the average velocity for each interval. We can even imagine taking the length of the interval to zero, in which case we would speak of taking a "limit." The concept of the "limit" is essential to the development of the calculus.

If we define the average velocity that the body travels over a distance D by

$$<V> = \frac{D}{T},$$

The roller coaster at Coney Island, near New York City, where the law of conservation of energy is illustrated daily.

Communications satellite.

Spaceman in training experiencing G forces.

where T is the time required to traverse the distance, then the instantaneous velocity is defined as

$$V_{inst} = \lim_{D,T \to 0} \frac{D}{T},$$

and the notation "lim" means that we understand that we are going to the limit in which the length D and the time T required to traverse it are approaching zero. This process is the heart of the differential calculus.

The reader will have probably noted one important point already. As D, the length of the interval, gets smaller, so too does T, the time it takes to cross the interval. This means that when we take the limit we have discussed, we are dividing one quantity which is becoming small by another quantity which is becoming small. The result of such a division need not be small.

Let's convince ourselves of this. Consider the two quantities x and 3x in the limit that x approaches zero. Obviously

$$\lim_{x \to 0} x = 0$$

and

$$\lim_{x \to 0} 3x = 0.$$

However, it is also true that

$$\lim_{x \to 0} \frac{3x}{x} = \lim_{x \to 0} 3 = 3$$

so that the quotient of two numbers, each of which is approaching zero, can itself be a perfectly finite number.

Once we have understood the concept of instantaneous velocity and the idea of the limit that underlies it, we can see how Zeno's paradox can be resolved. It is true that after crossing half the room, I must then cross half of what is left. But it is also true that *it only takes me half as long to do it.* As the intervals to be crossed get smaller and smaller, the time required to cross them does, too. In the limit that the interval approaches zero, the time needed to cross it also approaches zero, but the quotient of distance/time stays finite, and is, in fact, the velocity at which I am traveling. Zeno saw that space was infinitely divisible, but didn't see that time was also!

Let us approach the idea of a limit from a somewhat different angle. Let's go back to the example of the two-hour drive to Washington. Let's also suppose that the driver decides to travel the hundred miles by starting out slowly and gradually accelerating until he has reached the end. His distance-time graph would then look like the curve labeled Q in Fig. 7.7.

This situation is a little difficult. Up to this time we have talked about velocity as a distance divided by a time, but this concept requires that there be some distance over which the time can be measured and in which the velocity does not change. There is no such interval in our example.

The way we would apply the idea of a limit to this problem would then be to break up the interval into a series of sub-intervals, each of which would have its own average

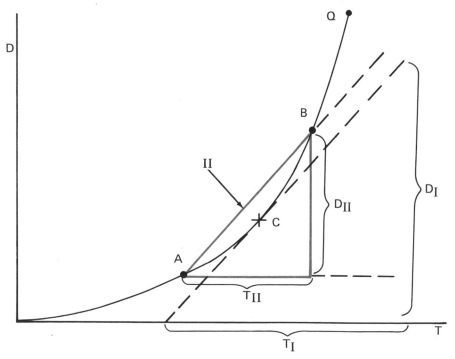

Figure 7.7.

velocity. The average velocity corresponding to each interval would, of course, be different from interval to interval. The curve labeled II in Fig. 7.7 is an example of such an average velocity between A and B. If we wanted to we could define the average velocity as

$$< V_{AB} > = \frac{D_{II}}{T_{II}} ,$$

as shown in the figure.

Now let us ask ourselves what would happen as we took the distance interval D_{II} to zero. Obviously, points A and B on the curve would move closer together. Suppose they converge at C in the limit. The average velocity for the interval between A and B would approach the value $\frac{D_I}{T_I}$, which is the slope of the line that is tangent to curve I at point C. The slope of this line is what we defined earlier as the instantaneous velocity at C!

This procedure illustrates one very important point about the calculus. We have initially a rather complicated problem (here the motion of a car with non-uniform velocity). We replace this complicated problem by a series of simpler problems, each appropriate to a small interval. In our example, this corresponded to taking the average velocities in the intervals in curve II, which amounts to replacing the original problem by one in which the velocity is constant over each interval, but changes from one interval to another. Finally, we take the limit as these intervals go to zero, recapturing our original problem. The study of finding ways to break problems down to infinitesimal intervals is called the DIFFERENTIAL CALCULUS.

So far we have discussed the idea of the limit only in terms of the motion of objects. If that were its only use, it would remain an interesting (but not very widely applicable) idea in the history of science. It has, however, a wide number of other uses which make it of much more importance than that, and we shall discuss one of them now as an example.

We saw that Newton had discovered the law of universal gravitation, which stated that between two point

151

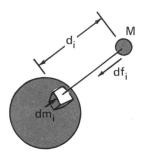

Figure 7.8.

bodies of mass m_1 and m_2 a distance d apart there exists a gravitational attraction given by

$$F = \frac{Gm_1\, m_2}{d^2}.$$

However, we know that most bodies in the universe are not point masses but have a finite extent. For example, the earth is approximately spherical in shape, and is thousands of miles across. How can we go about finding the gravitational attraction exerted by the earth on a small body near it, such as a satellite?

Without the idea of a limit, it would be difficult to know how to proceed with this problem. With the idea of a limit, however, it becomes quite simple. Imagine that we break up the earth into small chunks of matter (see Fig. 7.8). Then each chunk will exert a force on the small body near the earth, and the force will be given by

$$df_i = \frac{GM\, dm_i}{d^2},$$

where we are referring to the force due to the small chunk i as df_i, the mass of chunk i as dm_i, and the distance from chunk i to the body of mass M as d_i. Now we could approximate the force on the body by adding up the forces exerted on it by each of the little chunks. This would give

$$F = \sum_{\text{all } i} G \frac{M dm_i}{d_i^2}.$$

It is also obvious that if we went to the limit where the size of each small chunk goes to zero (while the number of such chunks becomes infinite), we would have a correct description of the force exerted on the body by the entire earth. As an aside, the sum over chunks which we wrote down above is called an integral in the limit that the size of the chunks goes to zero, and the sum is usually written

$$F = \int \frac{GM\, dm_i}{d_i^2}.$$

152

where the symbol $\int$ is called an integration sign. (It's sort of a capital S, which should remind you of "sum.")

The study of how to carry out such summations is the main subject matter of the INTEGRAL CALCULUS, and is beyond the scope of this book. Nevertheless, it should be clear that this approach provides a way of attacking problems in physics that is very powerful indeed.

The essential use of the calculus, then, can be thought of as follows: a complicated problem is broken up into pieces, each of which is simple. In the above example, the complicated problem of the gravitational attraction of the earth was broken down into a series of simpler problems, having to do with the force between two points. This breaking down process involves the differential calculus. The simple problem is then solved. In our example, the simpler solution just involved writing down Newton's law of gravitation for two point masses. Finally, the solutions to the simple problem are added up for all of the pieces, and the final answer obtained. This building up process involves the integral calculus.

What the calculus gives us, then, is a way of solving very complicated problems in terms of the solutions to very simple problems. Virtually all of the progress that has been made in the sciences since Newton have depended for their techniques on some form of the calculus as we have described it here.

E. The Newtonian World Picture

Up to this point, the consequences of the development of Newtonian mechanics had been primarily of interest to scientists. Physicists and astronomers and other scientists derived a great benefit from being able to use the new tools developed by Isaac Newton. It would be a mistake, however, to think that an intellectual revolution of this size could be contained in one area of human endeavor alone. In fact, the kind of outlook that Newton developed had a tremendous impact on all areas of life in the 18th century. Even the use of the term "Newtonian" world picture is evidence of this fact.

Many of the ideas that we now identify as "Newtonian" were not actually stated by Newton himself. It is not an unusual phenomenon to find the followers of a great man going far beyond the limits of what he was willing to do in order to carry through his ideas. Marx is reputed to have said "I am not a Marxist." In the same way, Newton might well have said "I am not a Newtonian."

Perhaps one example will illustrate this point. One of the great questions of 18th-century science and philosophy was the problem of "action at a distance." The problem can be stated in the following way. "Since the earth and the sun do not touch each other, how is it possible that a force can be transmitted across empty space between them?" Newton himself did not even try to answer this question. Here is his own reaction to this problem:

> Gravity must be caused by an Agent acting constantly according to certain Laws; but whether this Agent be material or immaterial, I have left to the Consideration of my Readers.

In other words, Newton felt that since he could not provide an answer, it was better to just leave an unanswered question.

Those who followed him, however, were not nearly so reticent. Many of them introduced a concept of the "ether" as some sort of material medium between the earth and the sun. The idea was that the sun pressed on the ether at one place and this force was transmitted through the ether which held the earth into its orbit. The concept of the ether proved to be as much a blind alley for physics in the 18th and 19th centuries as the concept of the epicycle was earlier.

Nevertheless, even granting that Newton's followers would probably push his ideas further than he was willing to himself, there is implicit in the work of Newton an entirely new way of looking at the world. This new way of looking at the world was something that was very much in consonance with other social changes that were going on, and particularly with the Industrial Revolution. It is a view of the world in which God at some point made a few laws— laws that men could learn—and in which everything in the universe proceeds in an orderly way according to these laws. There was no room in this world for special intervention by God or by any other supernatural force.

One rather interesting outgrowth of Newtonian science was an idea that was developed in the 19th century, the so-called "divine calculator." This concept had a great deal of influence on philosophical discussions of free will. Simply stated, it can be put as follows: "Since I know all of the laws that govern the universe, if at any given time I know the position and velocity of every piece of matter in the universe, then, using Newton's laws of motion, I can in principle calculate where those particles would be at any time in the future. The only problem is developing a calculator that is capable of following the motions of all the particles in the universe." This rather arrogant statement of scientific power was made repeatedly by philosophers in the 18th and 19th centuries. It illustrates, in its own way, how ridiculous ideas can be when they are carried to extremes. We might even call it Zeno's paradox of classical physics.

Just as Zeno overlooked a rather important point—the divisibility of time—the promulgators of the divine calculator idea also overlooked a very important point. They overlooked the fact that in order to measure the position of one particle at some point in the universe, I'd have to bounce another kind of particle off of it. This other kind of particle might be a light beam, for example. The question then arises as to the effect of the measurement itself on the position of the particle that is being measured. Just as Zeno's paradox, when properly thought out, could be said to contain within it the germ of the idea of the calculus, the idea of the divine calculator, when properly thought out, could be said to contain within it the germ of quantum mechanics and of the uncertainty principle. We shall return to this idea when we discuss the development of 20th-century physics.

SUGGESTED READING

Most of the books cited in Chapter 6 deal with the subject matter presented here as well. In addition, there is one more:

Pirsig, Robert. *Zen and the Art of Motorcycle Maintenance.* New York: Bantam Books, 1974. Contains an excellent discussion of Zeno's paradox in the last section.

QUESTIONS AND DISCUSSION IDEAS

1. Identify what agency exerts the centripetal force in the following cases:
   a. the earth moving around the sun
   b. the moon moving around the earth
   c. someone stuck to the wall in a "tilt-a-whirl" ride at a carnival

2. In aerial maneuvers, there are many different kinds of loops that an airplane can perform. For example, there is the "one g" loop in which the plane goes through a loop (so that the pilot is actually upside down at one point), but in which the pilot feels that he is sitting normally while the earth is turning around, Similarly, there are loops in which the pilot feels that he is hanging upside down at the top, and loops in which he feels that he is being pushed into the seat of the plane. From what you know about centrifugal force, discuss the difference between these situations.

3. In the same light as in Problem 2, discuss the technique used to train astronauts for conditions of weightlessness by flying them in airplanes.

4. Could the earth ever spin fast enough to throw us all off (at least in principle)? Write an equation that would tell you how fast a point at the surface of the earth would have to be moving for this to happen. Can you convert this into a speed of rotation?

5. In the text, we used something called "radian measure" to discuss angles. We saw that an angle of 2π in radian measure was actually $360°$ in degree measure. Given this fact, answer the following questions:
   a. How many radians in
      $30°$, $90°$, $180°$?
   b. How many degrees in
      1 radian, π radians, 4π radians?

6. Try your hand at writing a Platonic dialogue in which Socrates and Isaac Newton discusss circular motion with each other.

7. Why did we say that the solid line in Fig. 7.5 included a stop for lunch? Which portion of the curve corresponds to the lunch break?

156

8. Prove by means of a graph the accuracy of the state-
    ment made in the text that the slope of the line tangent
    to a distance-time curve at point C is the same as the
    instantaneous velocity at C.

9. Suppose that we were to cross a room in the following
    way: We cross the first half in one second, half of
    what's left in one second, half of the remainder in still
    another second, and so forth. Would we ever reach the
    other side? Why or why not? Compare your reason-
    ing here to Zeno's.

10. Describe what motion similar to that in Problem 9
    would look like to someone watching us cross the
    room.

11. What if we crossed the room as follows: The first half
    is crossed in one second, half of the remainder of the
    room in 3/4 seconds, half of what's then left in 9/16
    (= 3/4 of 3/4 sec) seconds, and so forth. Would we get
    across? Why or why not? What would *this* motion
    look like?

12. We have already alluded to one use of the calculus
    which Newton made—the calculations of planetary or-
    bits.
    a. By means of a graph or figure, show that an ellip-
        tical orbit can be approximated as a series of arcs
        of circles.
    b. Given the answer to 12a, discuss in general terms
        how Newton might have gone from the simple
        solution to the problem of motion in a circle to
        the more complicated problem of motion in an
        ellipse.

13. Calculate the energy needed to perform the following
    tasks:
    a. lift a 100-pound weight 10 feet
    b. lift a 10-pound weight 100 feet
    c. lift a 1000-pound weight 1 foot
    d. lift a 1-pound weight 1000 feet

14. If each of the tasks in Problem 13 were completed in
    10 seconds, how much power would be expended in
    each case? Give the answer in foot-pounds/second and
    in horsepower.

15. Two blocks of wood of equal mass are sitting on the ground. There is a compressed spring between them. At a given instant, the spring is released, causing the blocks to fly apart.

   a. What is the momentum of the system before the spring is released?

   b. What is the momentum of the system after the spring is released?

   c. It is observed that one of the blocks is traveling vertically after the spring is released. In what direction is the other block traveling (you may neglect the momentum of the spring itself). Why?

CHAPTER
VIII

ELECTRICITY AND MAGNETISM: THE SCIENTIFIC METHOD IN ACTION

"I don't care for dictators much, but I think this whole doggone country ought to be run by.. eeeee-lectricity."
Woodie Guthrie

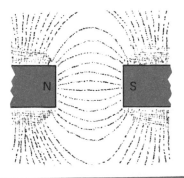

A. New Worlds to Conquer

With the development of the scientific method and its application to the problems of astronomy and mechanics, men had at their disposal a powerful new way of getting information about the world around them. It is only natural that this new tool should be turned to the investigation of other kinds of physical phenomena which had been known for a long time but had never been understood. In this chapter, we shall see how the scientific method was applied in order to gain understanding of the kinds of things we now associate with electricity and magnetism. The fact that it took less than two centuries to come to an understanding of these new phenomena, whereas it took over a millenium to come to the science of mechanics is not hard to understand. The first step in any endeavor is always the hardest. Once people have learned *how* to do something, whether that something is curing a disease or learning more about the physical world, it is much easier for them to apply the method they have developed to new situations. The difficult step was the development of the scientific method itself. That this method should have been developed by people investigating the motion of material objects and heavenly bodies is not too surprising either, since these are phenomena that are immediately available in everyday experiences. But once the *method* had been developed other less common experiences could be investigated as well. Electricity and magnetism certainly fall into this category.

Both electricity and magnetism had been known from very ancient times. There are records suggesting that as early as 2000 B.C., the Chinese were building implements that were the forerunners of the modern compass. Thales of Miletus discussed the "lodestone" (a rock which has magnetic properties). He argued that since the lodestone moved, it must possess a "soul." (The reader may find it interesting to compare Thales' comments on lodestones with the transitional stage of development of Piaget.) A folklore developed in ancient Greece around the magnet or lodestone. Greek sailors believed that in the Mediterranean there was an island made completely out of magnetic material. They believed that if a ship with iron nails in it were to sail near this island, the nails would be pulled out and the ship would fall apart. This was used as a rationale for building ships with wooden pegs, without the use of iron nails. (There are, however, more practical reasons for this choice.)

The Greeks also knew about the phenomenon of electricity, or, more precisely, what we would call today "static" electricity. They knew that if an amber rod were rubbed with a piece of fur, small bits of paper would be attracted to the rod. You have had direct experience with static electricity, too. For example, you have probably noticed that after combing your hair on a particularly dry day, bits of paper or other material may stick to the comb. You may also have received a shock when touching a door knob after walking across a rug. These are both manifestations of static electricity.

While electricity remained a curiosity without much practical importance, magnetism was eventually utilized for a number of purposes. In the early 1200s the fact that a sliver of lodestone would always align itself in the north-south direction was used to build the first rudimentary compasses for use aboard ships. As ocean commerce grew in importance, so too did the development of the magnet. In fact, magnetism entered a phase very similar to the development of science in the early Ionian period. A large group of men worked on the development of better magnets. These men were not scientists in the modern sense, but men of the technical tradition in the sense that we have been using that term. But, like other men of the technical tradition, they did not restrict themselves to an attempt to build a better compass. In the 1600s, William Gilbert studied the properties of the lodestone. He was very interested in the question of why a lodestone always pointed itself north. He made a large sphere out of magnetic material and observed that if he put a small compass anywhere on the sphere it would always point in a particular direction. From this, he concluded that the earth itself must act as one giant magnet, or lodestone. We now know this to be the case, of course, but it is interesting to realize that the first man to come to this conclusion was not a philosopher interested in discovering the properties of the earth, but a navigator who was interested in understanding how a compass worked so that he could build a better one.

After the excitement caused by the development of the Newtonian system of mechanics, the late 18th century saw a revival of interest in the properties of electricity. The men who began this investigation called themselves "electricians." They were really starting from scratch in a brand new area that had never been investigated systematically before. It was a period during which they groped for new ideas, wrestled with new experimental techniques, and, in general, tried to catalog all of the properties of electricity that they could. The reader will probably recognize this period as the first fact-gathering stage in the scientific method, which we have discussed before. The only difference is that in this case the fact gathering was done quickly, and it concerned phenomena that were not necessarily in everyone's everyday sphere of experience.

We have already discussed the phenomenon of static electricity as it was known to the Greeks. In addition to knowing that if a piece of amber were rubbed with fur, it would attract small pieces of paper or other material lying around it, they also knew that once these small bits of material had touched the amber they would be pushed away, or repelled, from the rod to which they had been initially attracted. People talked about an "electrical charge" being induced in the paper.

In the 1730s, an Englishman named Stephen Gray did a series of very interesting experiments on static electricity. First, he put identical charges on two pieces of oak, one piece being hollow and the other solid. He found that the two objects behaved in precisely the same way in the presence of other electrical charges. From this, he concluded that whatever electricity was, it did not reside in the center of the object but only in the surface.

He also discovered that the way in which a body was attracted to a charged piece of amber depended not so much on the properties of the body (like color or taste or weight) but on something else. He discovered, for example, that some objects (like human hair) were very much affected by the piece of electrified amber. On the other hand, other types of material (like a piece of glass) were not. The first kind of material are what we call "conductors" while the second type are what we call "insulators." At the same time, the Frenchman Charles DuFay discovered that if a glass rod were rubbed instead of the amber, materials would behave in much the same way that they behaved toward amber. However, it was discovered that materials that had been touched

162

to the glass rod would be attracted by an amber rod, while pieces of paper that had touched the amber rod and been repelled would then be attracted by the glass rod. In other words, it was known at the end of the 1730s that the response to static electricity depended on some innate property of the material, and it was known that there were at least two kinds of static electricity—that associated with amber and that associated with glass.

Originally, people felt that the property of static electricity had something to do with some kind of "fluid" that flowed in and out of the amber and glass rods. It was thought (from DuFay's experiment) that there were two such fluids, and they were given the names "positive" and "negative." The explanation of DuFay's experiment was then quite simple. A piece of paper, upon touching the amber rod, was filled with negative electrical fluid. When the paper was then brought near the glass rod, which was filled with positive fluid, an attraction existed between the two. From the fact that the paper filled with negative fluid was repelled by the amber rod and attracted by the glass rod, it was possible to conclude that a general law existed governing the force between bits of "electrical fluid." The law, stated simply, was that like bits of fluid would tend to repel each other, while unlike bits of fluid would tend to attract each other.

At this time in the history of electricity, a rather amazing figure enters the stage. He is probably familiar to Americans as the author of *Poor Richard's Almanac*, as one of the Founding Fathers of the United States, and as one of the more engaging characters to have emerged on the American scene. His name is Benjamin Franklin.

In 1746, Franklin happened to attend a public lecture on the subject of electricity in Boston. He was fascinated with the topic and, upon his return to Philadelphia, he obtained from friends in London some of the apparatus that was in use by scientists in those days. The main piece of apparatus was something called a "Leyden jar." It was simply a large glass jar coated inside and out with tin foil. This jar had the property of being able to store "electrical fluid." That is, if amber rods were rubbed and then brought into contact with the tin foil, it was observed that bits of paper would then be attracted to the tin foil as well as to the amber. In the language of the time, it was said that electrical fluid was somehow stored in the jar and could be used in experi-

163

ments at a later date. In modern terminology, the Leyden jar is called a capacitor. Franklin's first project was to devise an apparatus in which the jar was spun around while in contact with a piece of buckskin. This had the effect of eliminating the tedious process of rubbing pieces of amber and bringing them into contact with the Leyden jar. The spinning process, and the constant friction of the Leyden jar against the buckskin, had the same effect in producing electricity as rubbing amber with a piece of fur. This apparatus was immediately adopted throughout the world as an easy and convenient way of producing electricity for experimental purposes. This is the sort of thing we associate with the native ingenuity of Franklin.

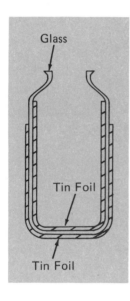

Once he had devised this technique for facilitating experimental investigations, Franklin turned to the problem of how it was that electrical fluid flowed into material bodies. It had been known for some time that if a wire were connected to one side of a Leyden jar and then brought around to somewhere near the other side, a spark would jump from the wire to the other piece of tin foil. After the Leyden jar had been "discharged" in this way, it no longer had a store of electrical fluid. Franklin noted two important facts about these electrical discharges. First, he noted the striking similarity between the electrical spark and the phenomenon of lightning. Second, he noticed that sparks seemed to jump more readily to and from pointed objects than they did to or from objects with more regular contours. In fact, he discovered that once electrical charge had been stored in the Leyden jar, it leaked continuously into the air. If one of the tin foil plates in the Leyden jar were connected to a sharply pointed piece of metal, the leakage took place much more quickly.

These investigations eventually led to Franklin's famous experiment which established the electrical nature of lightning. He proved this by flying a kite in a thunder storm in Philadelphia and observing that electrical sparks jumped off the end of the key suspended from the kite string when lightning struck. Once the electrical nature of lightning had been established in this way, Franklin again used his knowledge of science in a very practical way by developing the lightning rod, a pointed piece of metal which, when placed on the roof of a house, will attract lightning. These metal rods were then attached to metal wires which ran into the ground. Thus, when lightning would strike a house, it would be diverted into the ground rather than causing a fire in the house itself.

Although Franklin's experiments on lightning gained him a great deal of notoriety, his most lasting contribution to the science of electricity was the development of the single fluid theory of electricity. You will recall that it was known that there were two kinds of electrical charges, and it had been postulated that there were two kinds of electrical fluids, one to correspond to each kind of charge. What Franklin saw was that the same phenomenon could be explained in terms of a single fluid. He felt that the fluid that was normally "negative" was the one that actually flowed from one body to another, and that the presence of a "positive fluid" on an object could simply be interpreted as the absence of negative fluid.

The experiments of the amber rod and the glass rod would then be explained in the following ways: when an amber rod is rubbed with a piece of fur, a negative electrical fluid is pushed from the fur into the amber rod. This results in a negative charge in the rod. On the other hand, when a glass rod is rubbed, negative electrical fluid is drawn out of the rod. This means that the rod, which started out with equal amounts of positive and negative fluid, now has a deficit of negative fluid, and that is interpreted as a positive charge. However, the positive fluid does not actually flow at any time.

In our modern terminology, we say that what Franklin called the "negative fluid" is actually the flow of electrons, and the "positive fluid" is nothing more than the atoms from which the electrons making up the negative fluid have been stripped. These atoms are relatively immobile and do not move about very readily in material. Consequently, Franklin's explanation of the flow of electrical charge anticipated by over a hundred years the development of the modern electrical theory.

By the late 1780s, then, a great deal of qualitative knowledge about the behavior of electricity had been gathered by many different researchers. The situation was very similar to that which had obtained the mechanics before the time of Galileo, when a large number of facts about the behavior of

Figure 8.1.

material objects were known, but very little quantitative work had been done on them. For example, it was known that a force existed between two electrically charged objects. The precise mathematical description of this force, however, was not known. The next step in the science of electrostatics— the discovery of the exact mathematical form of the electrostatic force—was done by Joseph Priestley and later, in much more detail and with much more precision by Charles Augustin de Coulomb in 1789.

Coulomb used a variation of the Cavendish balance, which was described in the discussion of the gravitational constant (see Fig. 8.1). Two similar pith balls were charged with a known amount of electricity. For example, in Fig. 8.1, we show the balls being charged negatively. These pith balls are like two ends of a dumbbell which are suspended from a cord. Then two large positively charged objects are brought near the two small pith balls. Because of the electrical force between the large objects and the small pith balls, the dumbbell rotates until the force exerted by the twisting of the string exactly cancels the electrical force. In this way, Coulomb was able to measure quantitatively the magnitude of the electrical force between any two electrical charges.

He found that there was remarkable similarity between the electrical force between two charged objects and the gravitational force between two massive objects. This similarity can easily be seen by looking at the mathematical form of the force law which Coulomb discovered. He found that if there were two electrical charges, one of magnitude q and the other of magnitude Q, located a distance d apart, then the force between these two charged objects would be given by

$$F = \frac{kqQ}{d^2}$$

where k is a universal constant, much like the universal constant of gravitation G.

This constant, which we shall call for the moment the electrostatic constant, could easily be determined in an experiment of the type we have described above. For example, we could imagine doing the experiment with charges q_1 and Q_1 a distance d_1 apart. The force between these charges could be measured and, from the mathematical form of Coulomb's law, we could then deduce a value of a constant k for that particular experiment. We could then repeat this experiment with charges q_2 and Q_2 a distance d_2 apart, and determine the constant k for the second experiment. The point is that no matter how many different kinds of charges and distances we choose to do this experiment, the constant k always turns out to have the same numerical value. This is what is meant, in fact, by calling it a "constant."

With the work of Coulomb the first step in the scientific method as it applies to electrostatics has been completed. The various facts about electricity have been discovered, and the basic law governing the force between different kinds of electrical charges has been written in its mathematical form. If static electricity were the only subject of interest, the job would be completed at this point. The full knowledge about electricity would have been gained, and no new basic knowledge would be left. The only thing left to do would be to derive the various mathematical consequences of Coulomb's law—to carry out the second step of the scientific method.

Fortunately or unfortunately, depending on your point of view, this is only the beginning of the story. A number of important discoveries were made in the years immediately following Coulomb which lead to far more important consequences than the law of force between charges by itself.

B. Animal Electricity

The progress of science very rarely proceeds in straight lines. We have been discussing the acquisition of knowledge about electrostatics as if it were a uniform, continuous process in which no mistakes were made and in which no blind alleys were followed. Of course, this is not the way things actually happened. To illustrate this point, and to give the reader a little more feeling for progress in science as it actually exists, we shall discuss in this section the strange case of Luigi Galvani. Galvani was a physiologist at the University of Bologna. He had spent a good deal of time investigating the anatomy of frogs and experimenting on the nervous system of the frog. In 1780, he acquired a machine, much like the one invented by Franklin, for generating static electricity. He also acquired a Leyden jar. One day in the laboratory two experiments were being run simultaneously. On one side of the lab the electrical machine was running, producing sparks. At the other side of the laboratory, an assistant was dissecting a dead frog with a steel scalpel. The assistant noticed when he touched a certain part of the frog's leg with the steel scalpel at the same time that the machine was discharging a spark, the frog's leg contracted. This response, which we now call the galvanic response, was a totally new phenomenon in the study of living systems.

Galvani immediately began to perform experiments to try to understand the basic nature of the phenomenon he had discovered. He discovered, for example, that glass tubes held near the frog's leg would not produce the response, while iron tubes would do so. He discovered that electricity from lightning would cause the response as well as electricity from the generator. He also discovered that if he pierced the spinal column of the frog with a brass hook and then pressed the brass hook against a piece of iron or steel, the contractions would occur as if the electrical machine were there. From this experiment, he concluded that there must be something called "animal electricity" which resided in the living organism. He felt that the contractions he saw, and the methods by which the contractions could be induced, were simply artificial means of moving animal electricity around inside of the organism.

Today, of course, we know that this idea of animal electricity stored in the body is false. The correct explanation of the galvanic response was given by Alessandro Volta (after whom the volt is named). He realized that the nerves were simply pathways along which electrical current could flow, and that what Galvani had done was simply to put outside electrical current into the nervous system.

At the time of Galvani, however, the idea of animal electricity proved to be very powerful. Since this electricity was presumably stored in the body, Galvani believed that it was the contamination and pollution of this pure energy that caused the diseases of old age. It also happened that as this idea became more widely known among the populations of Europe at that time, a number of quacks and charlatans appeared who promised to cure any disease by use of the new wonderful phenomenon of "animal electricity."

This was a period when people did not know very much about electricity. They also did not know a great deal about anatomy. The fact that the two were connected seemed a very mystical and magical thing. You must remember that in 1780 we were still very much in the first stage of the development of the scientific method as it applies to electricity and magnetism. Consequently, it is not too surprising that people were willing to believe almost any idea that came along. They simply did not have facts at their disposal that would have allowed them to prove or disprove any idea that someone put forward. Thus, when someone came along and offered to cure arthritis or tetanus or any other disease by the use of electrical apparatus, there was no means by which he could be shown to be a quack.

It has often happened in the history of medicine that as new ideas containing a germ of truth become known, they have been exploited by unprincipled individuals. For example, in our own time, we are just beginning to learn some of the laws and regularities of human behavior. At the same time, a large number of people need and want treatment for emotional disturbances. In the proliferation of new kinds of "therapy" on the fringes of respectable psychiatry there is a striking resemblance to the practitioners of "animal electricity" at the end of the 18th century.

Science does indeed make progress, but some of the basic characteristics of human behavior have not seemed to change radically in the last few hundred years.

C. Unexpected Discoveries about Magnetism

We have seen how the phenomena of magnetism have been put to commercial use in navigation. At the same time, men who were primarily interested in developing a better compass had collected a good deal of empirical knowledge about the behavior of magnets. It is well known that every piece of lodestone had a "north pole" and a "south pole." The north pole, by convention, was taken to be the end of the lodestone, which pointed in a northerly direction.

In addition to the fact that a lodestone would turn itself to become aligned along the north-south direction, it was also known that if the north poles of two separate lodestones were brought near each other, they would tend to be repelled from each other. The same thing would happen if two south poles were brought near each other. On the other hand, if the north pole of one lodestone were brought near the south

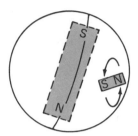

Figure 8.2.

pole of another, the force would be attractive—the magnets would tend to be drawn toward each other. Most of the readers have probably had first-hand experience with this sort of phenomenon.

If you will remember the conclusion of Gilbert that the earth itself was a gigantic magnet, then the behavior of a compass on the surface of the earth would be easy to understand. Consider the diagram in Fig. 8.2. If the earth is really a gigantic magnet, then it must have a north and a south pole. The north pole of our small magnet will then be attracted toward the south pole of the earth and repelled from the north pole of the earth. At the same time, the south pole of the small magnet would be attracted toward the north pole of the earth and repelled from the south pole. The net result will be that the magnet will be twisted into a north-south alignment.

One point in terminology has to be cleared up at this stage. Since the north magnetic pole of the earth must attract the south pole of the small magnet, it is clear that there is a possibility of confusion over what we actually call north and south. From Fig. 8.2, we see that this possible confusion is avoided by adopting the convention that the side of the small magnet that points toward the geographic north pole will be called the "north" pole of the magnet. This means that, technically speaking, the magnetic pole of the earth located in Greenland is the "south" pole of the earth's magnet even though in common usage it is referred to as the "north magnetic pole." It's pretty clear, though, that any attempt to get people to refer to something in Greenland as a "south pole" is doomed to failure, so we shall just have to live with this terminology as it stands.

Even as the controversy over animal electricity was raging, events were happening that would bring about a fundamental alteration in our ideas of what magnetism was. Alessandro Volta, in the course of his studies on the galvanic response, invented a device that was known first as the voltaic cell, but is today called the battery. He invented it for much the same reason that Benjamin Franklin invented the electrostatic generator. He wanted a convenient source of electrical charge for use in his experiments. The original cell was simply alternate layers of zinc and copper sandwiching pieces of wet pasteboard. When the zinc and copper plates were connected, a spark was observed to jump between them. Thus,

170

the zinc and copper plates must have had opposite electrical charges.

With this invention, it became possible to study not only stationary electrical charges, as had been done up to this period, but to study moving electrical charges as well. Because the battery provided a steady supply of electrical fluid, it was possible to connect one side of the battery to the other with a wire through which that fluid could flow. In modern terminology the flow of electrical fluid through a wire is called an electrical current. You will recall that the work of Franklin had shown that there was really only one type of electrical fluid, the type known as "negative." We know today that this fluid is nothing more than collections of tiny particles called electrons, each carrying a small negative charge. Thus, when one side of a battery is connected to the other, what actually happens is that electrons flow from the negative pole to the positive pole through the wire.

With the addition of the battery to the arsenal of scientific tools that were available to the experimenter, important new discoveries were not long in coming. In 1820, a Danish scientist by the name of Hans Christian Oersted made the fundamental discovery that proved the existence of a connection between the phenomenon of electricity and the phenomenon of magnetism.

Actually, Oersted had announced in 1807 that he wanted to investigate this problem. Unfortunately, he was not a very skillful experimenter. A student of his described him as "a man of genius but . . . he could not manipulate instruments." Oersted had set up a demonstration of the voltaic cell for a lecture in a class. He had connected a wire across the battery, so that a current was actually flowing in the wire. He noticed that when a compass was brought near the wire, the compass itself was deflected. This could only mean that in some way, a moving electrical charge (i.e., the charge flowing through the wire) produced a magnetic force in exactly the same way that a magnet produced such a force.

The importance of this discovery cannot be underestimated. We have seen in our study of mechanics that one of the most important realizations was that the physical processes that occurred on the earth and the physical processes that occurred in the motion of the planets were, in fact, one and the same. In much the same way, Oersted's discovery showed that the physical processes that occur in the phenomenon of electricity and the physical processes that occur in the phenomenon of magnetism, are, if not identical, at least interconnected and related in some fundamental way.

The organization of science is a very interesting subject. At any given time, most of the men who are working and carrying out research in a particular field are well acquainted with each other's work if not actually acquainted personally. It is not at all uncommon for a member of a faculty at a university to be better acquainted with the work of one of his colleagues on another continent than with the work of one of his colleagues who has his office down the hall. There is an "invisible college" in each field of science whose members correspond with each other, visit each other's laboratories, and, in general, keep themselves informed about the progress that is being made. It is not too surprising, then, that within a few months of Oersted's discovery of the connection between electricity and magnetism in Copenhagen, scientists all over Europe were investigating this new and exciting phenomenon. In particular, a young French physicist by the name of Andre Marie Ampère (after whom the "amp" is named) carried out a series of experiments that supplemented and widened the scope of Oersted's discovery.

The experiment conducted by Ampère was simple both in execution and in conception. He reasoned that if an electrical charge could affect a magnet, then a magnetic field ought to affect the current. He then set up an apparatus much like that shown in Fig. 8.3. Two long wires are made parallel to each other, and each wire is connected to its own battery. We have, then, a situation in which electrical current is flowing through each wire. The direction of the current in each wire can be reversed simply by reversing the connection at the battery (for example, if the one end of a wire is connected to the positive pole of the battery, the direction of the current can be reversed by connecting that end of the wire to the negative pole of the battery instead).

Ampère discovered that when the current in the two wires was running in the same direction, there was an attractive force between the two wires. That is, the two wires tended to be pulled toward each other. On the other hand, if the direction of the currents in one wire was opposite to the

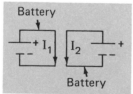

Figure 8.3.

direction in the other wire, the wires tended to be repelled, or pushed away from each other.

This is a simple statement of the results of Ampère's experiment. To interpret the experiment, however, it is probably easiest to make a short digression first and introduce the idea of an electric and magnetic field.

D. Technical Digression: The Idea of a Field

We saw in studying Newton's law of universal gravitation that if there was a mass M located at a particular point in space, then there would be a force on another mass, m, a distance d from the first mass, which was given by

$$F = \frac{G\,Mm}{d^2}.$$

What this means is that *if* I were to bring a small mass m to the position a distance d away from the large mass, and *if* I were to measure the force on that small mass, then I would find that the force was given by the above equation. It is sometimes useful, however, to talk about the points in space around the large mass M without actually carrying out the experiment explicit in Newton's law of gravitation. When we do this, we often speak of something that is termed a "gravitational field."

A gravitational field is defined as follows: It is the force that would be felt by a unit mass at a particular point in space, *if* that unit mass were to be brought to that point in space. For example, in the above paragraph we discussed the force between the large mass M and the small mass m. In order for the small mass m to be a "unit mass," it would have to be exactly one mass unit in whatever system of units we were using. For example, if we were measuring everything in English units, then m would be a unit mass if it were one pound-mass. If we were using metric units m would be a unit mass if it were one gram. If we were to bring this unit mass to the point a distance d from the large mass, it would experience a force given by

$$F = \frac{G\,M}{d^2},$$

where we have already set m = 1. This is called the gravitational field a distance d away from the large mass. Obviously, we can talk about the gravitational field at a distance d from the large mass without ever actually bringing small masses up to those points and measuring forces: We can simply use Newton's law of gravitation to calculate what the force would be if we did bring the masses up. We can imagine that a gravitational field exists all around the large mass M, even in the absence of other masses.

This gravitational field which we have defined can be interpreted in another way. Suppose that instead of bringing a unit mass up to a point a distance d away from the large mass, we brought a mass of size m. In this case, we would find a force given by the first equation in this section (F = GMm/(d²). If we divide both sides of this equation by m, we then find that we have

$$\frac{F}{m} = \frac{GM}{d^2}.$$

Comparing this equation to the one just above, we see that the force on a mass divided by the size of the mass is also equal to the gravitational field. Of course, this is implicit in our description of the gravitational field, but we mention it to emphasize the point that what is important in determining the gravitational field is the size of M, and that the gravitational field around M is completely independent of the size of the small mass m.

There is one more property of the gravitational field that we should notice. When we bring a small mass m up to a distance d away from the large mass, there is a force of a given size. But that force also points in a certain direction. That is, there is a force whose size is given by the equation we have just discussed, and whose direction is toward the large mass M. Thus, if we had a situation such as that pictured in Fig. 8.4, where points A and B are both the same distance from the large mass, the gravitational field is not the same at points A and B. The gravitational field is, in fact, one of those things, like the velocity, that has both magnitude and direction. It is a *vector*. It is possible, as in Fig. 8.4, for the gravitational field at two points to be equal in magnitude but to be different in direction.

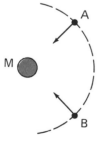

Figure 8.4.

174

In order to tie down the idea of a gravitational field, let us ask ourselves what the gravitational field would be at any point in space around the large mass M. If we imagined a sphere of radius d around the large mass, we know that the magnitude, or size, of the field at every point on that sphere would be the same, but the direction at every point would be slightly different. This is shown in Fig. 8.5. The arrows in this figure represent the gravitational field. The length of the arrow is supposed to be related to the size of the gravitational field, and the direction of the arrow shows the direction of field. If we now took a sphere of radius two times d (i.e., a sphere whose points are twice as far from the mass M as were the points on the original sphere), then, again we would have a situation in which all the points on the sphere would have the same magnitude field. Furthermore, we know from Newton's law of gravitation that the magnitude of the field at the second sphere will be only one-quarter as large as the magnitude at the first sphere.

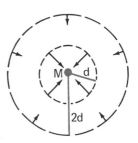

Figure 8.5.

It is obvious that we could carry out the procedure outlined above for a sphere of any size. This means that for each point in space around mass M, there is a gravitational field of a given size and a given direction. Furthermore, the magnitude of the field for every other point the same distance away from M will be equal to the magnitude of the field of the point in question, but will differ in direction. This is what we mean when we say that a mass "generates" a gravitational field in space around itself. It is important for the reader to realize that when I discuss a mass "generating" a gravitational field, I have done nothing more than restate Newton's law of gravitation in a slightly more obscure way. The statement that a gravitational field exists in space is no different from the statement that a force exists between two gravitating bodies. The statement that a gravitational field exists is simply a shorthand way of summarizing the results of a long series of hypothetical experiments in which masses are brought up to positions around the mass M and the forces on them measured.

It should now be obvious how one would go about defining an electrical field. We saw that Coulomb's law could be stated in the following way: If two charges q and Q are located at distance d apart, then a force exists between these two charges whose magnitude is given by

$$F = \frac{k\,q\,Q}{d^2}$$

and whose direction is such that the charges will be attracted if they are of unlike signs and repelled if there are of like signs. The electrical field around a charge Q is then defined as the force that would be felt by a unit *positive* charge at the point in question, or, equivalently, as the force per unit charge on a positive charge of any kind at that point. Going through the same mathematical steps on Coulomb's law as we did on Newton's law (dividing both sides of the above equation by q), we find that the expression for the electrical field at a point in space is given by

$$E = \frac{k\,Q}{d^2}\,.$$

We have chosen to define the electrical field in terms of the force that would be felt by a positive charge. Obviously, if a positive charge is attracted toward Q, then Q would have to be a negative charge. This means that a unit negative charge brought in the same position would be repelled instead of attracted. Thus, once we have defined the electrical field, we know the behavior of both positive and negative charges in that field. The only difference between the two is in the direction of the force that is exerted by Q.

In terms of fields, we can now restate Coulomb's law. Whereas before we had stated Coulomb's law in terms of actual forces between actual charges, we can now state it in terms of electrical fields around a single charge. In fact, stated in terms of fields, Coulomb's law is nothing more than the statement.

Around every electrical charge there is an electrical field whose magnitude is given by

$$\frac{k\,Q}{d^2}$$

and whose direction is such that like electrical charges will repel each other and unlike charges will attract each other.

The definition of a magnetic field can be made in an analogous way, but we have not yet defined what we mean by "the strength of a magnet." In other words, we do not have the analog for magnetism of the charge strength that we had for electricity.

We all know that some magnets are stronger than others. The magnets that are used to move old automobiles around in a junk yard are obviously of a different type from the small magnets that are used around the home and in children's toys. We know, however, that *any* magnet, when placed on the earth and allowed to move freely, will experience a torque that tends to twist it around and align it in a north-south direction. We can use this fact to define something that we shall call the "strength" of the magnet.

We could imagine performing the following type of experiment: A particular spot on the surface of the earth is chosen, and a magnet is placed there. By the same type of technique discussed in relation to the Cavendish experiment, we could measure the torque that is exerted on that particular magnet at that particular place on the surface of the earth. In the same way that we defined all masses relative to a standard mass, we could define the strength of any magnet in relation to one particular magnet, which we would use as our standard. The strength of a magnet is customarily denoted by the Greek letter "μ." It is usually called the "dipole strength" because each magnet has two poles, a north pole and a south pole.

Now, suppose that we take one of our magnets whose dipole strength is μ and place it in a particular region of space. Suppose that we observe that when the magnet is placed in this particular region of space, it tends to twist. The fact that the magnet tends to twist means that a torque is being exerted on it. Suppose, further, that we measure this torque and find that it has a particular value T. We can define that magnetic field B as the product of the magnetic strength of the magnet times the torque. In equation form, this is

$$B = \mu T.$$

We can regard this "magnetic field" in the same way as we regarded the electrical and gravitational field earlier. That is, it is a measure of the torque that would be felt by a particular magnet if that magnet were to be placed in the region where we believe the field exists. All other remarks concerning the

177

(a)

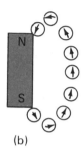

(b)

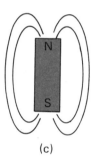

(c)

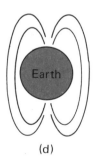

(d)

Figure 8.6.

existence of electrical and gravitational fields in space would apply to the magnetical field as well.

For example, we would say that a magnetic field exists around any magnet. This statement simply means that *if* we were to bring another magnet up, the second magnet would experience a torque due to the presence of the first.

To get some idea of what a magnetic field looks like, let us consider what would happen if we looked in a region of a standard bar magnet, which has a north and a south pole. If we were to lay a compass at any particular point around this magnet, we would find that the compass needle would turn in such a way that the north-seeking pole would point directly in the direction of the north pole of the magnet. If we laid a series of compasses down, beginning with one right at the north pole of our stationary magnet, we would find that these compasses would tend to line up in the way that is shown in Fig. 8.6a. If we then set down another magnet so that its north-seeking pole touched the south pole of the magnet in front of it, we could generate a line of magnets that would stretch from the north pole of the stationary magnet around to the south pole of the stationary magnet as shown in Fig. 8.6b.

We could, by laying down several lines of small compasses, generate a picture of the magnetic field around the large magnet that would look like that shown in Fig. 8.6c. This type of magnetic field is commonly referred to as the "dipole" field, because it is a type of field associated with magnets that have a single north pole and a single south pole. For reference, we show a sketch in Fig. 2.6d of a map of the magnetic field around the earth. We see that it, too, has the dipole form. A magnet laid anywhere on the surface of the earth will tend to line itself up along the lines of a dipole field, just as magnets laid around the stationary magnet tended to line themselves up along the lines of its magnetic field. It was the similarity between the field maps shown in Fig. 2.6c and Fig. 2.6d that led Gilbert to postulate that the earth was itself a gigantic magnet.

Now let us consider how we could state the findings of Oersted and Ampère in terms of fields. Oersted discovered that a compass (which is itself a small magnet) would be deflected near a wire that was carrying electrical current. This means that near the wire, a torque is exerted on a magnet.

By the above definition, this means that there must be a magnetic field there. Consequently, we can restate the findings of Oersted in the following way:

A moving electrical charge generates a magnetic field in the space surrounding it.

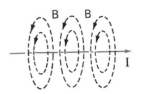

Figure 8.7.

In the case of Oersted's experiment, the moving electrical charge was simply the charge flowing through the wire. However, we now know that even a single charge moving along in free space will generate a magnetic field around itself. In this case, of course, the magnetic field will move along with the charge.

If we were to carry out the same experiment with the current-carrying wire that we carried out with the single magnet, we could map out the magnetic field around the wire. In Fig. 8.7, we show the results of such an experiment. In this case, the field lines would be circles with the wire at the center. This is an example of a magnetic field that is not of the dipole form. If the direction of the current in the wire is reversed, the direction of the field lines is reversed. There is an easy way to remember which way the field around a wire carrying a current will point, which is called the "righthand rule." If you put the thumb of your right hand in the direction of the current, then the fingers of your right hand will be curled in the same direction as the magnetic field.

Although the magnetic field around a single wire is not of the dipole form, it is very easy to construct a dipole magnet field from a wire. One simply has to wrap the wire into a coil as shown in Fig. 8.8. In this case, if a current were flowing through the wire in the direction as shown, the magnetic field line would have the form pictured. The reader will immediately see that this is a dipole field. The explanation as to why a single wire should have a circular field but a coiled wire should have a dipole field is left to the questions at the end of the chapter. We merely state that this principle, which allows us to produce something that looks very much like a permanent magnet from a coil of wire is the basic principle of the electromagnet.

Now let us turn our attention to the experiment of Ampère. He discovered that two wires that were carrying current would exert a force on each other. We know from

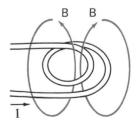

Figure 8.8.

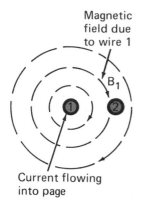

Magnetic field due to wire 1

B_1

Current flowing into page

Figure 8.9.

Oersted's experiment that around a single wire carrying current there is a magnetic field. In particular, there will be a magnetic field due to the first wire at the position of the second wire (see Fig. 8.9). This means that at the position of the second wire, we have a situation in which there is a magnetic field (produced by the first wire) and a moving charge (the moving charge which makes up the current flowing in the second wire). The existence of a force on the wire can then be interpreted as an effect of the magnetic field on the moving charge. In fact, we can restate Ampère's results in the following way:

> A magnetic field will exert a force on a charge moving through it.

As in the case of the magnetic field produced by the moving charge, this result holds both for a current (which is a collection of charges moving along) and for an isolated charge. In fact, something very similar to the right-hand rule, which we discussed above, holds for the direction of the force exerted on a charge. The rule can be stated in the following way: If you put the index finger of your right hand along the direction of motion of the charge, and the middle finger of your right hand along the direction of the magnetic field, then your thumb will point in the direction of the force. We leave as an exercise for the reader to verify that this rule will reproduce the results of Ampère's experiment.

We have seen that the introduction of a concept of the electric and magnetic fields allows us to restate many of the experimental results that have been derived up to this point. It is extremely important, however, that the reader realize that restating a result does not change it nor add to it in any way. The statement of Coulomb's law in terms of forces between particles and the statement of Coulomb's law in terms of the electrical field around a single particle are completely equivalent: simply the same experimental facts stated in alternative languages. We have also seen, however, that many experimental laws, such as Ampère's law, can be given a new depth of interpretation by talking in terms of field rather than in terms of force. In other words, we tend to understand Ampère's law better in terms of magnetic fields and the forces that magnetic fields exert on moving charges than we

180

Benjamin Franklin (1706-1790) investigating lightning.

A modern electronic device—this is a minaturized circuit for a computer. You can get some idea of the scale by looking at the Bicentennial quarter it is resting on.

could if we simply regarded it as a law concerning the forces between two wires. In addition, when we state the law in terms of the force between a moving charge and a magnetic field, we suggest other experiments that should be done and whose results we should be able to interpret in the same way. Thus, Ampère himself experimented only with long straight wires carrying current. If, however, we discuss his results in terms of magnetic and electrical fields, then we would predict that a moving charge in the vicinity of a stationary magnet should also have a force exerted on it. This suggests an experiment that could be done to test our understanding of Ampère's law. In fact, such experiments have been done, with results that we would expect—namely, that single electrical charges traveling in stationary magnetic fields are deflected regardless of the source of that field (that is, regardless of whether the field is created by a permanent magnetic or by a current-carrying wire).

With this understanding of the concepts of electrical and magnetic fields, we can now turn to some later discoveries that shed further light on the phenomena of electricity and magnetism, and that further emphasize the idea that these two seemingly very different things are in fact related at some fundamental level.

E. Induction and Displacement Currents: The Plot Thickens

We have seen that electrical fields can be generated by stationary electrical charges, and that magnetic fields can be generated either by permanent magnets or by moving electrical charges. This seems to be a rather one-sided state of affairs. Somehow, the fact that electrical charges can in some way produce magnetic fields leads one to suspect that it ought to be possible for magnets, in some way, to produce electrical fields. The proof that this supposition is indeed correct was given first by Michael Faraday.

Faraday was a rather interesting person in addition to being one of the prominent scientists of the 19th century. He was born into a humble background—his father was a blacksmith in northern England. At the age of 14, as was the custom in those days, he was apprenticed to learn a trade. In this case, he began working in the shop of a bookbinder.

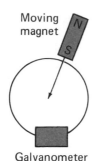

Galvanometer

Figure 8.10.

He quickly developed the habit of reading the books as he was binding them. Consequently, he became a self-educated man. It was during this time that he began to develop an interest in science and decided that he would like to become a scientist if possible. The obstacles set in his path by the English class system should not be underestimated by 20th-century reader. He broke into the field by convincing Sir Humphry Davy, a prominent physicist at that time, to hire him as an assistant. He convinced Davy that he would make a very good assistant by the simple stratagem of attending some lectures that Davy gave, taking extremely good lecture notes, and then binding them in a rather elegant way. These he presented to Davy as his calling card and as a proof of his interest in science.

It did not take Faraday long to move from being an assistant who helped Davy present his lectures to being a recognized scientist in his own right. Like other scientists of this period, he knew, of the discoveries of Ampère. He became convinced that it ought to be possible to produce an electrical field with magnets, just as it was possible to produce a magnetic field with electrical currents. Some of the experimental apparatus that he used to prove this point is shown in Fig. 8.10.

Suppose we have a large loop of wire that is not connected to a battery or any other source of electromagnetic charge. Suppose also that there is in this loop an instrument that can detect the presence of moving electrical charges (this type of instrument is called a galvanometer). Suppose now that we move a large permanent magnet around this loop in any way we please. What we observe is that as long as the magnet is moving, a charge will flow in the wire. When the magnet stops moving, the charge stops flowing.

The fact that the charge moved implies, by Newton's second law, that there must be a force acting on it. From our definitions, the presence of a force that can cause a charge to move would be equivalent to the existence of an electrical field. At the same time, the fact that a magnet is moving through the loop of wire means that the magnetic field near the wire is changing. Thus, one statement of the results of Faraday's experiment would be to say:

A changing magnetic field causes an electrical field in a conductor.

In the original language of the experiment, however, we would probably say something to the effect that a changing magnetic field induces a current in the wire. Thus, this process is sometimes called the process of "electromagnetic induction."

Faraday also discovered the laws governing the direction in which the current in the wire will flow. You will recall from our discussion of Oersted's discovery that when current is flowing in a particular direction in a wire, the direction of the magnetic field near the wire is given by the right-hand rule. Thus, in the Faraday experiment, when the current in the wire is flowing counterclockwise, the magnetic field due to the current flowing in the wire will be up and out of the page inside the loop (see Fig. 8.10). Conversely, if the current is flowing clockwise, the magnetic field in the loop due to the current flowing in the loop will be down, or into the page. Of course, when no current is flowing, there is no magnetic field in the loop due to anything happening in the wire.

Now suppose that we begin moving a magnet toward the loop of wire. Clearly, this will tend to increase the magnetic field in the loop. This increase is due solely to the motion of the magnet and has nothing to do with anything that is happening in the loop. What Faraday discovered was that as soon as the magnet began moving toward the loop, a current began flowing in the loop, and the direction of that current was such that the magnetic field created in the loop was opposed to the direction of change in the magnetic field due to the magnet itself. In other words, if we move the magnet in such a way as to make the magnetic field in the loop become larger in the upward (out of the page) direction in Fig. 8.10, then the current will start flowing in the loop in such a way as to create its own magnetic field, which will be downward (into the page). The current will always flow in such a way as to oppose the motion of the magnet. A simple way to remember this is just to recall that nature does not like things to change.

A series of experiments was undertaken by Faraday to make a quantitative statement of the law of induction. If the loop in Fig. 8.10 has an area A (for example, if the loop is a circle of radius R, then A would be equal to πR^2), and has length ℓ, (for a circle, ℓ would be $2\pi R$), then Faraday found that the relation between the change of magnetic field, ΔB, in

Michael Faraday (1791-1867).

a time, Δt, and the electrical field in the wire, E, is given by the following law:

$$\frac{\Delta (BA)}{\Delta t} = -\ell E.$$

Thus, not only had Faraday demonstrated that a changing magnetic field can cause an electrical field, just as a moving electrical charge can cause a magnetic field, but he had written down the precise mathematical form governing the change of the electric field.

With the discovery of induction, one more piece of the jigsaw puzzle of electromagnetism fell into place. It turned out that only one more piece needed to be supplied before the science of electricity and magnetism could be developed to the point that mechanics had been developed by Newton. This last piece of the puzzle was put into place by another English scientist—James Clerk Maxwell (1831–1879). We shall discuss the full significance of his work in the next section, but for the moment we shall confine our attention to the phenomenon of "displacement current," which he introduced.

Let us consider an experimental apparatus such as that shown in Fig. 8.11. A battery is connected to a switch through an apparatus consisting of two large metal plates with nothing in between them. Let us consider what will happen when the switch is closed. Closing the switch will cause negative electrical charge to start to flow onto the left-hand plate of the apparatus in the box. However, if a negative charge flows onto the left-hand plate, we know by Coulomb's law that the negative charges on the right-hand plate will be repelled. If the plates are close enough together, moving a negative charge onto the left-hand plate will cause some negative charge to move off the right-hand plate. Clearly, the only place that it can move is into the wire. Since the right-hand plate started out being electrically neutral—that is, it had equal amounts of positive and negative charges—pushing a negative charge away from it will result in an excess of positive charge on the right-hand plate. Thus, closing the switch and allowing negative charge to flow onto the left-hand plate will start the negative charge flowing off the right-hand plate and will result in the right-hand plate acquiring the net positive charge.

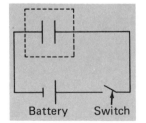

Battery   Switch

Figure 8.11.

When Maxwell began to think about this particular experience, he realized that there was a rather curious paradox with the laws of electricity and magnetism as they were known at that time. This paradox can be seen by asking a simple question: "Is there a magnetic field in the vicinity of the parallel plates?" From what we have seen so far, the answer would have to be no. After all, from the discoveries of Oersted, the only thing that can create a magnetic field is a moving electrical charge. Obviously, no electrical charge actually moved across from left to right in the above example. Therefore, there is no electrical current in the region between the two plates, and there can be no magnetic field.

On the other hand, suppose we enclose the entire apparatus with the plates in a large black box (denoted by the dotted line in Fig. 8.11). Then what we would see when the switch to the battery was closed would be a negative charge flowing in from the left and a negative charge flowing out from the right. Such a flow would quite rightly be interpreted as an electrical current. Consequently, we would use the discovery of Oersted to say that, since an electrical current flowed into the box on one side and out on the other, there would have to be a magnetic field in the region of the box.

Obviously, the magnetic field cannot both be there and not be there. This was the nature of Maxwell's paradox. He felt that it was unreasonable to suppose that the laws of nature would be such that the presence or absence of a magnetic field would depend on whether we knew what was inside of a black box. In order to resolve this dilemma, he asked himself what there was between the two plates. He knew that there were no electrical currents in the usual sense. But, he also realized that *something* was happening between the plates. Before the experiment starts (that is, before the switch is turned on) there is no electrical field between the plates. If we were to put a charge between the plates, it would not move. However, after the switch has been closed and negative charge has flowed onto the left-hand plate and off the right-hand plate, then the situation is different. If a small negative charge were placed between the plates, we know that it would be repelled by the left-hand side and attracted by the right-hand side. Thus, it would move toward the right. The fact that it moves means that there is an electrical field between the plates after the switch is closed, while there was no field between the plates before the switch is closed.

Maxwell then reasoned as follows: Since we have a situation in which an electrical field is changing, and in which a magnetic field ought to be present, it must follow that a changing electric field can generate a magnetic field, just as a changing magnetic field can generate an electrical field. Looked at this way, Maxwell's conclusions about the possibility of generating magnetic fields is just the other side of the coin of Faraday's law of induction.

Since a changing electrical field could generate a magnetic field, in the same way that an ordinary current could generate a magnetic field, Maxwell called the changing electrical field a "displacement current." He found that if an ordinary current "I" could create a particular magnetic field at a particular point in space, then an electrical field that changed by an amount ΔE in a time Δt would create a magnetic field at the same point. Thus, he felt that the true current at any point in space was given by the equation

$$I = I_N + I_D ,$$

where I_N is the normal current consisting of the flow of charged particles and I_D is the displacement current. The displacement current is given by

$$I_D = \left(\frac{1}{k} \right) \frac{\Delta E}{\Delta t} ,$$

where $\Delta E / \Delta t$ is the rate of change of the electrical field with time.

This new definition of an electrical current completely resolves the dilemma posed above. Between the plates in our apparatus, there is no flow of charge. This means that the normal current I_N is zero. This does not, however, mean that the total current is zero, since the electrical field is changing between the plates. Thus, even though there is no flow of charge, there is still a current, and, by our previously stated principle, a magnetic field will be generated. At the same time, in Oersted's experiment the displacement current is zero, since there is no changing electrical field anywhere, but the normal current is not zero since charge is flowing.

186

In this picture a magnetic field can be generated either by a normal current or by a displacement current or by some combination of the two.

With the introduction of the idea of displacement current, the science of electricity and magnetism was ready for its next big step forward.

James Clerk Maxwell (1831-1879).

F. Maxwell's Equations: Putting It All Together

Suppose that you are a scientist living in the last part of the 19th century and someone asks you to put together a list of everything known about electricity and magnetism. Suppose further that he asked you to do this in the language of electric and magnetic fields. You would probably come up with a list something like this:

1. Around every electrical charge there is an electrical field whose direction is determined by the requirement that like charges will repel each other and unlike charges will attract each other.
2. There is in nature no such thing as a magnet that has a north pole but no south pole or vice versa.
3. A magnetic field can be generated either by a moving charge or by a changing electrical field.
4. An electrical field can be generated by a changing magnetic field.

These four laws of nature, which were discovered over a period of more than a century of work, are today called Maxwell's equations. They play the same role in electricity and magnetism that Newton's laws of motion play in mechanics. They are the culmination of the first phase of the scientific method as it is applied to the field of electricity and magnetism. Like Newton's laws of motions, they are statements of experimental fact. They are not derived mathematically from anything else, but rather are generalizations and inductions of phenomena observed in the laboratory.

Obviously, all of these equations were not discovered by Maxwell. His contribution to the field was the introduction of the idea of displacement current and the mathematical working out of many of the consequences of the equations. This, too, is not unusual. Many of the laws that we saw

187

in mechanics were not derived by Newton. He was simply the one who saw the entire picture, put it all together, and worked out the consequences.

Since Maxwell's equations are *experimental* laws, there is always a possibility that they may be wrong. In other words, there is always the chance that when we look at some heretofore unexplored regions of the universe (for example, the world of very, very small distances) that we will find that Maxwell's equations would no longer hold. All we can say with certainty is that for the part of the universe that we have explored and for which we have good experimental data, we have not found any violation of Maxwell's equations. This is rather a striking statement, since our knowledge of the universe today extends from particles smaller than the atomic nucleus all the way up to enormous structures of super galaxies. In all of this vast range of sizes, no violation of Maxwell's equations has ever been found.

Many physicists have wondered about this. In particular, the second of Maxwell's equations—the statement that north and south magnetic poles always occur together, and are never found separately—is logically impossible to prove. It's a little like trying to prove that there is no such thing as a unicorn. All we can say is that we have not seen a unicorn (or an isolated magnetic pole) yet. It is always possible that the next round of experiments will find either a unicorn or an isolated magnetic pole.

Many physicists (the author included) have spent a good deal of time wondering why nature has chosen to construct the universe in such a way that there are isolated electrical charges but no isolated magnetic "charges." We have seen that the history of electricity and magnetism has taught us over and over again that there is a fundamental deep connection between electrical and magnetic phenomena. And yet, in this one case there seems to be a rather basic difference.

It is not that people have not searched for magnetic charges (called in the language of physics "magnetic monopoles"). Searches have been made for these hypothetical particles at accelerators, in chunks of ice around the north and south magnetic poles of the earth, and even in rocks brought back from the moon. There are good reasons to believe that if monopoles existed, they would be found in such places.

More recently, a group of scientists at the University of California at Berkeley has claimed to have found indirect evidence for a magnetic monopole by analyzing experiments in which apparatus was flown near the top of the atmosphere in a balloon. Other scientists have contested their results, and as of this writing the issue is not settled. All that we can say is that up to now no one has "brought a monopole back alive." It appears that the part of Maxwell's equations dealing with magnetic charges remains something of a puzzle.

As was the case with Newton's law of motion, when the correct basic equations for electricity and magnetism were written down it became possible to carry out a number of important calculations which had, up to this time, been too difficult for scientists to do. The immense technical advantages that have accrued to humanity because of the discoveries of electricity and magnetism cannot be underestimated. Try for a moment to imagine a society like ours without electricity. It is simply inconceivable. The entire industrial base of our civilization rests on the use of electricity—for power, for communication, for heating and lighting, to mention but a few examples.

For our purposes, however, we shall consider only one consequence of Maxwell's equations—derived by Maxwell, himself—whose discovery led to important new breakthroughs in basic science. This was the prediction of a phenomenon called the "electromagnetic wave."

In order to understand what electromagnetic radiation meant to Maxwell, we have to realize that his picture of the universe was rather different from ours. He believed in the physical existence of an entity called the "ether." This ether was supposed to be a substance that filled empty space. You can think of it as a kind of tenuous, jello-like material which existed between material objects. In Maxwell's view, this ether had a very complicated structure. He imagined it to be full of teeth and gears and wheels (much as Ptolemy thought of the solar system as being filled with epicycles). He interpreted the displacement currents, for example, as a stretching and spinning of certain parts in ether.

Now, if you think of empty space in this way, a rather interesting consequence follows. We all know that if we have a large piece of jello and we wiggle the jello at one point, pretty soon the jello will start to wiggle at another point. In the language of physics, we say that the wave has traveled from the first point to the second. We are all familiar with waves in many aspects of our life. For example, if you throw a rock into a pond, you observe a wave spreading out from that rock. If we imagine throwing a rock in the pond every second, we would see waves going out from that point every second. If there were a cork sitting in the water 10 feet away, then the cork would bob up and down every second— that is, every time a wave arrived at the position of the cork.

What Maxwell did was to take this intuition about the nature of ether very seriously. He felt that he had written down the correct equations describing phenomena of electromagnetism. Consequently, it should follow mathematically from his equations (just as elliptical planetary orbits followed mathematically from Newton's laws of motion) that waves ought to be propagated from one point to another. In our modern language, we would say that these waves would be propagated through a vacuum. In Maxwell's language, we would say that they were waves carried in the ether. Whatever the interpretation of the way that the waves travel is, however, it does follow from Maxwell's equations that such waves must exist. What he found was that if a charge is accelerated at one point in space, then at a later time another charge, located at a certain distance from the first charge, will begin to be accelerated as well.

Since the second charge moved, it follows that there must be a force on it, and hence there must be an electrical field at the position of the second charge. This means that the electrical influence caused by moving the first charge traveled through the intervening ether (or empty space) to the position of the second charge. Thus, the electrical influence moved from one point to another much as a wave moved from the rock to the cork in our above example. This movement of electrical influence is usually referred to as an electromagnetic wave.

The next question that Maxwell asked was how fast such waves would travel. It follows from the equations that the speed of electromagnetic radiation, commonly denoted by the letter c, is given by

$$c = \sqrt{k},$$

where k is the same constant that appears in Coulomb's law and in the other equations of electricity and magnetism. Of course, the numerical value of k had been known since the time of Coulomb. When Maxwell took the experimental value of a k in order to find what the speed of electromagnetic radiation was, he found that

$$C = 3 \times 10^8 \text{ m/sec}$$

$$300,000 \text{ km/sec}$$

or, in an English system of units,

$$C = 186,000 \text{ miles/sec.}$$

This number is probably familiar to many readers. This is, in fact, the value of the velocity of light. When Maxwell saw this result, he realized that it must be that these mysterious "electromagnetic waves," which had followed as mathematical consequences of his equations, must be identical with light itself. Thus, he had a situation in which the electromagnetic waves had been known all along, but had not been recognized as such.

Of course, the simple identification of the velocity of electromagnetic waves with the experimentally measured velocity of light was not a complete proof of the identity between the two. However, it was not long before further consequences of Maxwell's equations were worked out by Maxwell and other mathematicians, and the proof was made very solid indeed. For our purposes, the most important aspect of this stunning discovery is the fact that the speed of electromagnetic radiation depends only on the force between two stationary electrical charges. It can be determined by measuring the constant k which appears in Coulomb's law (and, the reader will recall, this constant can be measured by doing experiments in which the force between two charges is measured). Thus, we have the rather unique situation that the speed of light can be determined without ever measuring it directly. This, as we shall see, will have important consequences in the theory of relativity.

Maxwell had identified light as a species of electromagnetic radiation. In 1889 the German scientist Heinrich Hertz discovered another kind of electromagnetic radiation, which, although it traveled at the same speed as light, had a much longer wavelength (the successive crests). These waves were what are now called radio waves. The application of the discovery of Hertz to a practical situation was made by the Italian scientist Guglielmo Marconi in 1896. The consequences to all of us of these applications are, I think, obvious and we need not go into them any further.

We have seen, then, how the phenomena of electricity magnetism, lightning, and light itself have been unified by the work of many scientists over several centuries. The science of electricity and magnetism has gone from beginnings in legends and mythology to a set of mathematical equations summarizing precise experiments and experimental laws. From these mathematical statements, the important consequence of the existence of electromagnetic radiation, and the identification of that radiation with ordinary light, followed logically. In the years following Maxwell's original statement of the basic laws of electricity and magnetism, most of the important consequences of the equations have been worked out, and are now being applied in our everyday life. With this development, the acquisition of basic knowledge of electricity and magnetism and the application of the scientific method in this new field comes to an end.

G. Electrical Circuits

The electromagnetic phenomenon with which we are most familiar is probably the electrical circuit of the type we use every day in our home or car. In this section, we will discuss how such a circuit operates.

A circuit is simply a path made out of conducting material which allows electrons to move around it. In home wiring, this path is usually made of copper wires. Every circuit has two elements in it—an energy source and a load.

The energy source is the device that moves the electrons around the circuit, and hence the device that supplies them with kinetic energy. While supplying energy to electrons in the circuit, the energy source must itself draw on non-electrical sources of energy. For example, your car battery converts the chemical energy in the battery constituents to kinetic energy of electrons, while normal household power is derived from the burning of coal or the fissioning of uranium in the generating plant that is linked to your home. In either case, the function of the energy source is to take one kind of energy and convert it into another kind—electrical.

One way of thinking about the function of the energy source is to recall our discussion of the roller coaster in the previous chapter. In that discussion, we saw how a motor lifted the roller coaster up to a certain height, thereby doing work against the force of gravity and giving the roller coaster potential energy, which was then converted to kinetic energy in the fall. In the same way, we can think of the energy source as supplying the electrons in the circuit with potential energy, which is then used up in the movement around the circuit. Every energy source has a positive and a negative pole. In a battery, for example, these are the two terminals to which the car wiring system is attached. In household wiring, they correspond to the two prongs of the plug normally attached to appliances. In any case, the energy source works as shown in Fig. 8.12. Electrons enter the positive side of the source after having been through the circuit. It takes no energy to make them do this, because the electrons have a negative charge and would naturally move toward a positively charged object.

Inside the source itself, however, the situation is different. There the electrons find that they have to move from a positive to a negative object. They cannot do this on their own, any more than the roller coaster can lift itself up unaided. Something must do work against the electrical field inside the source, just as the motor had to do work against the gravitational field in the example of the roller coaster. The "something," of course, is just the outside source of energy.

We have mentioned the two most common energy sources for circuits—the battery and the generator. These sources, while similar in some ways, differ markedly in their operation. The battery is constructed so that one pole is always positive and the other always negative. Thus, the

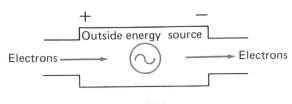

Figure 8.12.

current always comes out of the source at the same place and moves in the same direction through the circuit. Such systems are said to use direct current, or D.C.

A generator, on the other hand, produces electricity by spinning a loop of wire in a magnetic field, using the principle of induction to create current. In such a system, the current will flow one way in the loop for half of the turn and the other way for the other half. In such a system, the "positive" and "negative" poles of the power source would be reversed every half turn of the loop, and the current in the wire would then flow one way in the circuit for half of the time and the other way the other half of the time. Such a system is said to use alternating current, or A.C.

Electrical units are customarily stated in terms of the metric system. For our purposes, we can start by defining a unit of electrical charge. The total charge of 6.3×10^{18} electrons is called one coulomb of charge. When an energy source causes one coulomb of charge to flow past a point in the circuit each second, we say that it is delivering one ampere of current (usually abbreviated "amp"). Thus, the "amp" that we deal with in household wiring is simply a flow of electrons -6.3×10^{18} per second for each amp.

You will recall that in the metric system of units (see Appendix A), the unit of force was the Newton, and the Newton was defined to be the force that would accelerate one kilogram at one $meter/(sec)^2$. One Newton acting through one meter did one joule of work. The unit of potential energy in the electrical system is called the volt. A power source which expends one joule of work on one coulomb of charge gives that charge one volt of potential energy. In analogy with the gravitational potential, we usually say that the source "raises" one coulomb to one volt.

To get some idea of what these units mean, a normal household lighting circuit (A.C.) has a potential difference of about 110 volts between its terminals and may carry anything from one to 15 amps, depending on how many appliances are plugged in. A car battery maintains a potential difference of six or 12 volts between its terminals, and may deliver up to 50 amps for short times when the car is being started.

We have been careful to use the term "potential difference" in describing these voltages. The voltage represents

194

the difference in energy that the electrons have when they enter the source at the positive side and when they leave at the negative side. In analogy, the potential difference in the roller coaster example was the difference in energy of the roller coaster when it was on the ground and when it had been lifted up by the motor. Just as the roller coaster will not move unless there is a drop in potential between the point to which it is going and the point it is leaving, so too the electrons will not move unless there is a drop in potential between the point at which they enter the circuit and the point at which they come back to the source. This drop in potential, or potential difference, is often called the "voltage" of the source.

A source can be rated according to its voltage, or it can be rated according to the power it delivers. Since one watt of power corresponds to one joule per second, it follows that the power of a source is given by the formula

$$\text{watts} = \text{volts} \times \text{amps}.$$

Thus, a 100 volt source delivering 10 amps will also be delivering 1,000 watts, or one kilowatt (kw). If the source delivers 10 amps for one hour, it will deliver a total energy of 1 kilowatt hours. This unit—the kilowatt-hour—is what is usually used to compute your household electricity bill.

With this understanding of the energy source in a circuit, we can move on to discuss the other half of the circuit— the load. If the source is the device supplying energy to the circuit, the load is whatever uses that energy up. It may be a lightbulb, a toaster, a radio, or an electric range. Whatever it is, it is the ultimate user of the energy that enters the circuit.

In order to understand how a load affects the circuit, we have to introduce one more electrical concept—the concept of resistance. When the electrons leave the source, they are moving with some velocity. As they move through the wire, however, they begin to collide with atoms. These collisions result in a loss of energy by the electrons. The energy lost by the electrons has not disappeared, of course, but has simply been converted to kinetic energy of the atoms in the wire. As we shall see later, this kinetic energy is what we call heat, and this phenomenon explains why wires heat up when a current passes through them.

195

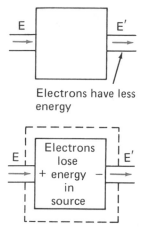

Figure 8.13.

Another way of describing the collisions between electrons and atoms is to say that the material resists the motion of the electrons. We can define this resistance quantitatively by noting that if we have a situation such as that shown in Fig. 8.13, where electrons enter a box with some energy E and leave the box with a lower energy E', the loss of energy is equivalent to having a source hidden in the box which has a positive pole on the high energy side and the negative one on the low energy side. The electrons entering this source would have to expend energy to climb the voltage "hill," and would come out with reduced energy just as surely as they would if they suffered collisions in the wire. Thus, there is a formal equivalence between energy loss due to resistance and energy loss due to having a source put in the circuit the "wrong way around." This means that we can talk about the energy loss associated with collisions as a voltage drop across the wire in which the resistance occurs.

This equivalence allowed the German physicist George Simon Ohm to define resistance in the following way: A material has a resistance of one ohm if there is a one-volt potential difference across it when one amp of current flows. In equation form, this definition takes the form

$$V = I \cdot R$$

where V is the voltage drop across the wire, I—the current, and R—the resistance in ohms. This equation is known as Ohm's law.

A circuit, then, must look like the drawing in Fig. 8.14. There is a source, which has a voltage V, emitting a current I into the circuit. This current then flows through the load, developing another voltage, oppositely directed to the source voltage, when it does so. For a steady situation, the voltage developed over the load will be just equal and opposite to the voltage developed at the source, so that the total potential drop for an electron that goes all the way around the circuit and through the source will be exactly zero. This, in turn, means that the current must be given by

$$I = \frac{V}{R}.$$

Some examples of the workings of familiar circuits are given in the problems.

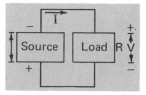

Figure 8.14.

196

SUGGESTED READING

Everitt, C.W.F. *James Clerk Maxwell: Physicist and Natural Philosopher.* New York: Scribner's Sons, 1976. A scientific biography of Maxwell by a contemporary physicist that contains some information about the man himself as well as his work.

Goodman, Nathan G. *A Benjamin Franklin Reader.* New York: Crowell, 1945. An excellent source on Franklin's scientific experiments.

Hall, A.R. *The Scientific Revolution 1500–1800.* Boston: Beacon Press, 1954. Has a good description of the early work in electricity and of the "animal electricity" episode.

Whittaker, Edmund. *A History of the Theories of Aether and Electricity.* New York: Humanities Press, 1973. A very complete, but very technical and mathematical, description of the history of electromagnetic theories. It has the best description that I have found of the way Maxwell thought about the world.

QUESTIONS AND DISCUSSION IDEAS

1. Using Coulomb's law, explain why electric current flows from the negative side of a battery, through whatever wires are in the circuit, and back to the positive side.

2. Using the right-hand rule, explain why the coil of wire shown in Fig. 8.8 produces a dipole field, even though a straight piece of wire does not.

3. Consider an arrangement as shown below, in which two wires carry current in opposite directions, but with the magnitude of the current being the same in both wires:

    a. What is the magnitude of the magnetic field midway between the wires? (HINT: You have to add the field from one wire to the field of the other.) You may leave your answer in terms of the field due to one wire alone.
    b. What will happen to a compass needle placed midway between the wires?
    c. Redo 3a and 3b in the case where the currents in the two wires flow in the same direction.

4. Consider the arrangement shown below, in which the pole of a magnet is placed above a loop of wire. It is

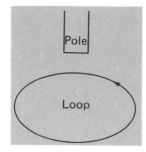

observed that when the magnet is lowered toward the loop, a current flows, a counterclockwise current flows in the loop.

    a. Which way will the current flow when the pole is withdrawn from the loop?

    b. Suppose the magnet is removed and then returned with its opposite pole pointing toward the loop. Now which way will the current flow when the magnet is lowered? When it is removed?

5.   How do you think the following men would have reacted to the idea that there was a relationship between electricity and magnetism?

    a. Aristotle

    b. Thales

    c. Benjamin Franklin

    d. Galileo

6.   I have the following pieces of equipment: some identical permanent magnets, and a coil of wire through which is flowing a current I. One side of the coil is painted red and one side of the magnet (the same pole on each magnet) is painted blue.

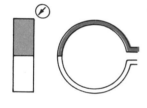

When I bring a small compass needle near the magnet, it points in the direction shown:

    a. If I put the blue end of two magnets near each other, will there be a force? If so, which way will it tend to push the magnet?

    b. The blue end of the magnet is moved toward the loop. After it has been sitting there for a while, will there be a force on it? In which direction?

    c. Now the magnet is jerked away quickly. Will the current I change? What general law tells you which way the extra current flows?

7.   Consider a situation like that pictured below, in which we have a gas made up of particles of equal mass, but where half of the particles are positively charged and half are negatively charged.

Suppose now that we begin running a negative charge onto the left-hand plate in the drawing. The plate is initially uncharged.

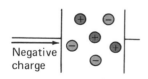

    a. Describe the motion of the particles in the gas when a little bit of negative charge has accumulated on the plate.

    b. Give an argument that shows that the right-hand plate will eventually acquire a positive charge.

    c. In analogy with Clerk Maxwell's discussion of displacement currents, show that we can regard this whole apparatus as a black box with current flowing through it.

    d. Is the current in 7c a displacement current, a real current, or a combination of the two? Why?

8. In the text, we mentioned that in modern terms, the electric current is actually made up of tiny negatively charged particles called electrons. Let's anticipate some of our present knowledge of atomic structure a little to see how this works. An atom (see drawing below) consists of a heavy, positively charged part called the nucleus surrounded by a swarm of small, negatively charged objects called electrons. The total charge is zero.

The total charge is zero

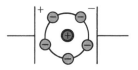

    a. Suppose two plates are brought near the atom, as shown, with one plate positive and one negative. Will there be a force on the atom? Why or why not?

Suppose now that for some reason one of the electrons which is usually on an atom is removed, as in the drawing below:

    b. What is the charge on the part of the atom (nucleus plus all but one electron) that is left? This kind of atom—with a missing electron—is called an ion.

    c. Suppose the ion and the loose electron are put between the charged plates in 8a. Will there be a force on the ion? On the electron? Which way will the force on each one point?

    d. Give an argument to show that the force on the ion and the electron are equal in magnitude, but opposite in sign.

9. Let us consider the consequences of the arrangement in the above problem. Given equal and opposite forces on the electron and the ion, and given the fact that the ion weighs several thousand times what the electron weighs,

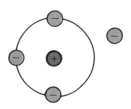

    a. Which particle will have the greater acceleration? How much greater?

       b. Discuss how we could generate an electric current from a collection of ionized atoms.

10. Let us consider the idea of the magnetic monopole. In analogy to the case of electrical charge, a monopole would experience a force

$$F = gB$$

in a magnetic field B (g is the size of the magnetic charge).

    a. Suppose a monopole were to enter the region near the earth. From what you know about the earth's magnetic field, which way would it be pushed?

    b. Hence make a guess as to why people have looked for monopoles in blocks of ice from Greenland.

11. Consider a normal household circuit that has 110 volts across its terminals and that is connected to a 100-watt lightbulb.

    a. If there is nothing else in the circuit, how much current will flow?

    b. What will the voltage across the bulb be?

    c. What is the resistance of the bulb?

    d. How long will the bulb burn before it uses one kw-hr? What will it cost to burn the bulb each hour at a rate of 4¢ per kilowatt hour?

12. Repeat Problem 11 for a toaster rated at 1 kw.

CHAPTER
IX

THE THEORY
OF RELATIVITY

There was a young lady named Bright
Who could travel much faster than light
She set out one day
In a relative way
And returned on the previous night

Anonymous

$$E = mc^2$$

A. The Speed of Light: Newton versus Maxwell

By the late 19th century, physics had come to the point where it seemed that its original mission—to explain the physical world—had been fulfilled. Most physicists felt that nothing new was going to turn up in their investigations of the world: that all that was left to do was to find the next decimal place in various physical constants and to work out in ever greater detail the consequences of the laws that had already been stated. But a few physicists realized that, contained within the structure of classical physics, were the seeds of its own destruction. In this section, we shall see how this "destruction" came about, and how it led to exciting new ideas and new areas of research in physics.

When we studied the law of falling bodies as it was enunciated by Galileo, we saw that one very important property of these laws was the rule of compound motion. This rule stated that motion of a projectile in the vertical direction was entirely independent of the motion of the projectile in a horizontal direction. As a consequence of this rule, we saw that a cannonball shot out from a cliff horizontally would strike the ground at the same time as a cannonball that had been simply dropped from the cliff, not shot from a cannon.

Another way of stating the same conclusion is to consider the following experiment: Suppose there is a man standing on the ground and another man on a railroad car moving past him at some constant velocity (for example, 20 mph). Suppose that at the time that the two men are standing next to each other (that is, at the time when the man on the railroad car passes the man on the ground) each man drops a ball. From the point of view of the man on the ground, he sees the ball that he drops fall straight down to his feet in a certain time T. He also sees the ball dropped by the man on the railroad car fall in a parabola, but he sees it hit the ground at precisely the same time as the ball that he dropped. On the other hand, the man on the railroad car sees the ball that he dropped fall directly to his feet, and he also sees the ball dropped by the man on the ground travel in a parabola and hit the ground at exactly the same time as his own ball. This fact follows because to the man in the train, the man on the ground appears to be moving backward (in much the same way that a telephone pole appears to be moving backward to someone sitting in a car).

Suppose that we ask each man, the one on the ground and the one on the train, to tell us what he found to be the law governing falling bodies. It is obvious that both men would tell us that the motion of projectiles obeyed the law of compound motion as stated by Galileo. However, if we asked each man to describe the experiment, each would give us a different description, corresponding to the fact that each man sees the ball that he dropped falling straight down and the ball that the other man dropped falling in a parabola. Thus, we have a situation in which two different men, standing in different places, give us different descriptions of a physical event, but give us the same law of physics to describe that event. This is an example of something called the "principle of relativity."

The principle of relativity can be stated in the following way:

The laws of physics are the same in all inertial frames of reference.

A frame of reference is simply the point of view from which a particular set of events is observed. For example, in the situation of the man on the ground and the man on the train, discussed above, the train would be one frame of reference and the ground would be another frame of reference. When we say that two frames of reference are inertial, we simply mean that they are moving with respect to each other at a constant velocity. In the above example, the frame on the train and the frame on the ground are inertial frames because they are moving with respect to each other at a constant velocity of 20 mph. The frames would not be inertial if they were being accelerated with respect to one another. For example, a man standing on a merry-go-round and undergoing circular motion is being accelerated, and so he is not standing in an inertial frame with respect to a man on the ground.

Actually, the principle of relativity as we have stated it—restricting its validity to inertial frames—is called the principle of special relativity. If we simply left out the requirement of inertial frames, and made the statement that the laws of physics are the same in *all* systems regardless of their motion, it would be called the principle of general relativity. We will discuss only special relativity at this time, but both principles—special and general—are believed to be true.

The principle of relativity, then, is built into Newtonian mechanics. It is discovered as a consequence of experimental laws and is most obvious when we talk about projectile motion. However, it can be shown to follow mathematically from Newton's laws of motion as well.

Let us ask ourselves what the consequences of the principle of relativity are. Let us suppose, for example, that we have two experimenters, each equipped with all the laboratory equipment that he desires, but each enclosed in a large room. We then ask each experimenter to determine what the laws of physics are in his own room. Each man would then busily set about reproducing the experiments of Galileo, finding the laws of falling bodies, and determining the law of universal gravitation—in short, retracing most of the history of physics that we have discussed up to this point. The principle of relativity tells us that, provided that the two rooms in which the men are enclosed are inertial frames, they will discover precisely the same laws of physics. This is true even if one of the rooms is on a spaceship traveling around at thousands of miles per hour and the other is stationary. To put the principle a different way, what we say is that it is impossible to tell by a series of experiments in a closed room whether a system in which you are standing is moving at a constant velocity or is stationary.

In fact, when you think about it, this principle does not seem so strange. After all, what do we mean by saying that something is stationary and something else is moving? To us, standing on the surface of the earth, it appears that we are stationary and everything moves around us. We saw that this was the principal difficulty that was encountered by early astronomers in explaining the motion of the heavens. However, to someone in a spaceship near the sun we are anything but stationary. We are spinning around as the earth rotates and we are moving from one point in space to another as the earth moves around its orbit. But even the observer at the sun would not be stationary if observed from the center of the galaxy, since the sun itself makes a grand circuit around the galaxy every few billion years. To an observer in another galaxy, it appears that our galaxy is moving with respect to him. It is difficult from our modern point of view to see how one could define such a thing as a "stationary point." In fact, what we mean by saying that something is moving is that it is moving with respect to something else, which we choose to consider stationary. For example, if we consider the man on the ground in the above discussion to be stationary, then the man on the train is said to be moving. We could equally well, however, consider that the man on the train was stationary and the man on the ground was moving.

A somewhat facetious comment on the principle of relativity could be made by saying that instead of asking "When does this train get to Vienna?" an observer on the train is perfectly justified in asking "When does Vienna get to this train?"

Although this conclusion—that there is no such thing as a stationary or privileged observer—seems obvious to us, it was far from obvious to a classical physicist. In fact, the concept of the ether was one manifestation of the idea that somehow, somewhere, there had to be a "God's eye" frame of reference. It was felt that there must be both an absolute time and an absolute space in the universe. Events such as the rotation of the earth or the movement of the earth around the sun were understood to refer to motion with respect to this fixed, immutable, "God's eye" frame of reference. The only thing that was thought to be stationary with respect to this immovable frame of reference was the ether itself. Hence, the motion around the sun was regarded by classical physicists as the earth swimming through a sea of ether. This concept came back to haunt them later, as we shall see.

If, however, the reader will go back and review our discussion of the derivation of Newton's laws of motion and of the laws of electricity and magnetism, he will quickly discover that nowhere in that derivation was it necessary to assume anything about the existence of absolute space or absolute time, or the existence of anything like a "God's eye" frame of reference, or even the existence of the ether. All of these were simply fixed ideas—ideas that existed in the mind of a scientist. They were not proven, and they were not questioned. In this respect, they were very similar to the fixed idea of circular motion that dominated the thinking of the Greeks.

205

There was another important consequence of the principle of relativity that we can see by going back to our example of the two men, each locked in his own room with his scientific apparatus. We said that the principle of relativity stated that the laws of physics would be the same in all frames of reference. Suppose we ask each man to discover for us the law governing the force between two charged objects. Clearly, if the principle of relativity is true, these men would rederive Coulomb's law, each independently of the other. Each would then write down an equation describing the force between two charged objects, which would simply be

$$F = \frac{kq_1q_2}{d_2}$$

The constant k which appears in this equation was discussed earlier. If the principle of relativity is true for electricity as well as for mechanics, then it follows that this constant k cannot depend on the motion of one frame of reference with respect to the other. Thus, a man on a moving railroad car would find the same value in the constant k as a man on the ground.

There is a hidden assumption in the statement that was made above. The statement that both men would find the same force between two charges assumes that the principle of relativity, which we derive only for the case of mechanics, applies to electricity and magnetism as well. It is not necessarily true that this should be the case. It could be, for example, that the principle of relativity is true for mechanics, so that the laws of mechanics are the same in all frames of reference, but that the laws of electricity and magnetism would not be the same in all frames of reference. It is quite possible in principle that the two men in the above example would measure different forces between charged bodies. On the other hand, it is also possible that they would find identical forces. We have come a long way from the time of the Greeks when such questions would be decided by thought alone. We know how important it is to decide questions of this type by experiment.

For reasons that will become obvious later, it is not so easy to get direct experimental confirmation of the principle of relativity for electricity and magnetism. Therefore, we shall take an alternate course. We shall *assume* that the principle of relativity holds for electricity and magnetism as well as for mechanics. We shall then find out what the consequences of this assumption are, which of these consequences can be tested experimentally, and what experimental evidence there is to support predictions associated with the assumption. If enough of the predictions turn out to be verified experimentally, and if none of the predictions turns out to be wrong, then we shall be justified in saying that we have proved that the principle of relativity is true for electricity and magnetism experimentally, even though we do not test it directly.

But now comes the hooker. We saw in our discussion of Maxwell's equations that once the equations were written down, certain mathematical consequences followed from them. In particular, we saw that the speed of light was related very simply to the constant k that appears in the above equation. We discussed at that time how the speed of light could thus be related to the force between two charged objects. Now we see an important consequence of that fact. If both of the observers in a hypothetical example find the same set of equations to describe electricity and magnetism—that is, if they find Maxwell's equations—then it must follow that both observers would claim that the speed of light was equal to c in their frame of reference. Once we have agreed that they will measure the same force between two charged bodies (a consequence that follows from the principle of relativity as embodied in Newton's laws of motion), then we cannot escape the conclusion that they must measure the same velocity of light (which, after all, simply follows from Maxwell's equations).

But this is a very paradoxical conclusion. How could two men, each sitting in a system that is moving with respect to the other, measure the same velocity for light?

To understand why this is paradoxical, consider the case of a man on a moving railroad car and a man on the ground. If the man on the moving railroad car throws a ball at the velocity of 10 mph with respect to himself (and hence with respect to the railroad car) and if the railroad car is moving at 10 mph, then the man on the ground will see the ball moving with a velocity of 20 mph—the velocity with respect to the car added to the velocity of the car itself. This is very straightforward, and is the only reasonable conclusion to draw. On the other hand, what we have said above is that if instead of throwing the ball, the man on the railroad car emits a beam of light (for example, by flashing a lightbulb), then he will measure a velocity of 186,000 miles per second for the speed of light, and the man on the ground will measure 186,000 miles per second. The man on the ground will *not* measure 186,000 miles per second plus 10 mph! At this point, something has to give. Either we accept this startling new conclusion about the addition of velocity, in which case we throw away the notion from classical Newtonian mechanics that velocities were to be added as they were in the case of the ball on the railroad car, or we say that this new conclusion is so patently false that there must have been an error in the logic that led to it. The only error we could find would be in Maxwell's equation itself. Hence, we must either conclude that Newtonian mechanics in the form that we have understood it up to this point is wrong, or Maxwell's equations in the form we have understood them up to this point are wrong. They cannot both be right. It is this basic contradiction between two important branches of physics that led eventually to the development of the theory of special relativity.

B. What's Wrong?

If there is a basic difference between the laws of mechanics and the laws of electricity and magnetism, then the first thing we must do is to try to understand and discover precisely where that contradiction arises. In the above example, the contradiction arose when we tried to compare two velocities—one in a moving frame and one in a stationary frame. Measuring a velocity involves measuring both the distance and the time. It is implicit in the framework of Newtonian mechanics that the distance measured in one frame (for example, the frame of the moving railroad car) will be the same as the distance measured in any other frame (for example, the frame of the man standing on the ground). The same statement could be made for measurements of time. It was this implicit assumption that allowed us to say that we would expect that the velocity of anything as measured in the frame of the railroad car would be equal to the velocity of that same thing measured by the man on the ground minus the velocity of the car itself.

But what do Maxwell's equations tell us about this basic assumption? If we assume for a moment that the velocity of light is indeed the same in all frames of reference, what does that tell us about the readings of a clock in two frames of reference?

A clock is simply a device that repeats the same motion over and over again. We could imagine making a clock like the one pictured in Fig. 9.1. The clock is a flashbulb, a mirror, and a photocell. A photocell is a device that registers the arrival of light. For example, the automatic doors in supermarkets are often activated by photocells. When you cross a beam of light, you change a light reading in one of these instruments, and this, in turn, triggers the door. In any case, we could imagine the clock could work as follows: first, the flashbulb flashes, the light travels from the flash-bulb to the mirror, a distance D, and back to the photocell. When the light has arrived at the photocell, the photocell triggers the flash, which then initiates the entire process again. Each time period during which the light travels up to the mirror and back is one unit of time. If we adjusted the mirror properly, we could make this one second (although the mirror will have to be a great distance away to do so). Whatever we choose as our unit of time, however, it is clear that this "clock" would do perfectly well for measuring time in principle.

Now, let us ask each of our two observers—one in the railroad car and one on the ground—to take one of these clocks along with him and to measure the time according to his clock. Suppose that we arrange things so that as the man in the train passes the man on the ground both flashbulbs go off simultaneously.

The man on the ground will see the light leave the flashbulb, travel up to the mirror, and come back to the photocell in his clock. Nothing will be different for him. On the other hand, let us ask how he sees the clock on the railroad car. He will see that flashbulb go off at the same time that his flashbulb goes off, but by the time the light from the flashbulb on the railroad car has gotten to the mirror on the railroad car, the mirror will have moved over a certain amount because of the motion of the railroad car

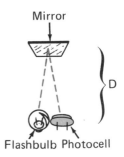

Mirror

Flashbulb Photocell

Figure 9.1.

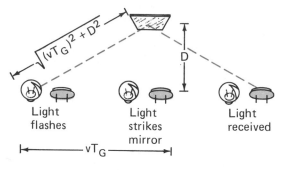

Figure 9.2.

itself. Thus, the man on the ground will see the light travel in an oblique line up to the mirror and then in an oblique line back down (see Fig. 9.2). If the velocity of light in the moving railroad car is the same as the velocity of the light on the ground, then the man on the ground would have to conclude that the clock on the railroad car was running slower.

Why?

The answer to this question is obvious. If the light in the clock on the railroad car must travel on the zig-zag path as shown above, then it is traveling a longer distance than the light in the clock on the ground. Since the light in both cases is traveling at the same velocity, this can only mean that it takes the light on the railroad car a longer time to get back to its photocell than it does for the light on the ground to get back to its photocell. Since we have defined the time that it takes for the light to get from the flashbulb to the mirror and back to the photocell as one unit of time the man on the ground can only conclude that a unit of time is longer on the moving platform than it is on the ground. Put another way, he says that the clock on the moving train is running slower than the clock on the ground, as seen by an observer on the ground.

This effect is called *time dilatation.* It is the origin of the famous "twin paradox." This paradox (which is not really a paradox at all) simply states that if identical twins were put on different spaceships and one spaceship moved very quickly while the other did not, then the twin in the fast moving ship would come back younger, since his clock had been running slower than the clock of the twin in the stationary ship. This conclusion seems outlandish at first sight. Indeed, it appeared

210

to be outlandish to physicists at the turn of the century, who were very reluctant to accept it. On the other hand, the only ingredients in the logic which led up to the conclusion were Maxwell's equations themselves. Thus, the only way that the conclusions could be wrong would be for Maxwell's equations themselves to be wrong.

What we have shown in the above example is that our very specialized clock (made up of a photocell, a mirror, and a flashbulb) will appear to be ticking slower if it is moving with respect to a particular person than would an identical clock sitting at rest with respect to that same observer. The question naturally arises as to whether this result is a consequence of the rather specialized kind of clock we have been talking about or whether it is a property of time itself. Let us suppose we wanted to know whether any clock would exhibit the same slowing down, or time dilatation, as we have found for our special clock. Take as an example an ordinary grandfather clock with a pendulum. Suppose we constructed one of our light clocks so that the flashbulb went off every time the pendulum went through one complete cycle. To us, sitting in the rest frame of both clocks, it would look as if there were a flash on the light clock and a tick on the pendulum clock, and these two events would take place simultaneously. Each time the light flashed, the pendulum clock would tick.

How would these events look to a man moving with respect to us? Obviously, from what we have just finished proving, he would see the flashes of the light clock occurring more slowly than the flashes on a clock that is at rest in his frame. Since I see the flashes of the light clock and the ticks of the pendulum clock as simultaneous in my frame, and since the two clocks are at the same place, then a man standing on the ground must also see them simultaneously. Thus, he would see both clocks slow down, and not just one.

This result would be true for any kind of clock you can imagine. For example, the modern standard for time uses the motion of an electron around an atom as its basis. The time it takes an electron to go around an atom corresponds to one "tick" of a normal clock. We could imagine carrying out the same experiment we described above for a grandfather clock with an atomic clock of this type. The processes that go on within the human body are also periodic and also depend ultimately on the motion of electrons around atoms.

211

There are many biological clocks operating in our body. The alpha rhythms of the brain could be an example of a clock. Thus, the conclusion in the twin paradox that one twin would be younger than the other follows from the fact that not only does a light clock appear to run slower in a moving frame, but a biological clock would appear to run slower as well. The time dilatation effect that we have derived is not some kind of artificial construct. It is a very real effect, which can be measured in nature.

Actually, it was a long time before direct experimental evidence for time dilatation was available. A number of indirect proofs were deduced from laboratory experiments. However, a few years ago a group of scientists from the University of Michigan carried out an experiment that proved once and for all that time dilatation really does exist. They took very accurate atomic clocks and flew them around the world on a commercial airline. By comparing the clocks before and after the flight (and, of course, leaving one clock stationary on the ground), they were able to measure the difference in time as measured by the clock that was moving in the airplane and the clock that was stationary on the ground. Thus, strange as it may seem, time dilation is a real effect. (See page 229 for a further discussion of this point.)

We see, then, that far from being "obvious," the Newtonian idea of adding velocities that we discussed above is incorrect. In fact, since the times as measured by two observers in two moving frames will be different from each other, one has to be very careful when defining what one means by velocity. For example, the velocity of the ball thrown by the man in the railroad car is easy to define in terms of the clock and measuring rod as carried by the man on the railroad car. However, the above example should have convinced us that one has to be a little careful about what the clocks and measuring rod of the man on the ground will be when *he* measures the velocity of the ball thrown from the car.

Before we start tearing down all of our clocks, however, we should try to get some idea of how big these effects of time dilatation would be at normal velocity. This will involve a mathematical derivation of the formula for time dilatation. (The reader who wishes to skip the mathematics can

Atomic clock.

Cooling Towers, Oconee.

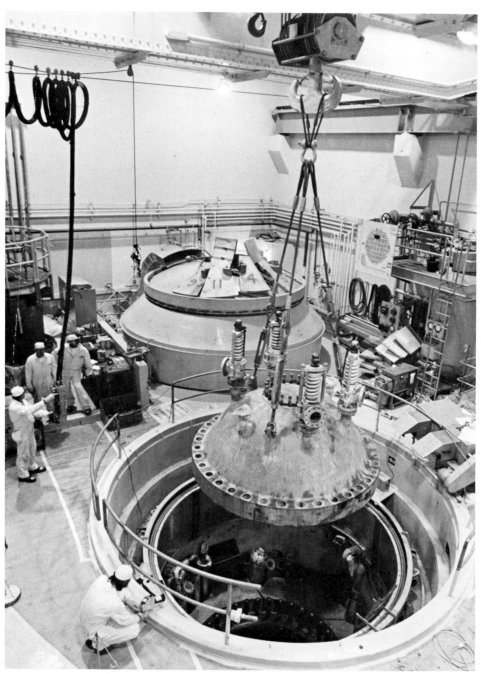

Annual fueling of nuclear unit, Humboldt Bay Power Plant, Eureka, California.

go ahead to the end without loss of continuity.) We discuss the situation in which there is a distance D between the flash-bulb and the mirror in the rest frame of the clock (the rest frame is defined to be that frame in which the clock is not moving). The time as measured in the rest frame of a clock is given the name "proper time." Then the time that it takes the light to travel up to the mirror (which is, of course, just half the time it takes for one "tick") is just

$$T_p = \frac{D}{C},$$

where we have called the time T_p to emphasize that it is the proper time for the clock.

How will the clock on the train look to a man standing on the ground? If we call T_G the time it takes for the light to get from the photocell on the train to the mirror on the train *as seen by the man on the ground,* then the man on the ground will see the mirror on the train move a distance vT_G while the light is traveling from photocell to mirror on the train. Some simple geometry (see Fig. 9.2) then tells us that as far as the man on the ground is concerned, the light on the train had to travel a distance

$$\sqrt{(vT_G)^2 + D^2}.$$

Since this light has to travel with velocity c, the time it takes the light to go this distance (which is what we have called T_G) is just

$$T_G = \frac{\sqrt{(vT_G)^2 + D^2}}{C}.$$

The next step is to solve this equation for T_G. If we square both sides and then subtract $(v^2/c^2) T_G^2$ from both sides, we get

$$T_G{}^2 - \frac{v^2}{C^2} T_G{}^2 = T_G{}^2 \left(1 - \frac{v^2}{C^2}\right) = \frac{D^2}{C^2}.$$

But from the definition of proper time,

$$\frac{D^2}{C^2} = T_p{}^2.$$

If we substitute this into the equation for T_G, and then take the square root of both sides and divide by $\sqrt{1 - v^2/c^2}$, we find

$$T_G = \frac{T_P}{\sqrt{1 - v^2/c^2}}.$$

This equation says mathematically what we said in words before—that the clock on the train, as seen by the man on the ground, will be ticking more slowly than the clock on the ground. This follows because the factor $\sqrt{1 - v^2/c^2}$ is less than one, and dividing by this factor will insure that T_G is always greater than T_P.

Problems are given at the end of the chapter that will give the reader a chance to work out some actual numbers for time dilatation. It is important to note that in the above equation v stands for the velocity of the moving clock and c for the velocity of light. Therefore, unless the clock is moving at a speed near the speed of light, the term v^2/c^2 will be very small and the two times will be essentially equal. For objects moving at normal speeds, the correction to our normal expectation that the two times should be the same would be very small. For example, a clock moving at 60 mph from the beginning of the universe would appear to have slowed down only one second compared to a clock sitting on the ground.

This explains why time dilatation was never discovered by Newtonian physicists. They simply could never have measured it because they dealt only with objects that were available to them in everyday experience. We do not have experience in everyday life of objects moving very close to the speed of light. Consequently, we cannot blame the Newtonian physicists for having missed this rather important effect. It was simply outside the reach both of their experience and their instruments.

This is a rather interesting conclusion. We have stated that a stationary observer will observe clocks in moving frames running slower than his own clock. How does a clock on the ground look to the man on the railroad car? Is it running slower or faster?

The first impulse is to answer that since his clock is running slower, he must see the clock on the ground running faster. This is not the case, of course. The way to see this is

214

to ask how the clock on the ground looks to the man on the train. To him, it looks as if his clock is perfectly normal. That is, the light travels straight up from the flashbulb to the mirror and straight back down to the photocell. He sees the clock on the ground doing something very different, however. When he sees the flashbulb go off, he sees the clock on the ground doing the same zig-zag pattern we discussed above, but in the reverse direction (see Fig. 9.3). Thus, by using the same argument given above, we would conclude that to an observer on the train, the clock on the ground appears to be running more slowly.

Which one is right?

Neither one is right in the sense that we use it in the sentence above. Each man describes the events as he sees them. Either description is a perfectly valid statement of what is actually happening in the physical world. There is no "right description" just as there is no "God'e eye view" of the universe. One of the basic tenets of relativity is that any observer can give an equally valid description of any event in the universe. Although a description may differ from that given by the other observers, this is simply a consequence of the fact that they are looking at it from different frames of reference. It does not imply that one is right and one is wrong.

This way of looking at relativity is something that is very congenial to the 20th-century mind, and, conversely, would have been anathema to someone like Isaac Newton. It is simply another example of how scientific and cultural ideas seem to move in similar directions.

Is there anything that the two observers would agree on? Yes, there is. For example, if we asked the man on the

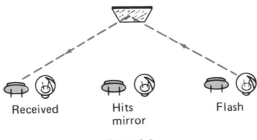

Received   Hits   Flash
                mirror

Figure 9.3.

215

train the question "What does the man on the ground see his clock doing?" he would give a perfectly sensible answer. His description of what he thinks the man on the ground should see and the description by the man on the ground of what he actually sees would coincide. The problem (and the paradox) arises from asking the man on the train "What do you see?" and then comparing his answer to what the man on the ground sees. In other words, the paradox arises because we are trying to compare the viewpoint of two different observers. These observers are in different frames of reference and will necessarily give different descriptions of what they see. However, if either one is asked what the other observer should be seeing, his answer will then coincide with that of the other observer. In this sense, two different observers can indeed give the same description of a physical process and the same answer to a physical question.

We have seen, then, that if Maxwell's equations are to be the same in all frames of reference, and if the principle of relativity is to be true, then it must follow that identical clocks will appear to read differently if one is in motion with respect to the other. The general rule that we have found is that a moving clock will always look as if it is running slower. Another way of stating this conclusion is to say that the proper time measured by a clock (that is, the time measured in the rest frame of the clock) will always be shorter than any other time that the clock will be seen to read.

Although this conclusion is startling, we have seen that it can be experimentally verified. This means that it is not inconceivable that the principle of relativity might be true. This, in turn, means that it makes sense for us to begin to ask what other consequences there are of the principle that might be equally as startling as the ones we discovered above. For example, we arrived at the conclusion about time dilatation by thinking about the example of a ball thrown on a moving railroad car. We reasoned that since the addition of velocities could not hold if Maxwell's equations were true, then something about the normal definitions of space and time must be changed. We have seen how the definition of time that we normally accept has to be changed. Let us now turn our attention to other things that seem just as normal to us now but will have to be changed if the principle of relativity is to be true.

C. Other Results of the Principle of Relativity

In this section, we shall study the consequences of the principle of relativity in four areas: length, simultaneity, mass, and energy. We shall also give a short account of some experimental verification of the results of this discussion.

1. *Length contraction*

Let us consider how we could use the type of reasoning that was developed in the above section to discuss the problem of measuring lengths. If we return to our example of two observers—one on the ground and one in the moving train—we can ask ourselves how they would go about measuring a length. Suppose, for example, that there was a measuring rod laid along the ground next to the railroad track. To an observer on the ground, the way to measure the length of the rod would be obvious. You would simply lay down meter sticks alongside of it. There is, however, another way that this length could be measured. The train is going by at a known speed (for example, 20 mph). If the observer on the ground has a clock, he can start the clock when the front of the train passes the beginning of the measuring rod, and he can stop the clock when the front of the train passes the end of the measuring rod. The length of the rod as he sees it will then be the velocity of the train times the time that he measured. If we call the time that he measured T_G (the G to remind us that this is the time as measured by an observer on the ground) then the length as measured on the ground, L_G, will be

$$L_G = VT_G.$$

For example, if the train is going 20 mph and it takes one hour for the train to get past the measuring rod, then we would say that the length of the rod was 20 miles. Now let us ask how this sequence of events appears to a man in the train. To the observer on the ground, the first event was the passing of the beginning of the measuring rod by the train, and the second event was the passing of the end of the measuring rod by the train. To the man on the train, however, the first event occurs when the beginning of the measuring rod passes him, and the second event occurs when the end of the measuring rod passes him. Of course, to him it appears that

217

the measuring rod is moving and he is stationary. Suppose that the man on the train also has a clock, and that he starts his clock at the first event (the passing of the beginning of the measuring rod) and stops his clock at the second event (the passing of the end of the measuring rod). He will find a time T_T has elapsed according to his clock. The subscript T reminds us that this is the time as measured by a man on the train.

If the train is moving at 20 mph with respect to an observer on the ground, it is obvious that to an observer on the train the ground is moving at 20 mph with respect to him, but in the opposite direction. That is, if to an observer on the ground it appears that the train is moving 20 mph toward the east, then to the observer on the train, it will appear that the observer on the ground is moving 20 mph to the west. This conclusion does not depend in any way on the addition of velocities that got us into so much trouble in the previous section.

It follows, then, that the observer on the train will conclude that the measuring rod is a length L_T long, and this length will be given by

$$L_T = vT_T,$$

where the quantity v is the same in this equation as it was in the previous equation.

The question that we want to answer is "How long does the measuring rod appear to be to the observer in the moving train? There are several measurements that were made in the above experiment. There was the measurement of the time interval between the two events as seen by the observer on the ground, and the measurement of the time interval between the two events as seen by the observer on the train. The proper frame—that is, the frame in which we would get the proper time—for the first measurement is obviously the ground frame, and the proper time for the second measurement is obviously the moving train. The proper frame for the measurement of the length with an ordinary meter stick is also obviously the the ground frame, since it is in this frame that the meter stick is at rest.

The important point to remember is that we are asking about a measurement that is being made *on the train.* This observer determines the length of the rod as he sees it by measuring the time interval between the passage of the front and the back of the rod as it goes by him. As far as he is concerned, the time between the two events as he sees them will be *shorter* than the time between those same two events as seen by another observer. This is a consequence of the fact that to an observer on the train, a clock in any moving system, including the ground system, will appear to be running slower than his own. This means that the time T_T, the time interval between the two events as measured by the observer on the train, must be less than the time T_G. We can determine just how much less this interval will be by dividing the first equation ($L_G = vT_C$) by the second equation ($L_T = vT_T$) to get:

$$\frac{L_T}{L_G} = \frac{T_T}{T_G} = \sqrt{1 - v^2/c^2} \ .$$

In other words, not only does an observer see clocks in a moving frame slowing down, he also sees lengths in a moving frame shortening in the direction of motion. This means, for example, that if he were looking at a basketball moving by with some velocity, he would not see a sphere (which is what we would see in the proper frame of the basketball) but would see instead a flattened pancake-like object. The dimension perpendicular to the direction of motion would be unchanged, but the dimension in the direction of the motion would appear shorter, hence the pancake-like appearance.

In thinking about this proof for the contraction of length, which is sometimes called the Fitzgerald contraction and sometimes called the Lorentz contraction (after two early 20th-century scientists who first investigated this effect), it is important to keep one distinction clearly in mind. In Section B of this chapter we showed that a consequence of the principle of relativity was that the time interval between two events appears shortest to the person in the proper frame for those events, and appears to him to be longer if he observes the events from a moving frame. The two events that we talked about were the flash of a flashbulb and the click of the photocell. Because of the greater distance that the light had to travel, it was obvious that it would take the second event a longer time to occur in a moving frame than it would in an identical apparatus on the ground. In the experiment we are discussing now, the two events are not the ticking of clocks but are instead the passage of the observer in the train past the front and rear ends of the measuring rod, respectively. (This point is covered more fully in Problem 16 at the end of this chapter.)

2. Simultaneity

We have seen that many of our ordinary concepts of space and time do not survive when we assume that the principle of relativity is valid for the laws of electricity and magnetism as well as the laws of mechanics. There is another concept that has to be modified as well. This is the idea that if two things happen at the same time—that is, are simultaneous—in one frame, they will then be simultaneous in every frame. In fact, it will turn out that this is true only if the two events happen at the same place; it is definitely not true if the two events happen to be separated by any distance at all.

There is a very easy way to illustrate this point. In fact, it is by thinking about the phenomena we are about to describe that physicists usually derive the equations that govern the behavior of space and time as seen by different observers. We know that the speed of light is the same in every direction. This means that if a flashbulb goes off in our frame we will see the light moving out in a large expending sphere, so that if we set up an apparatus like the one pictured in Fig. 9.4, in which a series of photocells were located around a sphere a certain distance from the given flashbulb (say, 10 feet),

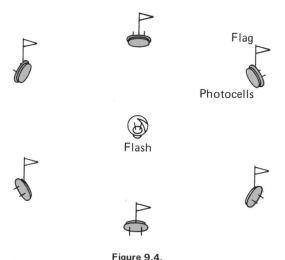

Flag

Photocells

Flash

Figure 9.4.

then a certain time after the flashbulb went off—the time that it would take light to travel the 10 feet in question—the light would get to the photocells. Suppose that at the location of each photocell we had a little yellow flag that would flip up when the photocell was activated. Then what we would see would be a sequence of events like this: the light flashes, a certain amount of time passes, and then all the yellow flags go up at once.

The fact that all of the yellow flags go up at once is simply the statement that the light arrived at every point on the sphere at the same time. In other words, the arrival of the light at the position of one photocell is simultaneous with the arrival of the light at the position of any other photocell *as seen by an observer in the same frame as the photocells.*

Although we have had very little experience in our life with making this kind of measurement with light, there is an analogy with water waves that will illustrate the above example. Suppose we throw a rock into a quiet pool of water. After the rock has hit the surface, we see an expanding ring of water corresponding to the wave moving out from the point where it started (the point where the rock hit the water) and moving away. If we had two corks, each located a foot away from the spot where the rock hit, we would see a sequence of events like this: first the rock hits the water, then we would see the expanding wave go out, and then we would see the corks bob up and down simultaneously as the wave gets to where they are. Since light is a wave itself, it is not unusual that a similar sequence of events should occur when a light beam is sent out.

Everything we have discussed up to this point concerns a measurement taken in a single frame of reference—the frame in which the photocells are at rest. Most of the developments that we have seen, however, have occurred when we have compared measurements taken in different frames of reference. Let us see what would happen if we asked the following question: "How would the events we have just described look to a moving observer?"

To answer this question, let us suppose that an apparatus identical to the one we have just described is mounted on a train. Let us suppose that a spherical array of photocells of radius 10 feet is moving along at some velocity (for example, 20 mph). Let us further suppose that on each of the moving photocells there is a red flag, and that this red flag will flip up when light activates that photocell, just as the yellow flags flipped up when light activated the photo- cells on the ground. Suppose that we arrange things so that the two arrays of photocells can actually pass through each other, so that the observer standing on the train at the center of his array can pass right next to the observer standing on the ground at the center of his array. Suppose that at the instant that the two observers are on top of each other, a flashbulb goes off. Let us now inquire as to what the observers should expect to see. We have already said what the observer on the ground will see. He will see a certain amount of time elapse, and then he will see all of his yellow flags go up simul- taneously. To an observer on the train, however, the events should be of the same form. After all, what he sees is a flash of light at the center of his sphere. Therefore, he ought to see a certain amount of time elapse (the time it takes light to travel from him to the photocells in his frame) and then he should see all of the red flags go up simultaneously. In other words, each observer should see the light going out in a sphere in his frame, and that sphere should be centered on him. The problem comes from the fact that the two observers are not at the same place when the light reaches the photocells (see Fig. 9.5). How can it be that each observer sees his sphere centered on himself? Is it possible for the light to be in two places at the same time?

The answer that comes to mind first is that somehow the man located in the same frame as the flashbulb should see the simultaneous events, and the other observer should not. Like many of the common sense answers that we have seen up to this point, however, this answer is wrong. The fact that light travels in a sphere is like the fact that it travels at the same velocity in every frame. It is a consequence of Maxwell's equations. The best way to think about this is to go back to the analogy of the rock and the water. Once the rock has hit the water, it makes no difference what it does afterwards. For example, if after it hit the water it suddenly speeded up or slowed down, it would not affect the wave; the wave travels outward according to the laws governing its be- havior. Once the rock has performed the function of getting the wave in motion, it makes no difference what the rock does. Thus, whether or not the observer sees a flashbulb

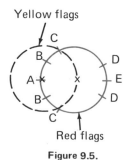

Yellow flags

Red flags

Figure 9.5.

moving by him, flashing at the instant it is opposite from him, and then moving on or whether he sees that same flashbulb stationary in his frame and flashing at the same instant makes no difference as far as what the wave of light does after it leaves the flashbulb. This means that the seeming paradox we have stated will not be resolved by recourse to the mechanics of the experiment. It is a genuine "paradox" of special relativity.

Of course, most of the "paradoxes" that we have discussed have not been paradoxes at all, but have simply been statements that events look different to observers in different frames. This "paradox" is no exception to that rule. Let us go back to the observer on the ground and ask how he sees the events we have just outlined. He will see the apparatus on the train moving toward him and at the precise instant when the observer on the train is next to him, the flashbulb goes off. He sees the light moving away from him in an ever-expanding sphere traveling at velocity c. He also sees the photocells at the rear of the apparatus on the train approaching him with whatever velocity the train has. Consequently, he sees the rear photocells coming up to meet the expanding sphere of light. Thus, he will see the light arrive at the photocell at point A in Fig. 9.5; hence, he will see the red flag associated with that photocell go up first. Shortly thereafter, the expanding sphere of light will have reached the points labeled B. Hence, he will see the red flags go up there. It is easy to extend this argument and to convince ourselves that what he will see will be the rearmost red flag go up first, then the next two, then the two after that, and so forth as the flags work their way around the apparatus on the train. At the time that he sees the light reach the photocells labeled C, another important event will have occurred. The light sphere will have expanded out until it has reached his own photocells. At that time, he will see the two red flags at point C go up, but he will also see all of the photocells in his own frame activated, and all of the yellow flags will go up at the same time. After all, as far as he is concerned the arrival of the light at *his* photocells constitutes a series of simultaneous events as he sees them. After all the yellow flags have gone up, the light sphere will continue to expand and will begin to overtake the photocells on the train that had moved past the photocells on the ground. Thus, a short time after he sees all of the yellow flags go up, he will see the two red flags at points labeled D go up, and this sequence will continue until the last red flag (at the point E) goes up.

Thus, the paradox has been resolved. Like the other paradoxes we have considered, it was not really a paradox at all, but simply a difference in the way two observers would see the same event or series of events. What we have seen is that a set of events appearing to be simultaneous in one frame of reference need not appear to be simultaneous in another frame. It is left as an exercise for the reader to describe in detail the events as they would be seen by an observer on the train. However, it is clear that he will see his flags (the red ones) go up at the same time in his frame, and so he must see the yellow flags go up at different times when he looks at the observer on the ground.

One amusing example of this conclusion is a famous problem—the "train and tunnel" paradox. It is stated as follows: There is a tunnel that has a length L located at a certain point. There is a train that, if it were brought to rest inside the tunnel, would have exactly the same length as the tunnel. The tunnel is equipped with two doors, which can be closed simultaneously by an observer standing on the ground.

The engineer on the train and the observer on the ground make a bet. The observer reasons as follows: Since the tunnel is of length L, when I see the train moving toward the tunnel at some velocity, the train will be shortened by the Lorentz contraction. Consequently, there will be some time when the train is completely inside the tunnel. If I close the doors of the tunnel at that time, I will be able to trap the train.

The engineer, on the other hand, reasons as follows: The tunnel is just as long as the train if the two are at rest. When I am moving toward the tunnel, however, it appears to me that I am stationary and the tunnel is moving toward me. Thus, I will see the tunnel contracted by the Lorentz-Fitzgerald effect. Consequently, there will be no time at which I will be completely inside the tunnel. Quite the contrary, when I am going through the tunnel I will actually stick out at both ends. Therefore, it will be impossible for the man on the ground to close the doors of the tunnel when I am inside it.

Thus, a bet is made. The engineer backs up the train and begins moving toward the tunnel, and the man on the ground sits with his lever to try to close the doors. The question is, who wins the bet? The answer is left to Problem 3 at the end of this chapter.

3. *Relativity and mass*

We have seen that many of our cherished notions of space and time and simultaneity have to be modified if the principle of relativity is true. It should come as no surprise, then, to realize that other notions will have to be modified as well. One of the most interesting effects is the modification in mass that occurs. It is not possible to give the kind of simple explanation of changes in mass that we gave for the changes in time and distance as perceived by different observers. On the other hand, the results are just as simple as the results for space and distance and time.

Just as one observed the moving observer's clock slow down and his measuring rods contract, one sees the moving observer suffer an increase in mass. In fact, the mass of a moving object as seen by a stationary observer is given by the formula

$$m = \frac{m_O}{\sqrt{1 - v^2/c^2}},$$

where v is the velocity of the object and m_O is the proper mass—the mass that would be measured by an observer in the same frame as the object.

There are some very interesting consequences of this result. For example, we remember from Newton's second law that the heavier an object becomes, the more difficult it is to accelerate it by applying an external force. This means that if we take an object that is now moving at some velocity below the speed of light and begin accelerating it, it will become heavier and heavier. Consequently, the closer its velocity approaches the speed of light, the larger the force that is needed to accelerate it. In order to get it *to* the speed of light where the mass becomes infinite, it is obvious that we would have to supply an infinite force. This is the origin of the well-known statement that according to the principle of relativity it is impossible for an object to travel faster than the speed of light. Actually, it is not at all impossible for an object to travel faster than light. Relativity just says that an object that is now traveling at less than the speed of light cannot be made to travel at or greater than the speed of light by applying an outside force.

While this may seem to be something of a play on words, consider that according to the argument we just gave, it is not only impossible for an object to be made to travel faster than light, it is also impossible for an object to be made to travel *at* the speed of light. Yet we do know of at least one thing that travels at the speed of light—light itself.

The reason that we make this distinction is to point out that there is nothing in the theory of relativity that says that if an object starts out traveling at the speed of light, it cannot continue to do so. All that it says is that no object now traveling at velocities less than c can ever attain the speed of light. Actually, the "photon" (which is the particle associated with light) has zero rest mass, so that when v=c in the above mass formula, we have a situation where the mathematical form corresponds to dividing zero by zero. From our discussion of limits, we know that this particular division can result in a finite value for the mass of the photon as seen by an observer stationary with respect to it.

More recently, scientists have speculated that there might be other particles that always move faster than the speed of light, just as photons move at the speed of light. These hypothetical particles are called "tachyons" (from the same Greek root that leads to the word tachometer), or "swiftly moving ones." As of this writing, however, such particles have not been found; they remain one of those interesting suggestions that might or might not be verified by experiment in the future.

4. *Energy and mass*

Probably the most famous consequence of the principle of relativity, and certainly the one that has the most direct effect on our everyday lives, is the discovery that mass and energy are equivalent. This discovery is embodied in Einstein's famous equation $E = mc^2$.

Like the statement that the mass of an object will look greater to a moving observer than to an observer in its rest frame, this equation is a mathematical consequence of the principle of relativity. However, the derivation of the equation from the principle involves the use of mathematics that is beyond the scope of this discussion. Therefore, the reader will simply have to take, on faith, the fact that arguments can be advanced to demonstrate this result just as they were advanced to demonstrate the results of time dilatation and length contraction.

But more important than an understanding of the derivation of the result is an understanding of what the result means. You will recall in our discussion of classical mechanics that we introduced the concept of energy—the ability to do work—and pointed out that there were many different forms of energy. For example, there are potential energy, kinetic energy, heat energy, and many others. We pointed out that the principle of conservation of energy that is so important in classical mechanics was simply a statement that any system had only a certain amount of energy. While it was possible for the system to shift that energy around—for example, from potential to kinetic energy and back again—it was not possible for the system to change the total amount of energy that it had. We used the analogy of the different kinds of energy as sinks and the shifting of energy as simply pouring water from one sink into another. You can pour the water and divide it among the available sinks in as many ways as you want, but you can never change the total amount of water.

What Einstein did when he derived his famous equation quoted above was to demonstrate to physicists that there was one more kind of energy—one more sink—than they had thought before. This extra category of energy, which is associated with special relativity, is called the "rest energy" of an object and is associated with its mass.

This means that it ought to be possible to convert the mass energy of an object to other forms of energy (for example, heat or kinetic energy), just as it is possible to convert the kinetic energy of an object to potential energy and vice versa. Since the speed of light, c, is a very large number, we can see by looking at the above equation that a very small amount of mass will give us a very large amount of energy. This was the basis of Einstein's famous letter to President Roosevelt in which he pointed out the possibility of using this principle both in the construction of weapons and the construction of power sources. It is the conversion of tiny amounts of mass in a uranium atom to energy which provides the basic power both of the atomic bomb and for the nuclear reactor.

On the other hand, the principle also states that it ought to be possible to convert other forms of energy (for example, kinetic or potential energy) into mass. It ought to be possible, in other words, to create new matter out of energy. When we talk about elementary particles in a later chapter, we shall see that this is done in a routine manner in modern accelerators. Thus, to modern scientists, the idea that mass and energy are two indistinguishable objects is no longer taken seriously. They are simply regarded as interchangeable forms of the same thing.

The consequences of this idea become important only at the level of the atomic nucleus. But, in principle, they are true in our everyday life as well. For example, a mouse trap that has been cocked has more energy than a mouse trap that has not been cocked. This is because compressing a spring is a way of changing the chemical energy in your muscles into potential energy in the spring. Thus, if we believe the Einstein equation, it must be true that a cocked mouse trap has more mass—must weigh more—than an uncocked mouse trap. Of course, the amount of extra mass is so small that it could never be detected by any experiment, but in principle this statement must be true. The important thing from our point of view is that, with relativity, a new classification of energy entered the world of science.

5. *Experimental proofs of the principle of relativity*

We have stressed repeatedly that any science and any scientific idea is only true or as valid as the experiments that were used to prove it. We have been spinning an elaborate web of strange new results and strange new ideas in our discussion of relativity, but up to this point we have said very little about whether or not there is any hard experimental evidence that would back up any of the assumptions that we have made. In this section, we will cite a few of the many experimental results that have been accumulated over the last 50 years which lead scientists today to accept the principle of relativity as one of the basic laws of nature. Like many other basic principles, it is not possible to test the principle of relativity itself directly. The logical steps are to (1) find out what the consequences of the principle are, and (2) find out whether these consequences can be verified and tested experimentally. We have already mentioned what I consider to be the most direct evidence for the truth of the principle of relativity, and this was the measurement by University of Michigan scientists of the slowing down of atomic clocks that had traveled in airplanes with respect to atomic clocks that had stayed on the ground. Such a result would be impossible to explain in any kind of classical way, and, since the slowing down of the clocks agreed to within experimental error with the predictions of relativity, there is little doubt in the mind of scientists that this constitutes an important corroboration of the theory. However, it should be pointed out that since the airplanes in question had to take off, achieve some air speed, and slow down (in fact, they had to do this many times in their flight around the world), the actual predictions of the slowing down of the clocks were not made according to special relativity but according to general relativity. Any "twin paradox" involving starting two clocks out at the same place, flying one around, and then bringing it back to the same place where it started necessarily involves an acceleration of one of the clocks, and hence involves more than just special relativity.

There are, however, other kinds of clocks in nature. We shall discuss later the fact that many elementary particles— tiny particles that live for very short times within the nucleus itself—have their own "clocks" built into them. These particles behave in such a way that after a certain amount of time has elapsed in their rest frame, they "decay," or come apart into other kinds of particles. It is a relatively simple experiment to determine how long a moving particle lasts before it decays. This is a way of measuring the time in a moving frame as seen by an observer in a stationary frame. It was this type of experiment that first provided evidence for time dilatation in nature.

The experiment involved a type of particle called the "mu meson." Like most elementary particles, it is designated by a Greek letter (in this case, the letter μ). We will discuss many of its properties later, but for our present discussion we will use only one of them. It turns out that the mu meson decays into other kinds of particles in a very short time—a millionth of a second or so—as seen by an observer in the rest frame of the particle. This means that if there were no time dilatation effect, the mu meson, even if it were traveling at the speed of light (or just under it) would be able to travel

Albert Michelson (1852-1931) who first showed that the ether didn't exist.

only about a few hundred yards before it decayed. Now mu mesons are regularly created by cosmic rays striking the earth's atmosphere. Typically, these reactions occur about six miles above the surface of the earth. If there were no such thing as time dilatation, no mu meson would ever be seen at sea level, since they would all decay high in the atmosphere. But it is a fact that many of these mesons are seen at sea level, which can only mean that the "clock" on the rapidly moving particle is slowed down as seen by an observer on the ground. Some examples of this type of thing are included in the problems.

Another famous historical experiment that actually preceded the theory of relativity was an experiment by two American scientists, Albert Michelson and Edward Morley, which was done in Cleveland in 1885. While this experiment did not directly test the idea of special relativity (indeed, these ideas were not even around at the time the experiment was done), they did test a very important classical concept—the concept of the ether. To the classical mind, the "ether frame" was the preferred frame of reference for all physical laws. All events that were observed were to be referred to this frame of reference, and an observer at rest in the ether was assumed to have the correct, or "God's eye," view of things. If such a frame exists, then it is obvious that the earth must be moving with respect to it, and that every point on the earth is constantly being buffeted by an "ether wind" as the earth moves on its daily rotation around its axis and its yearly revolution around the sun. Michelson and Morley set out to measure this "ether wind."

In order to understand the logic behind the experiments, let us consider a more mundane situation. Suppose I have a boat, and I know that that boat can go 5 mph on still water. I have another body of water, and I want to find out whether that water is moving or not. Suppose that for some reason or other I cannot actually see the water directly. Suppose that I then lay out a course on the bank of a known length—for example five miles. One way to find out whether the water is actually flowing (and if it is, how fast it is flowing) would be to take the boat and let it run over this course. If it took the boat precisely one hour to travel the five miles, then I would know that the water was not in motion, but was just like the lake on which we first tested the speed of the boat. On the other hand, if it took the boat two hours to go in one direction, we would know that the water was flowing in the opposite direction at the rate of 2-1/2 miles per hour. Similarly, if it took the boat less than an hour to make the return trip, we would have corroborating evidence that the water was in motion and was not flowing.

Another way we could check whether the water was flowing or not would be to run the boat across the body of water, rather than up and down the body of water. If we pointed the boat directly across the water and let it start, it would reach the bank directly opposite from us only if the water was not flowing. If the water were flowing, then the boat would be swept downstream slightly as it made the crossing and would reach the opposite bank somewhat downstream from the point at which it had originally been aimed. Either of these two experiments could be used to tell us whether the water was in motion or not.

What Michelson and Morley did was to set up an experiment in which they used light instead of a boat, and in which the hypothetical ether was supposed to play the role of water in our above examples. By using a complicated optical instrument called an interferometer, they were able to measure the difference in time that it took light to travel up and back to one mirror, and compare it to the deflection that light suffered traveling to another mirror and back—the second line of direction being perpendicular to the first. They could make a consistent check on their results by rotating the instrument so that, for example, if the first mirror were in the upstream-downstream direction in the first set of readings, it would then be in the cross-stream direction in the second set of readings. After extensive study, taking many months, they came to the conclusion that if there was an ether wind at the surface of the earth, its velocity had to be very small. Subsequent measurements (including some modern ones using lasers) have led to the conclusion that such an "ether wind" would have zero velocity at the surface of the earth. Of course, a zero velocity "ether wind" is equivalent to no "ether wind" at all.

Thus, the experiments of Michelson and Morley, while they did not bear directly on consequences of the principles of relativity, did point out that the idea that was prevalent in classical physics—the idea that there had to be some sort of

231

preferred frame of reference—could not stand the test of experiment. The absence of an "ether" cleared the way for new ideas and new thinking in physics, and hence this experiment can be thought of as being one of the crucial observational inputs into the theory of relativity.

D. Discussions and Conclusions

The principle of relativity states that the laws of physics must be the same as seen by every observer. The consequences of this principle, as we have seen, lead us to rather strange new ideas about a number of things. The question naturally arises, then, as to what we have left once these consequences are known.

What we have left, of course, are the laws of physics themselves. There is nothing in what we have done that tells us that, for example, Newton's laws must be wrong. What we *have* done is to redefine the concepts of space and time. Since Newton's laws are stated in terms of space and time, they may not look as simple as we had thought at first. For example, the second law involves the concept of acceleration which, in turn, involves distances divided by time. When we have accepted the principle of relativity, we still have the statement that force is equal to mass times acceleration. Although the mass of the same object as seen by two different observers will not be the same, both observers will agree that Newton's second law is valid in their frame of reference. Thus, the knowledge that has been acquired about the universe persists but, with the advent of relativity, we have to be very careful about stating who is observing an event, and taking careful account of the fact that two different observers may describe the same event in different ways.

Why didn't Newton see this? After all, he was a brilliant scientist. Why was it more than 200 years after his death before scientists realized the implication of the principle of relativity that was implicit in Newtonian mechanics? One reason, of course, was the fact that the electromagnetic theory of light was not known until 1865. We have seen the central role that the velocity of light and Maxwell's equations play in the derivation of relativistic physics. Until the time of Maxwell, there was really no reason to suppose that there was any contradition between the classical Newtonian idea of space, time, and the preferred ether frame and the laws of electricity and magnetism.

232

But even if this contradiction had been known in princi-ple, there is another important point that must be made. New-tonian physics was based on experimental observations that had been made on objects of "normal" size—that is, objects that could be manipulated in the laboratory or, at worst, seen in the sky. And, what is more important, the experiments had dealt with these objects moving at "normal" speeds. We have seen how difficult it would be to accelerate a normal ob-ject to anything near the speed of light. Clearly, the technical capabilities at the time of Newton and Galileo were not up to providing these scientists with objects traveling near c. Thus, the experimental basis of Newtonian physics was confined to normal objects moving at normal speeds.

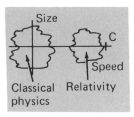

Figure 9.6.

We can visualize the situation as pictured in Fig. 9.6. Let the vertical axis represent the size of objects and the hori-zontal axis the speed at which they are moving. Normal every-day experience is clustered in the area shown. This region had been explored by Galileo and his predecessors, and the laws operating in this region were well known. These laws, in turn, formed the basis of Newtonian mechanics and of the Newtonian world picture.

When we begin talking about relativity, however, we are talking about the region on the right-hand side of the figure. This region involves normal size bodies, but it involves normal-size bodies moving at a velocity very close to the velocity of light. This region had *not* been explored experi-mentally. The contradictions and paradoxes that we have discussed have all resulted from our inclination to assume that the laws of physics that we discovered in one region (the region around the intersection of the two axes in the above diagram) must apply in all of the regions of the diagram. But this is clearly an unwarranted assumption. To use an analogy, we know that everyone speaks English in this country. This does not, however, mean that if we go to another region, we will find people speaking English. If we go to Europe, we may find that people speak English (if we visit Great Britain), but we could also find that people do not speak English (if we visit France). The question of what laws operate in any new region must remain open. There is no way of knowing when the laws that we have derived in one region will be appli-cable in a new region. There are examples when extension of this type can be made, and there are examples where they cannot be made.

233

Relativity is clearly a case where the extension cannot be made. The laws governing the motion of bodies moving near the speed of light are simply not the same as the laws governing the motion of bodies moving at normal velocities. Nor is there any reason why they should be the same. We have no right to expect that nature would be so kind as to present us with a world in which the laws we knew from everyday experience apply everywhere and equally to all systems, even those systems with which we have no direct experience.

There are, however, examples where extrapolations of this type can be made, just as it is possible to find English speakers in other parts of the world. One recent case involves the study of biological molecules. These molecules are very complex and highly ordered systems. There was a great deal of speculation in the 1930s that the laws governing biological molecules would prove to be different from the laws governing ordinary atoms. After all, we were extending our knowledge into regions of complexity that had never been explored before. As it turned out, however, the laws governing the simplest atoms also governed the most complex atoms. There are no new principles of physics involved in going from simple to complex systems. But it must be emphasized that there is no reason to expect that this would have been so. Had it turned out that new laws came in when systems became very complex, we would have had no right to be surprized.

Before we become too discouraged about the prospect of finding newer and newer laws of nature as we explore more and more of the world around us, one important point should be made. Although the laws of nature are different from the laws of Newtonian mechanics for particles moving near the velocity of light, it is also true that the laws of relativity reduce to the laws of Newtonian mechanics when we look at particles that are moving slowly. For example, we say that time in a moving frame is related to time in the proper frame by the equation

$$T_G = T_P / \sqrt{1 - v^2/c^2} \; .$$

The velocity v which appears in this equation is the velocity of the clock. When the velocity of the clock becomes very small compared to the velocity of light, the expression inside the square root in the above equation becomes approximately equal to 1. In other words, in the limit that the velocity of the object approaches zero, time measured in the moving frame and time measured in the stationary frame will become identical. The statement that these two times are identical is precisely the prediction of Newtonian mechanics. All of the other laws of relativistic motion that we have discovered have this same property. When the velocity of the objects in question becomes small, the results of relativistic mechanics became identical with the results of Newtonian mechanics.

In practice, this means that we can ignore relativity when we deal with normal objects moving at normal speeds. For example, no one who is calculating the flow of blood in a human artery would worry about the increase in mass due to the motion of the blood. It is just too small to be measured by any conceivable technique. Thus, provided that we keep in mind that Newtonian mechanics is only applicable in a certain range of velocities and sizes, we can go ahead and use it. In the same way, any new theory that might supercede the theory of relativity would have to reduce to the theory of relativity in the range of normal bodies moving near the velocity of light—the range for which the theory of relativity is known to be valid.

We shall see other examples of this idea of a hierarchy of physical pictures and physical theories as we proceed with our study of modern physics. For example, we shall find that the laws governing very small objects (the laws of quantum mechanics) are also not the same as the laws of Newtonian physics. On the other hand, when we take the laws of quantum mechanics and extend them to normal-sized objects, we find that the laws of Newtonian mechanics once again are recovered. Thus, each time we extend our knowledge away from our everyday experience in a new direction, we find new laws of physics. But when we take these new laws and extend them back into our normal, everyday sphere of experience, we find that we get results that are identical with the ones from which we started. This is a logically consistent and comforting result. It gives us some confidence that in extending our knowledge to new areas of the world around us, we will not be making radical breaks with the long history of scientific development.

SUGGESTED READING

All of the standard texts cited earlier contain sections on special relativity. In addition, the following may prove useful.

Clark, R. *Einstein: Life and Times.* New York: World, 1971. An exhaustive biography of Albert Einstein.

Einstein, Albert, and Infeld, Leopold. *The Evolution of Physics.* New York: Simon & Schuster, 1938. An informal and very readable account of relativity by the man responsible for it.

✓ Gamow, George. *Mr. Tomkins in Wonderland.* New York: Macmillan, 1940. An "Alice in Wonderland" account of the world as it would look if the velocity of light were smaller than it is. The book also deals with gravitation and quantum mechanics.

Jaffe, Bernard. *Michelson and the Speed of Light.* New York: Anchor Books, Doubleday, 1960. A biography of the man who pioneered the measurement of c and his contribution to relativity.

Russel, Bertrand. *ABC of Relativity.* New York: Harper, 1925. If you can get hold of a copy of this book, it remains one of the best discussions of relativity for the layman.

Trefil, James. "It's All Relative when you Travel Faster than Light." Smithsonian Magazine, Nov., 1976. A non-technical discussion of tachyons.

QUESTIONS AND DISCUSSION IDEAS

1. In the text, we showed that the time dilatation and length contraction effects depend on the factor

$$\sqrt{1 - v^2/c^2} \ .$$

   Calculate this factor for the following cases:
   a. a car moving at 60 mph (88 feet per second)
   b. a satellite moving at 25,000 mph
   c. an object traveling at half the speed of light

2. In each of the three cases in Problem 1, an observer on the ground measures a 10-second interval on his clock. What interval will be seen on the moving clock?

3. Resolve the paradox of the train and the tunnel posed in the section on Simultaneity.

4. Describe the sequence of red and yellow flags discussed in that section as it would be seen by a man on the train.

5. In each of the three cases in Problem 1, the moving ob-
    server is carrying a measuring rod 10 meters long. How
    long will this rod appear to an observer on the ground?

6. In light of the answers to Problems 2 and 5, comment
    on the possibility of detecting relativistic effects with
    experimental equipment available to the average per-
    son in each of the three cases.

7. Let us consider the problem of the mu meson discussed
    in the text. In its rest frame, the meson decays in
    0.000003 sec. Let us assume that the meson is created
    moving at a velocity almost equal to the speed of light.
    Show that if there were no time dilatation, it would be
    able to travel only 900 meters before decaying (the
    speed of light is about 300,000,000 meters per second).

8. Now suppose that the mu meson in Problem 7 was
    created moving at 0.99 times the speed of light, and
    suppose it was created in a collision 7,000 meters up
    (this is a little more than four miles).
    a. Calculate the factor $\sqrt{1 - v^2/c^2}$ for this meson.
    b. Hence, calculate the time interval between crea-
       tion and breakup which would be measured by
       an observer on the ground, remembering that in
       the rest frame of the particle, this time is .000003
       sec.
    c. Calculate how far something moving at almost the
       speed of light would travel in the time interval
       you worked out in b above.
    d. Hence, explain why this meson would be able to
       reach the ground, in spite of your answer in Prob-
       lem 7.
    e. Explain why the presence of mu mesons at sea
       level can be regarded as evidence for time dilata-
       tion.

9. Repeat the first four parts of Problem 8 for a meson
    moving at 0.9 times the speed of light.

10. Why is it the speed of light which appears in the equa-
    tions of relativity, and not some other speed (such as
    the speed of sound)?

11. State whether the following statements about relativity
    are true or false, and explain your answer.
    a. The principle of relativity requires that events
       look the same to all observers.

b. It requires that events look the same to all observers in the same frame.

c. Nothing can move faster than the speed of light.

12. The mass-energy equivalence in relativity is very important in nuclear reactors. Let's try to get some idea of how much mass correspond to how much energy. A 100-watt light bulb burning for one second uses 100 joules of energy (these units are defined in the appendix). Given that the speed of light is 300,000,000 meters/sec., how much mass would have to be converted to energy to keep the bulb burning for one second?

13. In the same spirit as the previous problem, suppose we were to lift a car weighing 2,000 kilograms (about two tons) up a distance of ten meters (about ten yards). This would correspond to 20,000 joules of energy. How much mass would have to be converted to energy to lift the car?

14. How much more massive than its rest mass would an object be if it were moving at
   a. 60 mph
   b. 25,000 mph
   c. .9 time the speed of light
   d. .99 times the speed of light

HINT: You may want to consult the answers to Problems 1, 8, and 9.

15. From your answers to Problem 14, comment on the problem of faster-than-light travel.

16. Consider the following objection to the proof of length contraction discussed in the Section: To the observer on the train, the light in the clock on the ground has to travel farther, so the clock on the ground will click fewer times. Therefore, the time seen on the ground will actually be shorter than the time measured on the train and, instead of contracting, the rod will appear to expand. Is this argument right or wrong? Why?

CHAPTER
X

SCIENTIFIC IDEAS IN THE 19th CENTURY

"Once or twice I have been provoked and have asked how many of them (literary intellectuals) could describe the Second Law of Thermodynamics.   The response was cold: it was also negative.   Yet I was asking something which is about the scientific equivalent of 'Have you read a work of Shakespeare's?' "

C. P. Snow, The Two Cultures and
the Scientific Revolution

A. The Atom in the 19th Century: Chemistry, Heat, and Entropy

We have already discussed how the 19th-century physicist had a very comfortable, perhaps even arrogant, view of the world. He felt that all of the important questions in physics had already been answered by the work of Newton and Maxwell, and that all that remained was an orderly working out, in greater and greater detail, of the laws of nature. We have also seen how, from a purely theoretical point of view, there was an inherent problem in the work of Newton and Maxwell. This problem, when explored by Albert Einstein, led us to totally new concepts of space and time, and to the special theory of relativity. But long before Einstein had formulated this theory, impressive evidence of another type was starting to accumulate that would also eventually lead to new laws of nature and an understanding of heretofore unexplored regions of the world. This evidence concerned the existence of tiny particles in nature—particles that are called atoms.

We have already discussed the fact that certain Greek philosophers had arrived at the idea that there must be some smallest particle of matter, to which they gave the name "atom." The method by which they arrived at this conclusion, and the motivation they had for thinking about this question were rather different from those of modern thinkers. Their primary goal was to find a compromise philosophical system, and their primary tool of investigation was pure reason. By the beginning of the 19th century, however, people were thinking about the structure of matter for rather different reasons. Throughout the Middle Ages, a steady build-up of knowledge about the chemical properties of matter had taken place. During the 18th century, scientists began to realize that there were two different kinds of material in the world. There were materials that could be "taken apart" into other kinds of materials, and there were materials that could not be taken apart in any way. For example, if wood were burned, we would get gases, ash, and other miscellaneous materials. Therefore, wood would be one of the first kind of materials we mentioned. To these materials, we now give the name "compound," or, in the case of a complex substance like wood, a "mixture of compounds."

240

On the other hand, there were a number of chemical substances that could not be broken down into other substances. For example, gold could not be broken down no matter what chemical processes or tests it was subjected to. This latter type of chemical substance we now call an "element." In 1796, the French chemist Antoine Lavoisier published a list of known chemical elements. These numbered 26 at the time of publication. Thus, it seemed that there were in nature 26 chemical substances that were in some sense more "fundamental" than other chemical substances. Any known substance could be broken down by one means or another until finally it was reduced to some combination of these 26 elements. From that point on, it could be reduced no further.

In addition, there were rather puzzling regularities that could be observed in the amounts of these chemical elements that would result from the reduction of any chemical compound, no matter what its composition. If water were broken down into its chemical elements, quantities of the two gases, hydrogen and oxygen, would be the result. No other elements were ever found in pure water. Not only was every kind of water made up of these two elements, but it was always made up of precisely the same *amounts* of these two elements. In other words, no matter where the water came from, whether from a glacier, from the ocean, or from rain, it was always approximately 88% oxygen by weight and 12% hydrogen by weight. This particular proportion of the two elements never varied. The same law was found to hold true for any other chemical compound that could be prepared or obtained. This was given the name "the law of definite proportions." When it was discovered, there was no understanding of why this law should be true. It was simply known that it held at the experimental level. In a sense, the law of definite proportions is rather similar to Kepler's laws of planetary motion. They are statements of facts that are found to be true in nature, but these facts are not understood in any basic way.

In 1808, an English chemist named John Dalton published a book called *New System of Chemical Philosophy* in which he proposed a rather clever explanation of the law of definite proportions—an explanation that succeeded at the same time in casting new light on the existence of elements and on other chemical regularities that were known at the time. Dalton's argument was as follows: Suppose that for each kind of element known (and there were 26 known at the time) there was a tiny particle of indivisible matter called an atom. Suppose further that the atoms corresponding to each chemical element were of different sizes and had different weights. Suppose that these atoms had the property that when two atoms of different elements were brought together, there was some mechanism by which they could stick together. If all of these facts were true, then the chemical phenomenon that we have just discussed would be easy to understand.

For example, the creation of chemical compounds would simply correspond to a process in which different kinds of atoms were brought together and combined. The decomposition of chemical compounds would correspond to a process in which a series of atoms that had been held together were taken apart. Chemical processes that we see in the laboratory correspond to building up and tearing down structures built of atoms—sort of a tiny tinker toy game.

The law of definite proportions was also very easy to understand. In Dalton's picture, water would look something like the structure we show in Fig. 10.1. It would consist of one heavy atom (oxygen) to which were attached two lighter hydrogen atoms. This combination of atoms was called a molecule. When the water was decomposed into its elements, we would always find the oxygen and hydrogen in the ratio of one oxygen atom for every two hydrogen atoms. Since the oxygen atom is 16 times as heavy as the hydrogen atom, the law of definite proportions is explained handily. Similar arguments could be made for any other of the chemical compounds for which the law of definite proportions was known to hold.

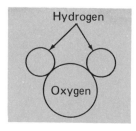

Hydrogen

Oxygen

Figure 10.1.

Thus, with John Dalton, the modern idea of the atom was born. It is very important to realize how different this idea is from the Greek atom proposed by Democritus. Democritus never considered whether or not any evidence of an experimental or observational nature could be collected that would support his ideas. He was purely interested in the philosophical impact of the idea of atoms. Dalton, on the other hand, introduced the concept of atoms purely and simply to explain observed chemical facts. He was not particularly interested in the philosophical impact of the idea of atoms, but rather in explaining known chemical phenomena. Furthermore, the main thrust of his arguments was to show how the idea of atoms was forced upon us by observation—by experiment. This attitude of dependence on observation should be familiar to the reader by this point. We saw it begin with Tycho Brahe, and it had reached its high point during the height of the Newtonian era in physics.

To Dalton, the Greek name of "atom" was particularly appropriate. He believed that atoms were indeed indivisible lumps of matter. In fact, the illustrations that he gave in his books of chemical compounds resembled nothing more than the pyramidal stacks of cannonballs one sees in old fortresses. To Dalton, the atom was simply a large, round, featureless, uniform ball of material. Later on, the followers of Dalton would try to imagine that these round, featureless atoms had hooks attached to them, to explain the fact that atoms could be stuck together. However, Dalton himself never advanced arguments of this type. He was content with showing how the processes of chemistry and the combinations and decompositions of matter could be explained in terms of a few simple building blocks called atoms. With this explanation, men advanced for the first time to a true scientific understanding of the basic structure of matter.

B. Technical Digression: The Laws of Thermodynamics from an Atomic Point of View

Once we have in mind the idea that pieces of matter that appear smooth and continuous to our senses are composed of tiny particles, it becomes possible to understand yet another area of physical phenomenon. This is the area dealing with properties of heat. The science of the study of heat and thermal effects is called *thermodynamics.*

If we could look inside of a block of material, we would see that the atoms or molecules of which it is composed are constantly in motion. Depending on the kind of material it is (i.e., depending on whether it is a solid, a liquid, or a gas), this motion can be anything from small vibrations about a fixed point (appropriate for a solid) to random wanderings through a large volume (appropriate for a gas). In fact, this is one way of classifying materials into the categories of solid, liquid, and gas. Let us take a block of ice

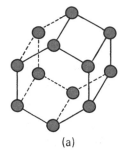

(a)

(b)

(c)

Figure 10.2.

and trace its progress from a solid crystal to steam. When the ice is solid, the water molecules are arranged in a regular crystalline lattice such as that shown in Fig. 10.2a. You can visualize this state of matter as sort of a giant tinker toy construction, in which the atoms are held fairly rigidly in place by forces that we shall discuss in detail later. This means that each atom is constrained to remain in the neighborhood of its original location, and is not free, in general, to go wandering about in the lattice. If we try to move one particular atom in the lattice, the series of interlocking bonds that ties all the atoms together makes it impossible to move a single atom without moving all of the other atoms as well. This, in turn, means that if we push on a solid at some point, the solid will push back. Thus, it is not possible to move the top of a table downward simply by exerting a small force on it. The force that we exert when we push down is exactly cancelled by the force exerted by the chemical bonds in the table top. This property—the ability to generate forces that can counteract externally imposed forces—is the basic characteristic of a solid.

But the atoms are *not* like tinker toys in every detail. In fact, although each molecule in an ice crystal is constrained to be in one general area, it is not stationary. It moves around, performing small vibrations and oscillations about its point in the lattice. You could think of this in our tinker toy analogy as a situation in which the things that hold the molecules together and keep them separated by the correct distance are not solid sticks, as they would be with tinker toys, but instead are fairly stiff springs. In such a situation the molecules could jiggle around a little bit, but they could never get very far from the point where they would have been had the springs been solid, stiff sticks. The points about which the molecules oscillate are called "lattice points." This is the picture that we now have of the internal structure of the crystalline solids.

Suppose that we start to heat the ice. We know, of course, that the ice will melt. But what does this mean on the atomic picture? As we start heating the ice, we would observe that the small oscillations around the lattice points that we discussed above would begin to become larger—i.e., the water molecules would begin to exhibit larger and larger motions around their lattice point. Eventually, if we supplied

244

Isaac Newton (1642-1727) investigating the theory of color.

Albert Einstein (1879-1955), the man who developed the theory of relativity, investigates some interesting applications of Newton's Laws.

enough energy, the oscillations would become so large that the water molecules would be able to tear themselves away from the lattice point (break the springs in the above analogy) and go wandering off by themselves. When they had done this, the material would look similar to that pictured in Fig. 10.2b. The molecules would be fairly closely packed together, but they would no longer be locked into place. Each molecule would be free to wander about to any part of the material that it wanted to. However, the molecules would still be fairly close together in this state of matter, which we call a liquid.

If we continued to heat the material, the motions of the molecules would become more and more violent. Eventually, those molecules that were nearest the surface of the material would be able to get out of the material and go off by themselves into the air. If we continued heating the material, pretty soon all of the molecules would have escaped, and we would have a state of matter something like that pictured in Fig. 10.2c. The molecules would be rather widely spaced, and would be moving around more or less randomly in a large volume. This type of material is called a gas, and since we started with ice, this particular gas is what we would call steam.

This orderly picture of the progression of materials from solid to liquid to gas is a simple consequence of Dalton's idea of the atom. One important ingredient in this picture is the idea that when we heat a material, the atoms move faster than they had previously. If we think in terms of energy, we know that an atom when it is moving quickly must have more kinetic energy than that same atom when it is moving slowly. Thus, we would have to conclude that a material that has been heated and in which the atoms are moving quickly has more energy than a material that has not been heated. This leads us, in turn, to conclude that heat itself must be a form of energy.

The realization of this connection between the speed of atoms in an object and the heat stored in the object is one of the fundamental points in the science of thermodynamics. We normally think of measuring heat in terms of temperature, but it is obvious from the above discussion that high temperature simply corresponds to a high level of kinetic energy in the atoms, with a similar statement being true for

Snowflakes.

245

low temperature. Thus, once we know about atoms, the temperature of any object has a simple interpretation.

Even the subjective feeling of heat when we touch something at a high temperature can be explained. When the fast moving atoms in the object collide with the atoms in our hand, the atoms in our hand move faster. The temperature in our hand increases, and this is interpreted as heat.

It is important to realize that this connection between heat and atomic motion holds no matter how the heat is applied to an object. For example, we could supply the heat with a flame, in which case the atoms in the object would be given increased velocity by the rapidly moving atoms in the hot gas. Alternatively, we could rub the object with something else, imparting kinetic energy through collisions of moving atoms with the more or less stationary atoms in the object. In either case, the result is the same. The atoms in the object move faster, and we say that the temperature increases. The latter example, in which the temperature is increased by contact, is an example of heat generated by friction.

The realization that energy and heat are in some way equivalent is often called the "First Law of Thermodynamics." We shall state it in the following way:

Heat is a form of energy, and energy is conserved.

Actually it is not very difficult to find direct experimental consequences and confirmations of this idea. For example, we know that if we lift a heavy weight off the floor that we have given the weight potential energy. Suppose we rigged up an apparatus like that shown in Fig. 10.3 in which the weight we had lifted was tied by a rope to a shaft which turned a set of paddlewheels. If this shaft were to be placed in a large trough of water, and the weight were to be dropped, the paddlewheels would be turned while the weight was falling. We know that the turning of the paddlewheels would cause some of the water molecules to be set into more rapid motion than they had been before the wheels started turning. If we had a sensitive enough thermometer, we could even measure the difference in temperature in the water before the weight was dropped and after it was dropped. We would discover that the temperature of the water had risen. This is, of course, only what we should have expected on the basis

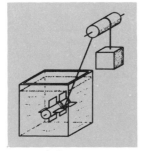

Figure 10.3.

of our picture of heat and temperature as kinetic motion of atoms. It was experiments of this type that led physicists to enunciate the first law of thermodynamics and to work out the exact numerical equivalence between ordinary energy (that is, potential and kinetic energy of large objects) and heat. For our purposes, however, we shall simply keep in mind that when an object is heated, the kinetic energy of its molecules increases. In this way, we stress the idea that heat is just one more kind of energy—like kinetic or potential energy—which must be taken into account when we work out energy balances in physical situations.

A second important point that we can understand when we consider the transition from ice to steam as seen from the atomic point of view is that ice is a highly ordered, very regular kind of material, and steam is a rather chaotic, unordered kind of material. In the process of changing the material from ice to steam we have increased the amount of disorder in the material. Physicists actually have a technical definition of the quantity that we have referred to loosely as "the amount of disorder." A measure of the "amount of disorder" in a material is given by a quantity called the "entropy."

We all know that there are general processes that go on in nature that have to do with the idea of order. For example, it is a great deal easier to scramble eggs than to unscramble them. It is almost always easier in the real world to go from a system with a high degree of order to a system with a lower degree of order. I am sure that the reader can think of many examples. It is easier to knock down a tower of blocks than to build it. Geological processes wear down mountains (a relatively ordered system) to piles of sand and rubble (a relatively unordered system). The examples could go on endlessly. In fact, the idea of order can be given a mathematical sense; it can be shown that no matter what we do in the universe, the total amount of disorder in the universe must either stay constant or increase. It can never decrease. This is a statement of something that is called the "Second Law of Thermodynamics." A precise statement is:

In every physical process, the total change in the entropy of the universe must be equal to or greater than zero.

247

In our example of the melting ice, we saw a particular system go from being highly ordered to unordered. This would be an example of a situation in which the entropy in the water molecules had been increased. But what about the reverse process, in which the steam is condensed as water and then melted as ice? Isn't this a violation of the second law of thermodynamics in that the entropy of the ice-water-steam has actually decreased?

This example is not a contradiction of the law as we have stated it, because although the entropy of the ice-water-steam system has, in fact, decreased, the system is not the entire universe. We know that when steam condenses, it gives off heat. Similarly, when water freezes, it gives off heat. This heat must be absorbed somewhere. For example, it may be absorbed by the immediate surroundings of the steam or the water. When these materials absorb heat, their atoms will begin to move faster, as we have discussed above. This means that although the entropy is decreasing in the water and ice, it is increasing in the surroundings of the material. Thus, if we add up the total change in entropy of both the steam-water-ice and its surroundings, it is reasonable to expect that it would not be negative. At best, the second law tells us that if we added up all the entropy changes attendant on changing a particular system *and its surroundings,* the change in entropy would be greater than zero or (at best) zero. Thus, according to the second law of thermodynamics, the universe might eventually "run down" like a clock, and the eventual end of everything will be a totally disordered, uniform blob of gas spread out throughout all space.

Actually, many cosmologists do not hold to this particular bleak view of the future of the universe. They argue that even though at the present time we have discovered a particular law of nature—the law of increase of entropy—it is not necessary that these laws must hold at all times in the future. We come back again to the statement that we have made so many times before. The laws of physics are only as good as the experiments that are used to establish them. Obviously, we cannot do experiments 20 billion years in the future, and we have no real way of knowing which of the laws that we now believe to be true will still be true then. There are a large number of theoretical futures of the universe in which at some time in the future the universe stops expanding and begins to contract. During the period of contraction, according to these models (and it is important to emphasize that they are just guesses at things that could happen), the second law of thermodynamics could be reversed. While entropy always increases during expansions, in some scenarios the entropy decreases during contraction. In this way, the quantity of order in the universe is undergoing a constant, cyclic change, and the universe renews itself every 20 or 30 billion years. I mention these "far out" theories to emphasize the fact that all laws of physics are contingent on experimental verification, and that it is always possible for clever theoreticians to imagine situations in which the laws as we know them now do not hold.

There is another interesting concept that follows from the atomic picture of solids. We have discussed the idea that even at normal temperatures the atoms in solids are not stationary, but instead perform small fluctuations about a fixed point (called the lattice point) in the solid. If we add heat to the solid, these fluctuations become larger. Conversely, if we cool the solid, the fluctuations become smaller. It is possible to imagine in principle that we could cool the solid enough so that the atoms of which it is composed became absolutely stationary. (Actually, although 19th-century physicists thought of things this way, the laws of quantum mechanics that we shall discuss in the next section tell us that atoms are never *absolutely* stationary, but have some smallest energy, called the "zero point energy" below which they cannot go.) Clearly, once we had cooled it that far, it would not be possible to cool it any further. In other words, once we have extracted *all* of the kinetic energy of the atoms that compose a material, it is not possible to extract any more kinetic energy. The point at which all atomic motions would stop is called "absolute zero." It occurs at about $-459°F$ ($-273°C$). In fact, there is a temperature scale (called the Kelvin scale) in which absolute zero is the zero of temperature. On this scale, room temperature would be about $300°K$.

Although you might think that it would be possible in principle to cool a body to absolute zero, in fact it is not possible. There is a "Third Law of Thermodynamics," which states:

It is impossible, by a finite number of operations,

to reduce the temperature of any object to absolute zero.

The basis of this law lies in the fact that as the temperature is lowered, the system becomes more and more ordered, so that its entropy (which is a measure of *dis*order) approaches zero. The third law, as we have stated it, then follows as a mathematical consequence of this fact. There is an entire area of physics research, called low temperature physics, which is devoted to the study of properties of materials near absolute zero.

In summary, then, the three laws of thermodynamics tell us about the behavior of heat and thermal phenomena. They play the same role in the science of heat that Newton's laws play in the science of mechanics, and that Maxwell's equations play in the science of electricity and magnetism. They are the basic, fundamental laws of nature that govern this particular area, and from these three laws all of the behavior of thermal phenomena can be derived.

C. Implications of Atomic Structure

The idea that the atom should be a simple, featureless piece of material, which is implicit in the works of the Greek philosophers and was also held by Dalton, proved to be one of those ideas that is just too simple and beautiful to be true. It was not long before people began to realize that there must be more to the atom than this.

In fact, we have already discussed some of the evidence that would point to this conclusion. With the study of the science of electricity, it quickly became obvious that it was possible to take ordinary material such as copper and zinc and sulphuric acid—none of which had an electric charge— and produce an electric current by building a battery. In other words, it was possible to take material in which there was no electric charge at the beginning and produce an electric charge at the end. If we look at this process on the atomic level, it is also obvious that this implies that there must be structure inside the atom. Why? Well, suppose we start with a group of electrically neutral atoms. This means that no atom has a net negative charge. We combine these atoms in a certain way (by building our battery) and suddenly find that we can extract from this group of atoms something that has an electrical charge. Where did this

250

electrical charge come from? The only place it could have come from was from inside the atoms themselves. This necessarily implies that inside the atom there must be something with a negative charge. And, since the atom is itself electrically neutral this implies in turn that there must be something with a positive charge. Thus, far from being a featureless, undifferentiated piece of matter, the atom must be arranged in such a way that inside of it are regions of positive and negative charge. It must have structure.

The actual form that this structure took did not become obvious for a long time. However, in 1897, the English physicist J. J. Thomson discovered a particle that he called the "electron." This is a particle much smaller than the atom, and it is a particle that carries a negative electrical charge. In fact, it is the movement of electrons in a wire that we perceive as flow of negative charge, and that we call an electric current. The discovery of the electron had important implications for the study of the atom. The reason is that once the carrier of the negative charge in electric current has been identified, it is obvious that this carrier of electric charge must also exist inside the atom. If it didn't, it would be impossible to extract electric current from neutral atoms. Thus, we know that inside the atom there is at least one kind of particle—the electron.

By the turn of the century, then, physicists were beginning to ask questions about the structure of the atom. From the point of view of the Greek philosophers, questions like these would have been meaningless. By definition, the atom was indivisible and could have no further substructure. However, the evidence for the substructure, which we have discussed above, was overwhelming. The picture of the atom that emerged at the end of the 19th century was called the "raisin bun" atom. People imagined the atom as a large blob of positive matter in which electrons were embedded in much the same way as raisins are embedded in a piece of bakery. We now know this picture to be wrong but it was the first attempt to come up with some sort of picture of what the atom must look like. Once we have established firmly the idea that there must be more to the atom than a single, indivisible, homogeneous piece of matter, the question of the structure of the atom and of the existence of subatomic particles becomes one of the most important questions in science. After all, we are dealing here with the question of the basic structure of matter, and surely this is one of the most interesting questions that human beings can ask about the world around them.

When we are trying to discover the nature of objects that we have never seen before (e.g., subatomic particles), it is natural to begin by trying to see whether or not it is possible to identify them with things with which we are already familiar. Thus, it was only natural that the first question that physicists asked about subatomic particles was whether they were particles or waves.

Both of these categories of objects are familiar to us in our everyday life. A billiard ball rolling along a table, a baseball being thrown through the air, and a rock rolling down a hill are all examples of particles in motion. We think of a particle as a lump of matter, having a definite position and a definite velocity at any given time. We can trace its movement from one point to another; at any given instant in time, we can say precisely where it is.

Waves are just as familiar in our everyday experience. We have all seen waves at the ocean or the beach. However, we have probably not thought too much about some of the detailed properties of waves. We shall have to do so before we can understand the questions that were asked at the beginning of the 19th century about the nature of subatomic particles.

D. Technical Digression: Waves and Particles

Let us consider a wave coming into a beach. We know that the wave moves with a certain speed. If we wanted to, we could measure the speed simply by positioning a friend a known distance out in the water and measuring the time it took a wave to go from where he was to the shore. On the other hand, we also know that if there were a little piece of cork floating on the water, it would not move with the same velocity as the wave. It would simply bob up and down. Now the cork is moving in the same way as the part of the water that surrounds it. In other words, the particles of water themselves bob up and down at the same time that the wave moves by. This is an important property of a wave.

252

The wave may travel *on* the water, but the wave is *not* traveling water. The wave is something totally different from the medium in which it travels.

To fix this idea more firmly in our mind, consider taking a long rope that is suspended from a wall and flipping one end. We all know that if we do this, a pulse will travel down the rope. The pulse can travel quite fast, but we know that each piece of rope simply goes up and down. Again, the wave is traveling at one velocity, and the motion of the wave is quite different from the motion of the particles making up the material on which the wave is traveling.

Let us begin our discussion of waves, then, by reviewing some of the simple properties that they have. In Fig. 10.4a, we have drawn a continuous wave. This is a type of wave that you would observe if you were standing at the beach and watching the surf come in. We call it a continuous wave because the wave's crests follow one another regularly, and there appears to be no particular end point to the wave. A continuous wave is different from the type of pulse we discussed in the example of the rope above, in that the pulse has a definite beginning and a definite end. The particular wave form that we have drawn is smooth. In Fig. 10.4b we show a type of wave that is not smooth. This type of wave could not be carried in water but is often seen in electrical circuits. It is called a square wave. The point of drawing these different kinds of waves is to emphasize the complete generality of the discussion that follows. We shall use water waves as an example, but *any* other kind of wave will have these same properties. It does not matter whether it is a wave on a rope, a wave in water, or an electromagnetic wave such as light.

Let us begin by introducing some definitions.

Wavelength:

The wavelength is the distance between points a and b or points c and d in Fig. 10.4a. It is defined to be the distance between one point on the wave and the corresponding point on the following wave. For example, at point e in the figure, the wave disturbance is zero (which simply means that at point e the level of the water is at the same level as it would have been had the wave not been there), and as we move from left to right the level

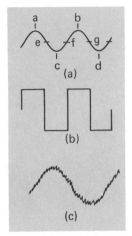

Figure 10.4.

of the water is going from a higher to a lower value. At point f, the level of the water is again zero, but at this point the level of the water is going from a lower to a higher value; f is not the point that corresponds to e; g is the point that corresponds to e. Thus, the wavelength of the wave in the above figure would be defined to be the distance between e and f. Wavelength is commonly denoted by the Greek letter λ.

Frequency:

The frequency is defined as the number of waves that pass a certain point in a certain time interval. For example, if we stood at one point and watched this wave go by and observed that five crests passed us in a second, we would say that the wave had a frequency of five cycles per second. Another way of thinking about this quantity is to think of it as the number of times that the water rises and falls at a given point per second.

Period:

This is the time it takes for a wave to go through one complete cycle. Obviously, the period is related to the frequency by the equation

$$\text{Period} = \frac{1}{\text{frequency.}}$$

For example, a wave having a frequency of five cycles per minute would have to complete each cycle in one-fifth of a minute.

Displacement:

The displacement of a wave is the amount that the medium in which the wave moves is raised or depressed over the level it would have if the wave were not present. In Fig. 10.4a, for example, the displacement is zero at points f and g, positive at points a and b, and negative at points c and d.

Amplitude:

The amplitude is the maximum displacement (positive or negative) of the wave. In Fig. 10.4a it would be the vertical distance between points b and f.

Node:

The node is the point at which the level of the wave is the same as the level that the water would have were there no wave at all. In Fig. 10.4a a node is at point e. In a traveling wave such as the surf coming into a beach, the node moves along with the wave. In standing waves, the node does not move.

With these basic definitions of the properties of waves, we can now move on to discuss another property exhibited by waves—the property of interference. We shall see that this particular property will allow us to distinguish between particles and waves. It is a property that waves alone possess.

To begin our discussion of interference, let us consider the example pictured in Fig. 10.5. A pulse, labeled A, is traveling on a rope toward the left, and a pulse, labeled B, is traveling on a rope toward the right. When we begin observing the rope, the pulses are well separated, and the rope is flat in between the two pulses as shown. One way of thinking about the progress of the wave along the rope is to imagine that each little bit of rope is receiving a "signal" from the wave, and that this signal tells that little bit of rope how to move. For example, at the point labeled 1 in the pulse labeled A, the rope is receiving a "signal" telling it to rise up from its equilibrium position by a distance that we have labeled d_A. Similar statements could be made about

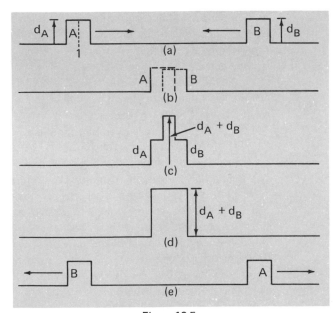

Figure 10.5.

any other point along the rope. The question that naturally arises, then, concerns what will happen as the two pulses begin to approach each other. For example, in Fig. 10.5b the dotted lines represent the positions that each pulse would take if the other pulse were not present. To find out what the rope actually does when the two pulses overlap, we have to be just a little more careful.

The phenomenon of interference arises when two waves, or pulses, overlap in the sense that we have discussed above. When this overlap occurs, a piece of rope in the region of overlap will be receiving two "signals"—one from pulse A and one from pulse B. As far as that single piece of rope is concerned, however, it is receiving only one signal. The signal it receives is the sum of the signal from pulse A and pulse B. Thus, in Fig. 10.5b, the rope in the overlap region is receiving a signal to lift up a distance d_A from pulse A and a distance d_B from pulse B; consequently, it rises a distance $d = d_A + d_B$. If we carry out this analysis for each point in the region of overlap, the rope will actually take the configuration shown in Fig. 10.5c.

As the pulses move through each other, the region in which the rope is receiving two "signals" increases, until the waves are completely overlapped. This situation is shown in Fig. 10.5d. However, nothing new happens when this overlap occurs. We can find the actual displacement of the rope simply by adding the "signals" received by each piece of rope in the entire region of overlap.

As the waves pass through each other, they eventually cease overlapping and then move on in their respective directions, as if nothing had happened. This is shown in Fig. 10.5e. It is left as an exercise to the reader to show this.

If it had happened that the pulses in the above example had been of opposite direction (i.e., if A had been as shown, but B had been down, the situation would have looked different but the rope would have obeyed the same law. In Fig. 10.6, we trace what happens as two pulses of equal opposite amplitude pass through each other. In Fig. 10.6a, the pulses are well separated and moving toward each other. There is no problem about what the rope will do at any point. In Fig. 10.6b, the pulses have started to overlap. In the region of overlap, the pieces of rope are again receiving two "signals." This time, however, one of the signals is telling the rope to move up and the other is telling the rope to move

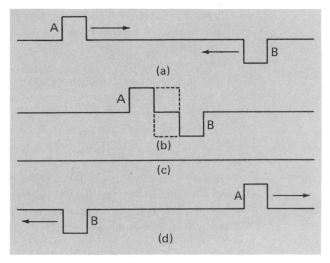

Figure 10.6.

down. Consequently, the distance that the rope actually moves will be less than the distance it would have moved if either wave had been in position alone. This is because the actual distance is given by the formula $d = d_A - d_B$ as opposed to the formula given above in which the two distances were added instead of subtracted. If the two waves are exactly identical in shape, there will be *no* displacement of the rope in the region of overlap. When the total overlap occurs, as in Fig. 10.6c, every piece of rope in the region of overlap is receiving signals telling it to move up from pulse A and signals telling it to move down the same distance from pulse B. Consequently, the pieces of rope do not move at all. When this happens, of course, we have the situation shown in which the rope is absolutely flat. In other words, when total overlap occurs, it looks as if there were no waves present at all. However, as the waves pass through each other, we again get the situation as shown in Fig. 10.6d of each wave moving along independent of the other.

Although we have demonstrated this process with pulses, it is equally true of continuous waves. The principle that we can use to discover the shape of the rope in the region of overlap is sometimes called "the principle of superposition." It can be stated as follows:

When two waves overlap, the actual displacement
of the surface on which the waves are moving is

given by the algebraic sum of the displacement that each wave would have if the other were not present.

What this means is that the presence of two waves at a particular point in space (e.g., the presence of two waves at any point in the region of overlap on the rope in the examples discussed above) causes the surface on which the waves are moving to be distorted in a way that is different from what it would be if each wave were there separately. This, in turn, means that the actual surface distortion is the result of some sort of combination of the effects of the two waves. The technical term for this kind of effect is "interference." We say that waves interfere with each other when they overlap.

It is important to contrast this kind of behavior, which we observe in waves, to what we would expect from particles. If two bullets collided with each other, we could never have a situation in which there were no bullets present. Even if they splintered into pieces, there would still be as many pieces in the region of overlap as there had been moving toward the region of overlap. Consequently, the properties of interference are unique to waves and do not apply to particles.

The type of interference we discussed in the first example above, in which both waves had amplitudes in the same direction, is often called constructive, or positive, interference. Conversely, the type of interference we discussed in the second example, in which one wave tends to cancel out another wave, is often called destructive, or negative, interference.

So far, we have only concerned ourselves with interference between pulses on a rope. Of course, the reader will realize that what we have been discussing is not restricted to this example. Two "tidal waves" on water would produce exactly the same kind of effect as those discussed above. However, from the point of view of understanding the structure of the atom, it is much more interesting to apply the principles described above to the case of continuous waves.

Suppose that we had a surface of water that was undisturbed initially. Then we could imagine putting a motor or a vibrator or something else in the water that would cause a wave to go out. In cross section, this wave would look like the one shown in Fig. 10.7a. The wave would move out continuously from the center, and each point on the surface of the water would bob up and down each time one cycle of the wave went past it. Of course, there can be no interference with just one wave. But suppose that somewhere else in the water, we had another motor that was also making waves. Suppose that if this machine were present alone, the wave would look like that shown in Fig. 10.7b. It is very reasonable to ask the question, "What will the surface of the water look like when the two machines are turned on simultaneously?" To answer this question, we have to invoke the principle of superposition, which we stated above. The surface of the water at any point will be the algebraic sum of the displacement that the first wave would have if the second wave were not there, plus the displacement that the second wave would have if the first wave were not there. Thus, in the above example where the high points of the wave in Fig. 10.7b are exactly at the same position as the high points of the wave in Fig. 10.7a, we would get a wave like that shown in Fig. 10.7c. It would be precisely twice as high—in technical terms, it would have twice the displacement—as either wave would by itself. This is an example of positive interference with continuous waves.

Suppose, however, that the situation was not like that discussed above. Suppose that when we turned on the first motor, we got a wave such as that shown in Fig. 10.8a, but when we turned on the second motor, we got a wave such as that shown in Fig. 10.8b. Suppose, in other words, that where the wave from the first motor was at its maximum, the wave from the second motor was at its minimum. In this case, the surface of the water would look like that shown in Fig. 10.8c. There would be no distortion at all. The reason for this is obvious. Each piece of the surface is receiving a "signal" from the wave in Fig. 10.8a to lift up a certain amount, and a "signal" from the wave in Fig. 10.8b to push down by the exact same amount. Consequently, each point in the surface does exactly nothing. The surface will be flat, and there will be no wave evident at all. This is an example of destructive interference for the case of continuous waves.

In dealing with waves on water, constructive interference will lead to waves whose height is higher than the

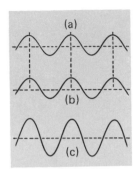

Figure 10.7.

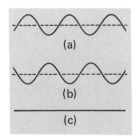

Figure 10.8.

259

waves that are interfering with each other, while destructive interference will lead to waves whose height is less than those that are interfering with each other. If, on the other hand, we were dealing with a wave like light, the perception we would have would be somewhat different. The amplitude of the light wave tells us how bright the light appears. Consequently, constructive interference of light corresponds to very bright areas in the region of overlap of two light waves, while destructive interference corresponds to dark areas.

Now let us apply our knowledge of the interference properties of continuous waves to a specific experimental setup. We shall give an example that concerns water waves, but obviously the same results would obtain no matter what kind of other wave we used. Suppose that we have two motors that move paddles up and down, thereby making waves in the water. We shall label these motors 1 and 2 in Fig. 10.9. If either one of these motors is turned on by itself, the result will be a concentric outward moving series of circular waves in the water. The question that we wish to address, however, is the question of what happens when we turn both motors on at the same time. Suppose at some point in the water there is some sort of a pier so that we can walk up and down a line, which is indicated in Fig. 10.9. We could then observe the type of wave that existed anywhere along that pier, simply by looking down in the water.

Let us begin by standing at point A, which is midway between the two motors and a long distance away. If we look down at the water at point A, what would we see? To answer this question we shall begin by asking a simpler question. Suppose that motor 1 were turned on, but motor 2 were not. After the motor had been running for awhile, so that the original waves were well beyond the pier, we would see a wave such as that shown in Fig. 10.10 at point A. It would be a single wave, of amplitude A, and it would be characteristic of one of the motors acting by itself. If, on the other hand, we turned off motor 1 and turned on motor 2, we would see exactly the same thing. To understand what we would see if both motors were turned on at the same time, we'd have to apply the principle of superposition.

Point A is the same distance from motor 1 as it is

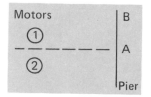

Figure 10.9.

from motor 2. Consequently, if the two motors are synchronous (by which we simply mean that when motor 1 is producing a crest in a wave, motor 2 is also producing a crest, and the same statement is true for troughs) then a crest of a wave that starts out from motor 1 at the same time as a corresponding crest from motor 2 will arrive at point A at the same time. We would then have a situation like that shown in Fig. 10.11. The wave from motor 1 (labeled 1) and the wave from motor 2 (labeled 2) arrive at point A in such a way that a crest from 1 is superimposed on a crest from 2, and a trough from 1 is superimposed on a trough from 2. By the principle of superposition, this means that the resultant wave will look like the one shown. That is, it will be exactly twice as big as the wave from 1 alone or the wave from 2 alone. There is a technical term that is used to describe this situation. When two waves arrive at a point in such a way that the crest of one arrives at the same time as the crest of another, and the trough of one arrives at the same time as the trough of another, we say that the waves are "in phase." Obviously, the statement that two waves are in phase is equivalent to the statement that there will be constructive, or positive, interference at the point in question. The result of our experiment, then, will be to see a wave at point A that is twice as large as that from either motor.

Suppose now that we begin walking up the pier until we arrive at point B. Let us suppose that at point B the distance from motor 1 to the point is exactly one-half a wavelength shorter than the distance from motor 2 to that same point. If this is the case, then the waves from 1 and 2 will be arriving as shown in Fig. 10.12. Even though the waves started out in phase from the synchronous motors, the waves from motor 2 had to travel farther to get to the point of observation than did the waves from motor 1. Consequently, the two waves do not arrive in phase. In fact, it is clear from the figure that the resultant disturbance in the

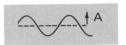

Figure 10.10.

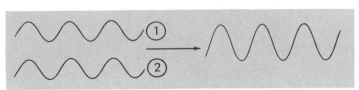

Figure 10.11.

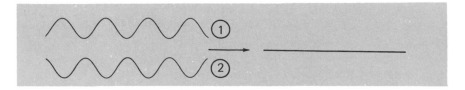

Figure 10.12.

water at point B will be such that there will be no disturbance at all. In other words, at point B we have complete
destructive interference between the two waves. Just as we
saw in our example of waves on a rope, destructive interference between two waves corresponds to the situation
we would get if there were no waves at all at that particular
point. Consequently, if we walked from position A on
the pier to position B, we would start by observing a wave
that was twice as big as the original wave and end by
observing no wave at all. We would see a still spot in the
water. If we were looking at light waves instead of water
waves, we would see a bright spot at position A and a dark
spot at position B.

Let us continue walking up the pier. Suppose that at
position C, the distance from motor 1 to C is precisely one
wavelength less than the distance from motor 2 to C. In
this case, we would have a situation such as that shown in
Fig. 10.13. Again, the wave from motor 2 would have to
travel farther to get to the point of observation, but it would
arrive at the point of observation exactly one wavelength
behind the wave from motor 1. This means that the first
crest from motor 2 would arrive at exactly the same time as
the second crest from motor 1, the second crest from motor
2 would arrive at the same time as the third crest from
motor 1, etc. If the motors had been running for a long
time, the difference of one wavelength does not matter.
Consequently, applying our principle of superposition, we
would get a wave as shown. Once more, we would get a
wave that would be larger than the wave from either of the
motors. In other words, we would see a large wave at point
A, no wave at point B, and a large wave at point C. If we
were observing light, we would see a light spot at point A,
a dark spot at point B, and a light spot at point C. It is left
to the reader to continue the walk up the pier to find out
what happens farther along.

262

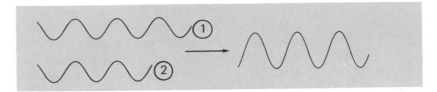

Figure 10.13.

This discussion of interference patterns in water waves would be interesting but not important were it not for one important fact. The property of interference, as we have said, is one that is shared by all waves. In particular, as we have hinted, light itself exhibits this property. Therefore, light was classified as a wave by classical physicists.

If you review our discussion of interference, you will see that it would apply to any kind of wave, whether it was in water or in something else. It would even apply to electromagnetic waves, since they have the same properties as any other wave, even though they don't travel in an easily visualizable medium. All that is necessary for interference to occur is the existence of two or more sources of waves.

Actually, you have been looking at interference patterns in light all of your life without knowing it. One of the most common ways to produce several light "sources" is to have a beam of light fall on a piece of material that has holes in it. Light falling on an ordinary window screen would be an example of such a situation. If we view the screen from the other side, each hole plays the role of a light source, and is analogous to one of the motors in our water wave example. Since each hole is illuminated by the same initial beam of light, the "sources" of the waves on the other side of the screen would all be synchronous.

Have you ever looked at a street lamp through a screen at night? If you have, you probably remember seeing a pattern something like the one shown in Fig. 10.14. There is a central bright spot and then a "cross" made up of alternating light and dark bands.

The alternation of light and dark should immediately remind you of the kind of interference pattern that we saw in water waves in our two-motor example. The bright spots correspond to places where constructive interference is occurring with the light waves coming from the various holes in the screen, and the dark spots correspond to places where

Bright spots

Figure 10.14. A light viewed through a screen.

destructive interference is occurring. When two or more waves from different sources come together to form a pattern in which we see light and dark bands, we often refer to this as a "diffraction pattern" and speak of the waves exhibiting diffraction. Diffraction can only occur when interference can occur, and interference can only occur for waves. There are many other examples of such interference patterns being seen with light, so it should not be too surprising to learn that the classical physicist felt that he had ample proof that light, because it exhibited interference, is a wave.

This same sort of criterion can be used to classify new things as well. One way of deciding whether a particular object in nature is a particle or a wave is to try to do experiments to see whether these things exhibit diffraction or not. For example, if we were dealing with light, we could produce diffraction with an apparatus similar to that pictured in Fig. 10.15. A beam of light from some source is shown on a screen. In the screen we cut two tiny slits, which allow the light to go through. These two slits act as "sources" of light to observers on the other side of the screen, and play the same role in the experiment that we are about to describe as the two motors played in the experiment involving water waves discussed above. Suppose we put another screen a certain distance back from the large screen. What would we see if we look at that screen?

Obviously, if we look at point A on the screen, we would see a bright spot. This is the point that is equidistant from the two slits in the first screen, and, consequently, the waves from the two slits arrive at this point in phase. If we look up the screen, we eventually see some point B, at which the distance from the top slit is half a wavelength less than the distance from the bottom slit. At this point, we have destructive interference, and consequently we see a dark spot. We could continue this discussion, but it is obvious that what we will see on the screen will be a series of light and dark bands corresponding to areas of constructive and destructive interference. The most important of these light areas for our subsequent discussion is the first one we discussed at point A. No matter how we arrange the slits, there will always be a bright spot at the point that is equidistant from the two of them.

Now suppose that instead of waves coming into the slit, we had particles. For example, suppose we had exactly

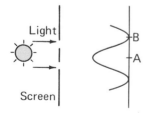

Figure 10.15.

264

the same situation—a screen with two holes cut in it—but we were shooting bullets at the screen rather than shining light on it. Suppose, further, that we set up an apparatus in back of the screen to catch the bullets that come through, as shown in Fig. 10.16. We could imagine, for example, having large drums filled with sand to catch the bullets as they come in. These drums would be stacked up on top of each other along the plane of observation. At the end of the experiment, we could simply empty each drum and count the bullets that were in it.

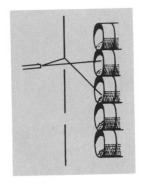

Figure 10.16.

Obviously, most of the bullets that get through the screen will be those that come through the slits without touching anything. These will go primarily into the collectors immediately behind the slits. What this means is that after the experiment is over, we would find that the two collecting drums located directly in back of the two slits would contain most of the bullets. However, there is some small chance that a bullet would hit the side of a slit, ricochet, and wind up in one of the other collecting bins. Consequently, there will be a few bullets scattered around throughout the other bins. If we made a graph of the number of bullets in each bin, we would probably get something like that pictured in Fig. 10.17. The histograms in Fig. 10.17a corresponds to the number of bullets that would appear in each collecting bin if only one slit were open. That is, the graph labeled U would correspond to the bullets that would appear in the collecting bin if only the upper slit were open, and the graph labeled L would correspond to the bullets that would appear in the bins if only the bottom slit were left open. Obviously, if both slits are left open, the result will simply be the sum of what would happen if one slit were left open plus what would happen if the other slit were left open. There is no such thing as interference between particles. This means that we would get a distribution of the number of bullets like the histogram in Fig. 10.17b. There would be a peak in back of each slit, and in general there would be a rather small number of bullets that would appear at the point midway between the two slits at the screen of observation. This is to be contrasted to Fig. 10.15 in which we showed the intensity of light from the first experiment that we considered. The intensity of light is such that it is brightest at the point that is midway between the two slits, and then has alternating light and dark bands. Comparing these two figures, it is

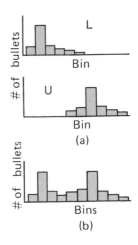

Figure 10.17.

obvious that in an experiment of the type we have described, waves and particles would behave in a very different way.

It is this difference between the behavior of waves and particles that formed the crux of one of the great puzzles in physics at the turn of the century. The question people asked was: "What are atoms and subatomic particles?" Since they were familiar only with objects that are readily available to the senses, this question was often phrased in the form "Are subatomic objects particles or are they waves?" The way that this question could be answered, of course, was to ask the question "Do subatomic objects exhibit diffraction?" Obviously, if they do exhibit diffraction, they must be waves. If they do not exhibit diffraction, they must be particles.

Of the subatomic objects that we have discussed so far, it is obvious that light (which we call subatomic since it is emitted from atoms) must be a wave, since there are many, many examples in nature of light exhibiting diffraction. At the same time, it is obvious that electrons must be particles. This follows from the fact that they have weight and mass and can be deflected from their course by things like magnetic fields.

With this understanding of the problem as it was perceived by physicists in the early 20th century, we can turn, after one more digression, to a series of puzzling experiments that were done over a period of about 30 years, and that eventually led to a totally new way of looking at the world—the science of quantum mechanics.

E. Technical Digression: Superposition and Standing Waves

In the last section, we discussed the principle of superposition as it applies to traveling waves and the way in which this principle might be used in distinguishing between waves and particles. The principle of superposition, however, is much more general than that. In this section, we shall discuss how the principle may be applied in problems involving standing waves. In addition to providing a useful introduction to classical ideas about black body radiation (which we shall discuss in the next chapter), the principle of superposition for standing waves, under the name "harmonic analysis," has practical applications in fields as different as music and oceanography.

By definition, a standing wave is a wave whose nodes do not move. It is a wave that is in some sense contained. For example, if we think about a wave in water the only way we could get a standing wave would be to have the water held in a vessel, such as a bathtub. In this case, we could get a situation such as that illustrated in Fig. 10.18, in which the water at each edge of the tub goes up and down with time, but the water in the center remains stationary. This stationary point is the non-traveling node.

Another example of a standing wave might be the vibrations of a string. In Fig. 10.19 we show a time sequence that we might see on a vibrating string. The string moves up and down between the two extreme configurations shown. For the case we have shown, the string has its ends tied down, and the fixed nodes are at the end points of the string.

Are these the only waves that could be present in the tub or on the string? Of course not. In the case of the string, for example, it is obvious that we could get any wave provided only that the wave has a wavelength that would allow the end points to remain fixed. In Fig. 10.20 we show several examples of other waves that could be put on the string.

From these drawings, it is easy to see that a standing wave may exist on the string provided that its wavelength is one-half the length of the string (see Fig. 10.19), equal to the length of the string (upper wave in Fig. 10.20), three halves the length of the string (lower wave in Fig. 10.20), etc. We can summarize this result by saying that a wave may be fit onto a string of length L provided that the wavelength satisfies the equation

$$\lambda = \frac{n}{2} L,$$

where n is any integer. The case n = 1 would be the wave shown in Fig. 10.19, while n = 2 and n = 3 are shown in Fig. 10.20. Sketching the waves for larger n, and going through the same exercise for the water waves, is left to the problems at the end of the chapter.

The wave shown in Fig. 10.19 is actually the longest wavelength that can exist as a standing wave on the string. It is called the "fundamental" mode of vibration on the string. From our general discussion of the properties of

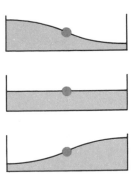

Figure 10.18.

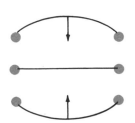

Figure 10.19. Vibrations of a string. The string moves from the top configuration to the bottom and then back again.

waves, we know that the fundamental, having the largest wavelength, will also have the lowest frequency of any wave that can exist on the string. If the string were part of a musical instrument, its vibration would set neighboring air molecules into motion, and the motion of air molecules would be perceived as a sound. The fundamental mode in this case would be the lowest pitch sound that the string could emit when plucked.

While it may be true that the waves we have discussed could exist on the string or in the bathtub, it is extremely unlikely that any one of them would exist alone. To understand the process by which a wave gets set up on a string or in water, let's consider a typical situation. A calm surface of water or a motionless string is disturbed in some random way. For example, we may drop a rock in the water or hit it with our hand. In a similar way, we may pluck the string or run a bow across it. In both cases, the surface on which the wave is to exist is distorted from its natural state by some outside force, and is then allowed to react in its own way. In the case of water, the molecules of water are set in motion by the force, and these molecules will collide with each other and with the walls of the container. Eventually, these collisions will cause the water to "settle down" and exhibit a regular motion. This motion need not be like any of the single standing waves we have discussed so far. In Fig. 10.21, we show an example of a steady state vibration of the surface of the water.

At first glance, it would appear to be difficult to say anything about such a complicated wave, but actually, we can say a great deal. Although the actual shape of the surface shown in Fig. 10.21 is not one of the simple waves that we have discussed up to this point, we can use the

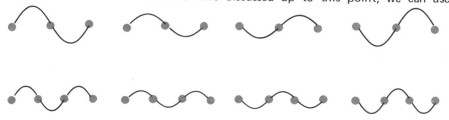

Figure 10.20. Two different waves on a string. Each wave moves successively through the configuration from left to right and then back again.

principle of superposition to describe it in terms of simple waves. As we have stated the principle, it says that whenever two or more waves overlap, the resulting displacement is simply the algebraic sum of the displacements each wave would have if it were present by itself. In other words, if many simple waves are present at a point, the resulting displacement may be quite complicated.

To see this, let us consider an example. Suppose that two waves are present on a string, one being the fundamental, and having wavelength $L/2$, and the other being of wavelength L. Suppose, further, that both waves have the same maximum displacement. What will the resulting surface look like? An answer is given in Fig. 10.22. The resulting wave does not look like either the fundamental or the harmonic, which were added together.

In the same way, we could imagine constructing more and more complicated shapes by adding together more and more harmonics. We could make the shapes even more complicated by varying the amplitude of each harmonic present. In the above example, this would correspond to adding together a fundamental wave with maximum displacement A and a harmonic with some other displacement (e.g., $A/2$ or $A/10$). Each of these additions will result in another complicated-looking surface, as the reader will discover for himself in the problems.

Thus, we see that adding together a number of harmonics of varying amplitudes can generate some very complicated-looking surfaces. The converse of this idea—that *any* complicated surface can be thought of as the sum of a number of components, each of which is simple—was proved in 1822 by Jean Fourier, a French mathematician. The process of decomposing the complicated surface into its simple constituent parts is referred to as harmonic analysis, or Fourier analysis, and each simple wave (i.e., each harmonic that goes into making a complicated wave) is called a Fourier component.

With the discussion of standing waves and harmonic analysis, we have at our disposal the important properties of waves as they were understood by the classical physicist. We can now turn to the problems that they were trying to solve at the beginning of this century with some understanding of the way in which they looked at the world.

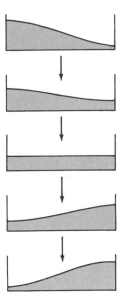

Figure 10.21.

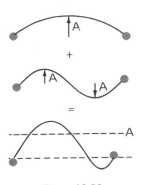

Figure 10.22.

SUGGESTED READING

All of the standard texts have sections on thermodynamics and on the particle-wave problem. The following may prove interesting:

Cardwell, D. S. L. (Ed.) *John Dalton and the Progress of Science.* Manchester, Manchester University Press, 1968. These are the proceedings of a conference held at Manchester. The book contains many essays and short papers by noted historians of science. There is no general theme, though, so it should be regarded as a book to use when checking on in-depth studies of narrow fields of inquiry.

Patterson, Elizabeth C. *John Dalton and the Atomic Theory.* New York: Doubleday, 1970. A biography of Dalton and a description of his work, which is quite thorough. It is readable and has many interesting photographs and plates.

Rosco, H., and Harden, A. *A New View of the Origin of Dalton's Atomic Theory.* London: Johnson Reprint, 1970. A reprint of some of Dalton's original papers and diaries connected with the atomic theory. It is particularly valuable because the originals were destroyed in a bombing raid in 1940.

QUESTIONS AND DISCUSSION IDEAS

1. Given the composition of the water molecule discussed in the first section of this chapter show that the numbers which were found for the Law of Definite Proportions for water follow.

2. In light of the Second Law, discuss the possibility of constructing a perpetual motion machine. Comment on the fact that the U.S. Patent Office requires a working model of such a machine in order to grant a patent.

3. In Fig. 10.5, draw the configuration of the rope at the following times:
   a. a short time after complete overlap
   b. when the back edge of wave A is even with the back edge of wave B
   c. a short time after (b)

4. Repeat Problem 3 for Fig. 10.6.

5. Define the wavelength for the waves shown in Fig. 10.4b and Fig. 10.4c.

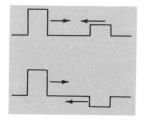

6. Carry out the kind of analysis done in the text for two waves approaching each other on a rope for the case shown at the upper left, where the waves are of different amplitude. Be sure to show the rope when there is partial as well as complete overlap.

7. Repeat Problem 6 for the case when the amplitudes of the two waves are in opposite directions.

8. Consider the example in Fig. 10.8, in which there are two motors making waves, and we observe the effects of these waves at the pier. Suppose that there is a string of corks floating in the water along the pier. What will the motion of the cork be
    a. at point A when only motor 1 is on
    b. at point A when only motor 2 is on
    c. at point A when both motors are on
    d. at point B when motor 1 is on
    e. at point B when motor 2 is on
    f. at point B when both motors are on

9. Consider an apparatus like that pictured below, in which a light is shown on a screen which has three slits in it. At the left, the light from the slits strikes another screen. The slits are evenly spaced in the first screen.
    a. Under what conditions will there be a maximum bright spot at point A directly in back of the middle slit?
    b. Is it possible to arrange the slits so that there is a dark spot at A?

10. Consider an experiment like that shown to the right, in which two synchronous motors are making waves in the water. An observer sits at point B. He finds that when motor 1 is on alone, there are 5½ wavelengths between B and the motor, and when motor 2 is on, there are 9 wavelengths between B and the motor. Both motors are turned on and allowed to run for a while
    a. What will he see at B? Why?
    b. He now walks down to A, equidistant from the two motors. What will he see there?
    c. How many null spots did he pass on the way from A to B? Why?

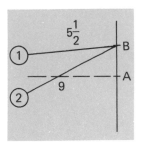

11. Suppose that in the previous problem, there were a line of corks along the path from B to A. Describe the observed motion of the corks that the observer will see if he walks down from B to A.

12. In the "two-slit" experiment described in the text, would moving the slits farther away from each other affect the fact that there was a bright spot on the screen midway between them? Why or why not?

13. According to the Second Law of Thermodynamics, entropy must always increase. But in a refrigerator, objects are cooled down (thereby increasing the internal order and decreasing the entropy). Doesn't this violate the Second Law?

14. To get some idea of how cold absolute zero is, look up the following temperatures:
   a. freezing point of water
   b. temperature of "dry ice" (carbon dioxide)
   c. freezing point of mercury
   d. boiling point of nitrogen

15. Draw the analogue of Fig. 10.19 for a string on which a wave with n = 5 exists. Repeat for n = 7, n = 10.

16. We saw in the text that the waves which could exist on a string were those that were consistent with the condition that a node exists at each point where the string is tied down. The analogous condition for water in a container is that the position of the maximum displacement must coincide with the walls.
   a. Using this information, show that the equation $\lambda = nL/2$ holds for water waves as well as for waves on a string
   b. Draw the standing waves in a tub for n = 2, 5, 7, 10

17. If one wave has a wavelength twice as long as another, then it will have a frequency only half as large. Given this fact, draw the wave in Fig. 10.21 at several different instants in time throughout one cycle of the fundamental.

18. Repeat Problem 17 for the following cases:
   a. the maximum displacement of the harmonic is 1/2 that of the fundamental
   b. the maximum displacement of the harmonic is 1/10 that of the fundamental

19. Suppose that we had a situation like that shown in Fig. 10.15 in which there were three slits in the screen, rather than two. If the slits are equally spaced, and if the distance to the observation point behind the middle slit is such that light must travel on a path one wavelength longer to get there from the upper slit, and the light falling on the screen has amplitude A, what will be the amplitude of the wave directly behind the middle slit?

CHAPTER
XI

SPLITTING
THE ATOM

*"I thought science was where you found the answers.
All you're doing is telling us about problems."*
> Anonymous undergraduate, on being
> introduced to quantum mechanics

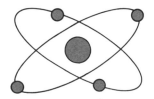

A. Introduction

In the beginning of this text, we talked about scientists as people with questions, rather than people with answers. The scientific method itself is simply a way to go about solving problems. Each new advance in our knowledge of the world—astronomy, mechanics, electricity, relativity—has been concerned with new and puzzling information.

The problem that we began discussing in the previous chapter is no exception. When men began asking questions about the basic nature of matter and about the structure of the atom, a whole new world was opened up. The first thing that happened was that the experimental facts that developed just didn't seem to fit in with the standard classical conception of the way the world should be. The first two decades of this century were a period of enormous confusion in physics. Each new fact seemed to lead to a logical contradiction. We shall examine a number of these "contradictions" now and see how the experimental facts about the atom force us to the realization that a totally new kind of physics would have to be developed to explain them. The new physics (called quantum mechanics) will be discussed in the next chapter.

The new experimental facts concerned themselves with two general categories: the interaction of light and atoms, and the structure of the atom itself. As these questions were investigated, they seemed to become more and more concerned with the fundamental nature of light and subatomic particles. As we trace the puzzling series of experiments that led to our modern ideas of the atom, it would probably be a good idea to try to keep in mind that our ultimate goal is an understanding of the atomic process, and that each new fact is one piece of a puzzle that can be assembled only when all of the facts are in.

It is also a good idea to try to remember how these experiments appeared to physicists at the turn of the century. To these men, everything had to be either a particle or a wave, and evidence that objects on the atomic scale had properties of both was interpreted as a logical contradiction, as a "wave-particle problem." After our experience with special relativity, however, we know that there is no reason why objects in a new region (in this case, the region of the very small) should behave like macroscopic objects with which we have direct experience. But the confusion of the classical physicist on this issue was so great that one can still find references to "wave-particle duality" as a philosophical problem in modern books.

B. Some Old Problems; Some New Problems

1. The Ultraviolet Catastrophe

The simple picture in which light was regarded as a wave because it exhibited interference effects and the atom and electron were regarded as particles because they were point-like quickly ran into trouble. The difficulty arose over theoretical and experimental consideration of something called a "black body" (the reason for the name will be made clear below).

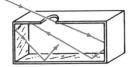

Figure 11.1.

Suppose we considered an apparatus like the one shown in Fig. 11.1, in which a box whose interior is completely mirrored has a small hole in one side. If we were to shine light into the hole, it would bounce around inside the box, and the chances that it would re-emerge through the hole are very small. Thus, the hole has the property that light entering it does not come back out. An object that absorbs all the light falling on it is said to be black, and hence the hole would be termed a "black body."

Adding energy to such a system by shining light through a hole is exactly analogous to adding energy to the water in the tub or the plucked string that were discussed in the previous chapter. If light is a wave, then we should get some sort of standing wave pattern in the box. But in considering such a system, the classical physicist had to face a very knotty problem for the first time—the problem of the interaction of light and matter in the form of atoms. Here, at the atomic level, were the two components of the classical world coming together. Imagine his consternation, then, when he discovered that if he applied ordinary classical ideas to this system, the answers turned out to be logically impossible!

To understand what classical physics predicts for the black body, it is easiest to imagine the possible light waves that could exist in the box as shown in Fig. 11.2. As was the case with waves in water and on a string, only certain

275

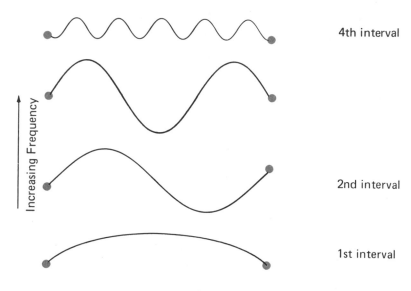

Figure 11.2.

frequencies of light will be able to exist in the box. Another way of saying this is to realize that in the case of light, frequency (or, equivalently, wavelength) is perceived as color. Thus, light of one frequency will be perceived by the human eye as blue, while light of another frequency will be seen as yellow, yet another as red, and so on. The fact that only certain frequencies of light are seen in the box, then, corresponds to the statement that only certain colors of light should be there once equilibrium has been established.

To the classical physicist, what happened in the light-box system was not a complicated process. The light striking the walls was absorbed by the atoms there. The light accelerated the electrons in the atoms, and these electrons, in turn, emitted their own light. In the classical picture of the atom, however, the electron would emit light at many frequencies, not just the frequency that was absorbed. Consequently, no matter what color light was initially introduced into the box, eventually it would be transmuted into the frequencies appropriate for that particular box. When we reached the state when the interaction of the light and matter was such that whenever light of a given frequency was absorbed at one point by one atom it was emitted at another point by another atom, we would say that equilibrium had

276

Ocean waves.

Ernest Rutherford (right) talks to J. J. Thomson, the discoverer of the electron, at Cambridge.

been established. This equilibrium would be somewhat analogous to a population at zero population growth. Some people would die and others would be born, but the total number of people at any given age would stay constant. In just such a way, the amount of light in the box at any given frequency would stay the same, although light of that frequency was constantly being absorbed and re-emitted.

One important distinction between light waves and the water and string waves we discussed in the previous chapter lies in the speed at which equilibrium can be established. Consider the case of a wave on water. Suppose we had started the water off in the fundamental mode of vibration. As time went on, the collisions of the water molecules with the wall would produce slight irregularities in the surface, which we would interpret as the presence of small amounts of higher harmonics. This is a relatively slow process in water, so that at any time after the water had been set in motion we would expect only a small number of harmonics to be present, and the energy associated with them would be small. Before a large number of harmonics could be generated in the water, the entire wave would be damped out by the natural viscous forces in the water.

In the case of the light box, however, the process for generating harmonics is the process by which atoms emit light, and this takes place very quickly. Consequently, a short time after the light enters the box, *all* of the possible frequencies necessary to establish equilibrium will be there. Thus, for the first time, the classical physicist had to start asking questions about the equilibrium of waves and particles.

One obvious question to ask in this situation has to do with the way that energy is apportioned among the various possible standing waves—how much energy does each standing wave carry? To answer this question, it is convenient to break the frequency levels up into intervals of equal size (see Fig. 11.2), and ask (a) how many frequencies can exist in each frequency interval, and (b) how much energy is associated with the frequencies in each energy interval?

Given the classical idea of light as a wave and the idea that atoms could emit light at any frequency, both of these questions could be answered by the classical physicist. Unfortunately, the answers turn out to be (a) the number of possible waves grows quickly, increasing as we go to higher

277

frequency intervals, and (b) the amount of energy in each frequency interval gets higher as we go up the scale in Fig. 11.2. These two answers together imply that the amount of energy associated with the waves in a box at equilibrium *must be infinite,* since there are an infinite number of frequency intervals (corresponding to higher and higher frequencies), each of which has an energy associated with it.

The higher frequency the light has, the more blue it appears to us. Violet is the highest frequency light we can see, and "light" outside the visible range beyond violet is called ultraviolet. Since the problem of infinite energy is associated with high frequencies, it was sometimes termed the "ultraviolet catastrophe." This name gives a pretty good idea of how the classical physicists felt about the whole situation.

The ultraviolet catastrophe was not something that was discovered suddenly at the beginning of the 20th century. Maxwell knew about it, and so did all other classical physicists. What they didn't know, however, was how to resolve it. In the face of this problem, the physicists at the end of the 19th century did what scientists always do when confronted with an unsolvable problem. They simply ignored it for the time being, and hoped that eventually someone would come along who would be able to solve it.

In the case of the ultraviolet catastrophe, that someone was Max Planck, a German physicist. He recognized that the key point to the entire argument given above lay in the question of how light, once it had been absorbed by an atom, was re-emitted. According to the classical theories that were prevalent at that time, light could be emitted at any frequency at all. What Planck did was to ask himself what would happen if this were not true. In effect, what he did was similar to what Copernicus had done 400 years earlier. He assumed something that would have been considered totally crazy by other scientists of his time, and asked what the consequences of this assumption would be. He then compared these consequences to known experimental observations to see whether the original assumptions were as crazy as they seemed at first. Whereas the classical physicists had thought that when an atom absorbed light of energy E it could re-emit that energy at any frequency whatsoever, Planck assumed that this was not so. He asked what would happen if instead of having the classical kind of light emission, we had something totally different. What if the light could only be emitted by the atom at a frequency determined by the equation

$$E = nh\nu.$$

where n is any integer from one on, h is a new constant of nature, now called Planck's constant (much more will be said about h later), and ν is the frequency of the light. In other words, light can be emitted at a frequency

$$\nu = E/h$$

or a frequency

$$\nu = \frac{E}{2h}$$

and so forth, but never at a frequency like

$$\nu = \frac{E}{(.716)h}.$$

What does this new assumption mean? One way of looking at it is to think of light as coming in basic units of energy $h\nu$. The energy of the light of a given frequency can be any number of these basic units. This is what we'd expect if light were a particle. After all, it is not unreasonable to talk about emitting one, two, or three billiard balls. If we think of light coming in "bundles" then these bundles were given a special name by Planck. They were called "quanta" of light and we say light energy is quantized.

It is interesting to note that although Planck introduced the idea of a quantum of light, and hence cast the first doubt on the identification of light as a wave, he never really accepted the idea himself. He believed that what he had uncovered with his work had to do with the nature of the atom itself, and had nothing to do with the nature of light. Whatever he believed about his assumption, however, Planck was able to show that it led to a resolution of the ultraviolet catastrophe.

To understand how this resolution came about, let us go back to our discussion of how harmonics could be generated by light in a mirrored box. We saw that the only requirement on the presence of an harmonic in a box was that the wave be of such a wavelength that it had a node at each wall of the box. What the assumption of quanta does is to add a second requirement. Not only must the wave have a node at each end of the box, but it must be a wave that can be emitted by atoms in the box. There are many waves that could fit in the box, but could not be emitted by atoms in the box. These waves will not appear among the harmonics. Similarly, there are many waves that could be emitted from the atoms in the box, but would not fit in the box because their nodes are in the wrong place. These, too, will not appear as harmonics in the box. Only those waves that satisfy *both* criteria—that the wavelength be such that there is a node at each end of the box, and that the wave be one that can be emitted by particles in the box—will actually appear as overtones. In this way, the number of possible harmonies is cut down dramatically. Although the actual working out of the results requires a great deal of mathematics, it should not seem unreasonable that by cutting out so many harmonies the total energy stored in the box goes from being infinite (as it would be for the classical atom) to being finite (as it is in fact).

The idea of light somehow existing in quanta could be checked experimentally very quickly. Experimental techniques that could measure the amount of energy in each harmonic inside of a black body had been in existence for a long time. To construct a black body, one had simply to build a large furnace with a very small hole in the side. Not only did Planck's assumption that light came in quanta eliminate the ultraviolet catastrophe, but it was quickly realized that the assumption explained experimental data on the amount of energy in various harmonics in laboratory black bodies as well (see Fig. 11.3). Thus, it represented a great step forward in our understanding of the interaction of light and matter.

However, this great step forward was taken with a certain cost. You will recall that it was well known that light always exhibited interference, and hence it was thought that light had to be a wave. However, if we begin thinking

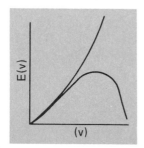

Figure 11.3. The amount of energy E(v) at frequency v in a black body. The upper curve is the classical prediction the lower the experimental result explained by Planck.

of light as a quantum, its wave-like nature becomes a little blurred. If it comes only at certain frequencies and if it comes in little bundles, then perhaps it has some properties of a particle as well. After all, we think of particles as little bundles, too. Although Planck himself refused to accept that his work cast doubt on the idea that light might be a wave, this refusal was not shared by other physicists of the time, and Planck's ideas very quickly were adapted to other problems that had been around for a long time. It is to these problems that we now turn our attention.

2. The Interaction of Light and Atoms: Is Light a Wave or Particle?

At about the same time that Planck was resolving the ultraviolet catastrophe, and postulating for the first time that light might exist in quanta, another phenomenon had appeared to puzzle physicists. This phenomenon is known as the "photoelectric effect," and is one that is now familiar in our everyday life. Basically, the phenomenon has to do with the fact that if light or other kinds of radiation are shown on certain metals, electrons are observed to come off from the metals. If the metal is connected to a wire, the movement of electrons away from the metal through the wire would be interpreted as an electric current. Thus, the photoelectric effect tells us that it is possible to convert light into an electric current.

You have had a great deal of experience with photoelectric effect in your life, although you may not know it. For example, "seeing-eye doors" operate according to the photoelectric effect. A beam of light is shown between two spots. On one side is a light that is producing the beam, and on the other side is a piece of metal that has the property that when the light hits it, an electric current will flow through a piece of wire attached to the metal. When someone walks through the beam of light, the light no longer falls on the piece of metal, and hence the current stops flowing in the wire. This sudden change in the current is then used as an electric signal to trigger a motor which opens the door. There are many other uses of this effect in modern machinery.

What was puzzling to physicists was not the fact that the electrons came out. After all, it was known that electrons existed inside of atoms. It was also known that light was

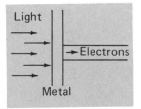

Figure 11.4.

an electromagnetic wave and hence had the ability to accelerate electrons. It was therefore not too surprising that light would have the ability to accelerate electrons enough to tear them out of the atom in which they normally resided and push them out of the metal. What was puzzling to physicists at the time was the fact that it did not take light very long to do this.

The process by which classical physicists imagined electrons to be expelled from the atom was a very gentle one. Each crest of the wave of light would come along and nudge the electron a little bit and gradually, over a period of time that was very long on the atomic scale—on the order of one minute—the electron would be lifted slowly out of the atom. It is a little bit like the process by which a piece of wood is washed onto a beach by a wave. It takes a long time for the wood to reach the beach, because each crest of the wave moves it only a short distance.

On the other hand, it was observed experimentally that the electrons were emitted from the metal as soon as the light was turned on. There is no discernible time delay between the time the light hits the metal and the time the electron came off the other side. This fact could not be reconciled with the idea of light as an electromagnetic wave.

The man who showed how this problem could be solved was Albert Einstein. His resolution was quite easy to understand in terms of the analogy given above of the waves washing a small piece of wood into shore. Granted that it would take the waves a long time to do the job, there is another way of getting that piece of wood to shore very quickly. For example, if we shot a cannonball at it, and the cannonball hit the wood, the wood would be knocked up onto the shore almost instantaneously. If we think of the wood as the electron in the atom, then the light would have to be the cannonball. This would explain how electrons could be emitted from atoms as quickly as they were observed to be emitted. On the other hand, it is very puzzling to try to understand this explanation if we think of light as a wave.

What Einstein had done was to take the hypothesis that Planck had made very seriously. Unlike Planck, he did not restrict himself to the idea that the emission of light in quanta was something that had to do with the structure of atoms and had nothing to do with the nature of light. What Einstein saw was that the emission of light in quanta could arise from one of *two* causes: it could arise because of the structure of the atom, or it could just as well arise because light itself exhibited particle-light properties, and light itself existed only in quanta. Although there is no way of choosing between these two options on the basis of the black body problem alone, Einstein realized that another problem—the problem of the prompt photoelectric effect—could be resolved if the second option were true.

In other words, both the black body problem and the photoelectric effect could be understood if it were assumed that light acted as a particle! This is a very disconcerting conclusion to draw, particularly if you share the 19th-century idea that light had to be either a particle or a wave. What Planck and Einstein had shown was that light, although it exhibited interference and other properties of waves, also exhibited properties of particles. It came in bundles called quanta. It could collide with another particle (the electron) and push it quickly out of an atom, just as one billiard ball can push another in a collision. Thus, the nature of light suddenly becomes very puzzling. It acts like a wave at times and like a particle at times. This fact is sometimes discussed even today in terms of a "wave-particle dilemma."

Actually, of course, it is not a dilemma at all. The dilemma arises simply because we have made the assumption that everything has to be either a wave or a particle. We made this assumption because everything in our everyday experience is either a wave or a particle. However, as we have pointed out in our discussions of special relativity, our everyday experience does not deal with everything that there is in the world. It does not deal with objects moving near the speed of light, and neither does it deal with objects as small as atoms. It should not surprise us, therefore, that categories which are appropriate to the description of objects in the everyday world of normal-sized objects are *not* appropriate to the description of objects whose size is very small. In fact, light is *neither* a particle *nor* a wave, but it is one type of subatomic object that happens to have some properties in common with waves and other properties in common with particles. It is an entirely new kind of beast.

283

We have seen the kind of problems that can arise when people insist on applying categories that are not appropriate in areas where they don't belong. For example, we saw that the Aristotelian insistence that motion be classified as either natural or violent was an actual hindrance to Galileo and his fellow scientists as they tried to understand the motion of projectiles. When the correct description of the law of compound motion was finally arrived at, there was no mention of natural and violent motion. In a sense, the question of whether the motion of a projectile is natural or violent has never been solved. What has happened is that it has been realized that this question is not particularly meaningful. In the same way, the question of whether elementary subatomic particles are "particles" or "waves" has never been resolved. What has been realized is that the properties of these subatomic objects are simply not the same as the properties of the normal everyday objects, and consequently it doesn't make sense to apply categories such as "particles" and "waves" to them. Just as Galileo was able to discover the laws that govern projectiles without ever referring to the question of natural or violent motion, scientists have been able to discover the laws governing atoms and subatomic objects without referring to the question of whether they are "really" waves or particles.

Einstein's explanation of the photoelectric effect also allows us to get a better idea of what Planck's constant is. In the problems at the end of the chapter, the reader is given a chance to convince himself that there is a connection between the frequency of a light wave and the color that we perceive. Since different colors of light correspond to different frequencies, from Planck's relation

$$E = h\nu$$

it follows that each different color corresponds to quanta of light of a different energy. In our cannonball analogy, this would correspond to saying that each color would correspond to a cannonball of different speed.

Suppose we did the following experiment: we shine light of a known color on a piece of metal, and measure the energy of the electrons that are emitted. From the energy of the electron, we can obtain information about the energy of the light, and from the color of the light, we can obtain its frequency. For this particular color, then, we could find an energy E_1 (the subscript means that this is the first color used) and a frequency ν_1. From Planck's relation, we also have

$$E_1/\nu_1 = h.$$

The significance of Planck's relation is this: if we now repeated the experiment with a different color of frequency ν_2 (and quantum energy E_2), we would find that

$$E_2/\nu_2 = h = E_1/\nu_1.$$

No matter how many colors we tried, we would always find that the ratio of the energy of the light to its frequency would be the same. We call the value that this ratio has "h", and find that its numerical value is

$$h = 6.6 \times 10^{-34} \text{ joule sec.}$$

This number is one of the basic physical constants, like Newton's constant of gravitation. In the problems at the end of the chapter, more concrete examples of its significance are given.

3. Discrete Spectra

At the end of the 19th century it became possible to study the emission of light by a single chemical element. The vapor lamp, the device that allowed this type of experiment, has become a commonplace in our lives today.

The vapor lamp works in the following way: small amounts of a chemical element in the form of a gas are placed in a sealed glass tube. An electric current is passed through the gas, heating the material in the tube. The light given off by the heated gas is then studied, and we can be sure that the light came from the particular element in the tube, since there is no other source of light in the apparatus.

Mercury and sodium vapor lamps are commonly used as street and highway lights in the United States. The yellow lamps used on expressways are usually sodium vapor,

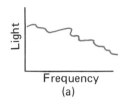

Frequency
(a)

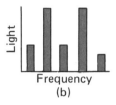

Frequency
(b)

Figure 11.5.

while the bright bluish-white street lamps used in large cities are mercury vapor.

From the point of view of the development of the atomic theory, however, what was important was not the practical uses of these devices, but what was learned about the light given off by chemical elements in isolation. Recalling that the classical physicist regarded the light being emitted from atoms as being due to the acceleration of electrons inside the atom, the most naive version of the "raisin bun" model might be expected to have light emitted at just about any frequency. Consequently, if we plotted the amount of light emitted as a function of frequency (which, you will recall, is related to the color of the light), we might expect to get something like what is shown in Fig. 11.5a. A plot of intensity versus frequency is called a "spectrum," and the spectrum in which light is emitted at all colors is called a "continuous spectrum."

Instead of finding a continuous spectrum being emitted from vapor lamps, however, a spectrum like that shown in Fig. 11.5b was found for each element. Only certain colors were present in the emission, and others were absent. A spectrum like this is called "discrete." Actually, we have already hinted at this result in our discussion of street lamps. The fact that sodium vapor lamps appear yellowish results from the fact that sodium emits a lot of light at a frequency corresponding to yellow, and only a little at other frequencies in the visible range. Thus, when we look at a sodium vapor lamp, our impression is of a yellow light.

When this result became known, the first reaction of classical physicists was to modify the raisin bun atom to account for it, just as Ptolemy modified the spheres of Eudoxus. The electrons in the atom were imagined to be connected together by a set of springs which only allowed certain types of motion. The springs were adjusted so that a discrete spectrum would result from electron motion in the atom. However, as the data became better and better, the springs got more and more complicated, and ultimately failed to match the data completely. By the early 1920s, the raisin bun atom (with springs) bore many similarities to the Ptolemaic version of the solar system. Both were exceedingly complicated and neither accounted for the main facts in a natural and simple way.

C. Splitting the Atom

We have seen that most of the problems that arose at the end of the 19th and beginning of the 20th century had to do with the interaction of light and atoms. It is only natural that given problems of this kind, people should begin to focus their attention on the nature of the atom itself. After all, the atom was half of the problem.

The atom as it was envisioned by Democritus and Dalton was a totally unstructured blob of matter. The existence of electricity and the subsequent discovery of the electron meant that this particular view of the atom could no longer be sustained. The atom was not an unstructured blob of material, but in fact had to be a rather highly complicated and complex system of many electrons with some sort of arrangement of positive electrical charge so that the overall charge of the atom was zero. The raisin bun atom was one attempt to understand how such systems could exist in nature. But no one had made any serious attempt to discover experimentally what the structure of the atom had to be. Most of the attempts to ascribe structure to the atom were rather indirect, depending on calculations that would match the hypothetical structure of the atom to actual observed emission of light. In 1911 Ernest Rutherford, a physicist at the University of Manchester who had already won the Nobel Prize for his work on the science of radioactivity (which we shall discuss later), proposed, in an off-hand sort of way, that direct studies of atomic structure might be useful. By the time his proposal had been realized, the scientific world was presented both with something very like our modern picture of the atom and with a brand new dilemma which matched any that we have discussed so far in complexity and puzzlement.

What Rutherford proposed was an experiment in which a thin sheet of gold foil would be bombarded by something called an alpha particle. The alpha particle was actually the nucleus of a helium atom, but for the purposes of our discussion we can regard it as a tiny subatomic "bullet," which can be produced in a laboratory and aimed at any target we wish. Later, when we discuss the periodic table and the modern structure of the atom, we shall be able to understand in more detail what Rutherford's "bullet" was. In terms of understanding atomic structure, however, we only have to know that the bullet was there.

Electrons

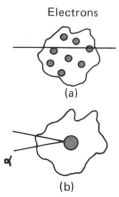

(a)

(b)

Figure 11.6.

When the experimental apparatus was set up, a very strange and puzzling result emerged. Most of the alpha particles went right through the gold foil. But a few—about one in a thousand—came bouncing back. There was something about the gold atoms inside the foil that turned the alpha particles around. There was nothing in the ideas about the gold atoms in the foil that could account for this strange behavior.

To understand the real impact of Rutherford's experiment, let us look at Fig. 11.6. In part (a), we have a representation of the standard raisin bun atom. If a bullet were fired at this atom, it would be a little bit like firing a cannonball at a cloud of smoke in which a few ping-pong balls were embedded. There would be nothing that would be heavy enough or massive enough to deflect the cannonball. Even if the cannonball hit one of the ping-pong balls, it would not be deflected by any measurable amount. In the same way, the alpha particle going through the raisin bun atom could not be deflected by the electrons or by the rather tenuous positive material in which they were embedded. The alpha particle had to go straight through if the raisin bun picture were correct.

On the other hand, suppose that the atom were not a raisin bun. Suppose that there was something at the center of the atom that was very massive and very hard. In this case we would have something like Fig. 11.6b. The cannonball which we shoot at the atom would miss the hard heavy center most of the time. But, when it did hit the center, it could bounce backward. This, then, was Rutherford's explanation of his experimental results. His experiments showed that at the very heart of the atom lies an unimaginably dense speck of matter, which we call the "nucleus." Almost all of the weight of the atom is contained in the nucleus. Around the atom are the electrons, which are very light and which carry the negative charge. The positive charge of the atom resides in the nucleus.

It is very important to get some idea of how empty an atom really is. We are used to thinking of atoms as small, compact, dense clumps of matter. In fact, they are not. For example, if the nucleus of a uranium atom were a bowling ball sitting in front of you, the electrons of the atom would be like 92 ping-pong balls scattered over 100 square miles of area. Everything except the bowling ball and those 92 ping-pong balls would be empty space.

The picture of the atom which Rutherford developed, then, was very much like our modern picture (see Fig. 11.7). At the center of the atom, containing most of its mass and all of its positive charge, sits the nucleus. Going around the nucleus in orbits are electrons as shown above. The collection of electrons is usually referred to as the "electron cloud" of the atom. This picture of the atom is rather like our picture of the solar system, in which the sun sits at the center, and the planets travel around in orbits. The main difference from the point of view of Rutherford was that instead of having a gravitational force such as the one that holds the earth and the sun together, the atom was held together by an electrical force which operated between the positively charged nucleus and the negatively charged electrons.

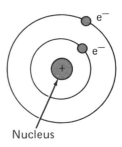

Nucleus

Figure 11.7.

The picture of the atom developed by Rutherford has become familiar to all of us. It can be seen anywhere from the symbol of the Atomic Energy Commission down to comic books read by grade school children. What is not so obvious, however, is that if we believe in classical physics the picture of the atom that Rutherford developed was impossible. Why? You will recall that it was a consequence of Maxwell's equations that if a charged object is accelerated, it must radiate light or some other form of electromagnetic radiation. Now you will also recall that a particle traveling in circular motion is being accelerated. This follows from Newton's first law. Indeed, we have discussed in some detail how the motion of the earth around the sun obeys Newton's laws of motion and, in particular, how the attraction of the sun constantly counteracts the tendency of the earth to move off in a straight line. The same argument can be carried over into the discussion of electrons going around an atom with a simple replacement of the electrical force, given by Coulomb's law, for the gravitational force given by Newton's law of gravitation.

If the electrons that are going around the atom are accelerated, then it must be true that they radiate. But light carries energy with it, and this energy can come from only one source—the electron itself. Thus, by the simple fact that the electrons are moving in an orbit, they must be losing energy. If they are losing energy, they will begin to spiral in toward the nucleus, just as a satellite going around the earth and losing energy to friction in the upper atmosphere eventually spirals in and crashes. It is a relatively simple matter to estimate how long an atom could live (i.e., how long it would take the electron to spiral into the nucleus). The number turns out to be considerably less than one second. Thus, once Rutherford had established beyond any doubt the fact that the nucleus existed in the atom, it became impossible to understand how any atom could exist at all according to classical ideas.

In other words, the minute that physicists began to probe the structure of the atom experimentally, it became clear that something was wrong with the way they were thinking about the world of the very small. The loss of the raisin bun atom was, after all, no more serious than the loss of epicycles. It was simply an idea that had been proved to be incorrect. On the other hand, it would appear that in some way the ideas of Maxwell's equations—the basic laws governing the phenomena of electricity and magnetism—had to be modified or corrected at the atomic level. Otherwise, atoms could not exist at all.

D. The Bohr Atom

The first successful attempt to "explain" the results of Rutherford—an attempt that also explained the balck body radiation, the photoelectric effect, and the problem of discrete spectra—was made by Niels Bohr, a young Danish physicist. The model of the atom that he proposed is now called the "Bohr atom," and is regarded as the first step toward the new science of quantum mechanics—the science dealing with the behavior of objects on the atomic scale. With some minor adjustments, it is still the model of the atom that we believe to be correct today.

What Bohr did was to make a totally unwarranted assumption about the behavior of electrons in atoms. We have seen how Planck did the same for the process of light emission when he assumed that the energy that the emitted light had could have only certain values. Bohr assumed that there was another set of variables describing the electron orbit that could have only certain values as well. If the electron has mass m and is moving with velocity v on an orbit of radius r, then the assumption that Bohr made can be written

$$mvr = nh,$$

when n is any integer and h is Planck's constant.

Why did Bohr choose this particular set of variables? The quantity mvr is called the "angular momentum" of the electron, and it can be shown that in classical mechanics this quantity is conserved if there are no outside forces acting on the system. Consequently, this particular set of variables is one of the "natural" quantities (like energy or momentum) to use in describing circular motion and Bohr's equation says that it is quantized.

Can this assumption be justified? No more than Planck's assumption could be or, for that matter, Copernicus'. It is simply one more example of a scientist taking a crazy idea, working out the consequences, and then showing by comparison to the data that the idea wasn't so crazy after all.

From our previous discussion of the mechanics of circular motion, it will be realized immediately that the assumption that Bohr made carries rather striking consequences for the motion of electrons around an atom. You will recall that we can determine the orbit of a particle by equating two forces—the force of attraction between the object in the orbit in the central body, and the centrifugal "force" that tends to throw the body out of its orbit, but that is, in reality, simply the tendency of the body to move in a straight line unless acted upon by a force. For a body like an electron being attracted toward a nucleus of charge Q, equating these two forces will give us

$$\frac{mv^2}{r} = \frac{kqQ}{r^2},$$

where Q is the charge of the nucleus, q is the charge of the electron, m is the mass of the electron, v is the velocity of the electron, and r is the distance between the nucleus and the electron. From this equation, we see that according to the ideas we have encountered so far it should be possible for the electron to be in an orbit around the nucleus at any distance r that it chooses. This follows from the fact that the above equation, once the value of r is known, can be solved for the value of v, the velocity of the electron. In other words, there is a velocity for the electron that is appropriate to an orbit at each distance from the nucleus. You have probably watched the injection of satellites into orbit around the earth. The same principle holds there. The NASA scientists know how high up they want the orbit of the satellite to be, and they can adjust the velocity of the satellite so that when it reaches that height it will have just the correct velocity so that the gravitational analog of the above equation (or a more complicated version of the equation) is satisfied and the spaceship stays in orbit. In any case, according to what we know now it ought to be possible for the electron to be in orbit anywhere around the nucleus; the only thing that would change from one orbit to another (other than the distance to the nucleus, of course) would be the speed at which the electron was moving.

What Bohr has done is to add another requirement on the orbit of the electron. Since this requirement says that

$$v = \frac{nh}{mr},$$

it is obvious that only certain values of the velocity are allowed for electrons. Thus, for example, an electron can be at that distance from the nucleus that is appropriate to a velocity of (h/mr), 3(h/mr), 10(h/mr) and so forth. It cannot, however, be at an orbit a distance from the nucleus that would be appropriate to a velocity of $1^1/2$ (h/mr), $7^5/8$ (h/mr) and so forth. It follows from the fact that angular momentum is quantized that the electron can exist only in certain well-defined orbits, at certain well-defined distances from the nucleus. It cannot exist anywhere in between. The orbits that satisfy both of these conditions are called "allowed orbits."

If we accept this basic hypothesis, then a number of very interesting consequences follow. However, it has to be emphasized that at the present stage of development the only justification for these hypotheses is the hope that they will explain the puzzles that we have been unable to explain in any other way.

How could electrons exist in a Bohr atom when they could not have existed in a classical atom that had a nucleus?

To answer this, we have to realize one other fact about orbits. This is the fact that to each orbit there corresponds not only a distance from the center and a velocity, but energy as well. The energy of an electron in orbit is made up of two parts: the kinetic energy associated with motion, and the potential energy that exists because it is a certain distance from the nucleus of opposite charge. Changing the distance between the electrons and the nucleus would change both of these energies. The velocity appropriate to the new orbit would not be the same as the velocity appropriate to the old orbit, and the potential energy would be different as well since the electron and nucleus would now be a different distance apart than they were originally. Consequently, it follows that electrons in different orbits will have different energies, as well as different velocities.

Now let us consider an electron that is at a distance from the nucleus satisfying both the balance of forces and the requirement that the momentum be quantized. Can such an electron radiate energy? Well, if it were to radiate a little bit of energy, it would have to move to a slightly different orbit. It would, in other words, have to move away from the orbit that is allowed to an orbit that is just a little bit closer to the nucleus. But the orbit to which it would have to move is *not* allowed. The quantization condition is not satisfied for this orbit. Thus, by assuming the quantization of angular momentum, we make it impossible for the electron to radiate light while it is moving in a circular orbit. This impossibility arises because in order to radiate light, it would be necessary for the electron to move to a region and an orbit in which it is not allowed to exist.

In other words, what Bohr has done is to make the tacit assumption that there are certain "safe" orbits in which the electron can go on forever without ever radiating. He has assumed that there is something operating at the level of the atom and the quantum that does not obey Maxwell's equations in the form that Maxwell wrote them down. He is assuming that it is possible for an atomic object to be accelerated without having it radiate, something that could not happen for a charged particle in our everyday experience.

The Bohr atom then, looks something like that shown in Fig. 11.8. The nucleus sits at the center (for the sake of simplicity, we are drawing the hydrogen atom in which the

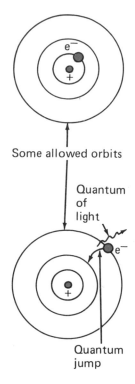

Some allowed orbits

Quantum
of
light

Quantum
jump

Figure 11.8.

nucleus is a single positively charged particle, and there is one electron), and there are a number of allowed orbits for the electron. At any given time, the electron can be in any one of these allowed orbits, but it is not allowed to be anywhere in between. Normally, the electron will be in the closest orbit to the nucleus, since this is the orbit of lowest energy.

Suppose that for some reason, however, the electron finds itself in one of the higher allowed orbits, as in Fig. 11.8b. This situation might come about because of a collision with another atom. In a collision it is possible that the atom we are looking at will gain energy, and it is possible that this energy gained will be used by the atom to lift the electron from the first orbit to a higher orbit. After the collision, the atom finds itself with an electron in a higher orbit, which we call "an excited state." What happens? Well, the electron wants to go back to the lowest possible state. It will do so by "jumping" from the excited states back down to the lowest state, which is called the "ground state." A jump of this type is given the name "quantum jump."

When the electron changes its orbit by making a quantum jump, however, it has also changed its energy. The energy of the higher orbit is not the same as the energy of the lower orbit. This means that the difference in energy between the two orbits has to go somewhere. What happens is that this energy is given off as radiation. If the energy level difference is appropriate then we will see visible light. Otherwise, the radiation will be either in the infrared or the ultraviolet.

We see immediately how the introduction of the quantization of angular momentum for electron orbits by Bohr explains many of the effects that have been puzzling us up to this time. For example, it is obvious that light can only be emitted at certain energies. Those energies at which light can be emitted correspond to the energy differences between the allowed orbits of the electrons. Thus, we have an explanation in terms of atomic structure for the hypothesis that Planck introduced. It also tells us why it is that atoms will emit only discrete spectra. They can only emit light whose energy corresponds to the difference between the allowed orbits, and, by Planck's formula, this means they can only emit certain well-defined frequencies of light. These frequencies, of course, correspond to the colors that are seen for each separate atom.

We have only drawn a picture of the hydrogen atom, in which there is one electron sitting in the allowed orbits. We have also only considered a simple case where the electron jumps from one orbit down to the next lower one. If the electron for some reason found itself in the fifth or sixth orbit, it would have a series of choices as to how to get back to the ground state. It could make the leap all at once, in which case it would emit light at a very high energy, or it could make a series of intermediate steps, in which case it would emit light of lower frequencies. The only requirement is that the total energy emitted as light be equal to the energy difference between the electron before and after it begins its journey.

If the atom we are interested in is not hydrogen, but some other atom, then the situation becomes more complicated, although the basic principles remain the same. If we are dealing with a heavier atom (for example, carbon or oxygen or uranium), then the nucleus will be heavier and will have a much higher positive charge. This means that the allowed orbit will be closer to the nucleus than the corresponding orbits would be for hydrogen. In addition, there would be many electrons scattered among the allowed orbits, rather than just one. We will discuss this point more fully when we talk about the Pauli principle.

With the work of Bohr we have a model of the atom that seems to work, and to explain and correlate a number of very puzzling experimental facts. However, it has to be emphasized that at this stage it is just a model. We know that it works, but we don't know why it works, and we don't know what the deeper principles are that lead to the principle of quantization of angular momentum. Somehow, we feel that it must be tied up with the nature of the electron, which up to this point we have always treated as a particle. We know that light (which we had always treated as a wave) exhibited some properties that we would think appropriate to a particle. It should not surprise us, therefore, if we find that the electron exhibits some properties of a wave as well. It is in the interplay between the particle-like and the wave-like properties of an electron that the explanation of a Bohr atom and, ultimately, the development of the science of quantum mechanics is to be found.

E. The Electron as a Wave

We began our discussion of the interaction of light and matter by pointing out that evidence existed to indicate that light, which we normally think of as a wave, actually had some of the properties of a particle. In 1925, Louis de Broglie, a young French physicist, asked what the consequences would be if the other side of this coin were found to be true as well. That is, he asked what the consequences would be if we assumed that electrons, normally thought of as particles, actually had some of the properties of waves. He went so far as to write down an equation, now called the de Broglie relation, which related the particle properties of electrons to their supposed wave properties. The equation takes the form

$$p = \frac{h}{\lambda},$$

where p is the momentum of the electron, λ is the wavelength of the wave associated with the electron, and h is the ubiquitous Planck's constant. At the time, physicists did not feel that this equation was a serious contribution to their science. In fact, several of them gave it the name "Comedie Francaise." The reason for this attitude is not very hard to see. We have discussed particles, and we have discussed waves. To the classical physicist, these two different kinds of things had to be distinct and separate. The implications of the particle-like nature of light had never been fully accepted by the physics community up to this point. So when de Broglie wrote down an *equation* that seemed to link these two properties, wave and particle, in an inexplicable way, it was not well received.

However, like all of the other "crazy" ideas that we have discussed, it turned out to be right. (No one has ever kept track of all the "crazy" ideas that turned out to be wrong.) In fact, it provided a very neat "explanation" of the Bohr atom, and the existence of the "safe orbit," which seemed so puzzling when we discussed them in previous sections.

Actually, although the de Broglie equation may seem very ad hoc and arbitrary as it has been presented here, there is a strong mathematical analogy between the de Broglie equation, which relates the momentum associated with an object to the wavelength associated with the same object, and the Planck equation, which relates the energy of an object to the frequency associated with that object. The de Broglie equation, then, is not so much a totally new departure in physics as an attempt to complete and unify a picture of subatomic objects that had been developing since the time of Planck.

But how does the de Broglie equation explain the Bohr atom? To understand this question, we have to go back to our discussion of the problem of fitting waves into a box. You will recall that in the case of the string, the requirement that a wave be able to exist on the string as a standing wave translated into the statement that the wave had to be such that it could have one node on each wall of the box, with any number of nodes (including zero) in between. This means, for example, that on a string of length ℓ we could fit waves and harmonics such as those shown in Fig. 11.9. On the other hand, we know that we could not have waves such as those pictured in Fig. 11.10. The reason that these waves could not exist on the string is because to have them on the string would require that one end of the string move. Since by hypothesis both ends of the string are tied down, this would be impossible.

Suppose we now ask ourselves a slightly different question. Suppose we asked how we could fit waves on a string where both ends were free, but on which we imposed a requirement that the wave be such that at the left-hand side of the string it is at exactly the same stage of its cycle as it is at the right-hand side of the string. For example, if the wave is a crest at the left-hand side, there must be a crest at the right-hand side. Similarly, if there is a trough at the left-hand side, there must be a trough at the right-hand side. We picture in Fig. 11.11 several examples of waves that can fit on a string subject to this requirement.

The mathematical requirement for fitting waves on a string subject to the condition stated above is that

$$\lambda = L/n,$$

where L is the length of the string, λ the wavelength of the wave and n is an integer. Examples of this type of fitting problem are to be found at the end of the chapter.

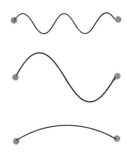

Figure 11.9.

Figure 11.10.

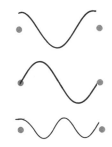

Figure 11.11.

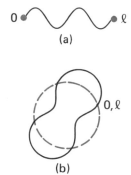

Figure 11.12.

De Broglie began by asking what the requirement would be on a wave that could be used to represent an electron going around in an orbit. Obviously, if we are to think of the electron as a wave instead of as a particle, we have to begin thinking about how we could fit a wave onto a circular orbit. But the answer to this question is obvious once we understand the problem of fitting a wave onto a string that we discussed above. Suppose we take one of the waves that can be fit onto a string of length ℓ, subject to the requirement that the wave be in the same part of the cycle at both ends of the string. Suppose we now take that string and bend it around into a circle. Where the two ends of the string join up, the wave (since it is in the same part of the cycle at both ends) will be continuous. For example, in Fig. 11.12 we show the case of a wave on a string (Fig. 11.12a), and of that string bent into a circle (Fig. 11.12b). Obviously, not every wave will fit on the string of length ℓ, and not every wave will be able to be continuous if it went around in an orbit. For example, in Fig. 11.13 we show a wave that could not be made to be continuous if we bent the string on which it was sitting into a circular orbit.

The requirement that de Broglie is stating here is sometimes called the requirement of "continuity" or "single valuedness." There are many ways to think of this. For example, if we took the wave in Fig. 11.13b, we observe that where the string comes around and meets itself, the wave has a different value than it had originally. In other words, at this point the deflection of the string would have to have two separate values. We could imagine this situation in the following way: Suppose we had a large circular string and we started a wave at one point. Once it had traveled around the string and it got back to the point where it began, there would be interference if the displacement did not have the value it had originally, at that point. The actual displacement to the rope at this point, once the wave had returned, would be given by the principle of superposition, and would be the sum of the wave that started and the wave that returned. Similarly, after the wave had had time to go around twice, the principle of superposition would be in operation at each point along the rope. Obviously, if the wave has had time to go around many, many times, at each point it is just as likely that the wave will have arrived with

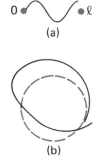

Figure 11.13.

a positive deflection as a negative deflection. The sum of all the deflections at each point, then, after a wave has traveled around many, many times, will be zero. In other words, the wave will "interfere itself out" if it is allowed to go around the orbit many times. The same argument would hold for any wave that did not satisfy the condition that it return to the point of origin in precisely the same stage of its cycle as it had at that point when it left. Thus, the requirement of continuity is equivalent ot the requirement that the wave exists in the orbit for a long period of time.

Once we have established the fact that a given wavelength can only be fit on an orbit of particular circumference, simple geometry tells us that a given wavelength can only exist at a certain distance from the nucleus. A wave of wavelength λ must fit on a "string" of length $2\pi r$. But de Broglie's relation tells us that once you know the wavelength of a wave, you also know its momentum, and hence the velocity of the particle with which the wave is associated.

In other words, if we described the electron as a wave, there are only certain orbits that are allowed. These are orbits on which we can fit the wave associated with the electron. The obvious question to ask is whether these orbits are the same as the "safe orbits" derived by Bohr from his assumption of quantization of momentum. Obviously, from the fact that we are presenting the discussion at all, the answer is yes. Much to the amazement of de Broglie and other physicists, if the electron were treated as a wave, in analogy to the treatment of light as a particle, the existence of the Bohr orbit could be understood simply in terms of the ability to fit the wave onto an orbit of given circumference.

One way of understanding the explanation of the Bohr atom that arises from the de Broglie relation is to say that the only orbits allowed are those for which the wave and particle descriptions of the electron are consistent with each other. There may, for example, be an orbit (other than an allowed orbit) for which the radius and velocity of the electron will allow the centrifugal and electrical forces to balance. These orbits would be "allowed" if the electron were a classical particle. If we took the momentum of an electron in this orbit, however, and used the de Broglie relation to get a wavelength, we would find that the resulting wave would

299

not fit on an orbit of circumference $2\pi r$. Hence, even though the particle properties of the electron would be all right for an orbit such as this one, they would not be consistent with the wave properties, which would demand that the electron wave fit on the orbit. The only orbits for which a calculation like the one outlined above would lead to a consistent result for both wave and particle properties turn out to be the Bohr orbits. In this way, the original guess of Niels Bohr is explained in terms of the use of subatomic objects having properties of both particles and waves to describe the structure of the atom.

Actually, if the de Broglie hypothesis had been only an idea, people would have been very slow indeed to accept it. However, we have already discussed how one would go about seeing if a particular object were a particle or a wave. You will remember that waves have the property that they can give diffraction patterns, while particles do not have this property. Once the idea that the electron might be a wave (or at least have wave-like properties) was stated, it was a relatively simple thing for experimental physicists to devise experiments to see whether electrons could ever give us a diffraction pattern. In fact, these experiments were done in 1927 in the United States for the first time. A beam of electron was directed at a crystal. From our point of view, the crystal can be thought of as simply a series of slits, much like the two slits that we discussed in the previous chapter. When the number of electrons falling at each point on a screen far away from the slit were counted, the resulting distribution of electrons looked exactly like an interference pattern, with the size of the peaks and the distance between the dark bands being precisely those that one would expect for a wave whose wavelength was given by the de Broglie formula. A sketch of this apparatus is given in Fig. 11.14.

Thus, as soon as the idea that an electron might have wave-like properties had been put forward, it was possible to test it experimentally. When these tests were done, the idea was found to be correct. We have come full circle. Not only does light have properties of a particle, but electrons have properties of waves. There are not two kinds of objects in the subatomic world after all. There is only one kind of object—the subatomic particle. This subatomic particle has properties of both waves and of normal particles. It is, however, a totally new object, and one that we have to study if

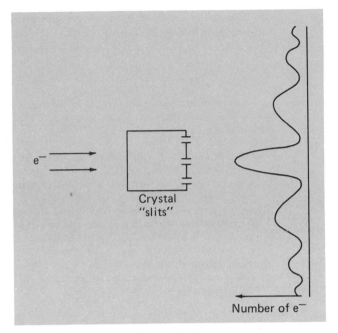

Crystal
"slits"

Number of e⁻

Figure 11.14.

we wish to understand the structure of the atom, and hence
the basic structure of matter.

Incidentally, the de Broglie relation implies that *any*
particle of matter that is moving (and therefore has a velocity
associated with it) also has a wavelength. In principle, it
applies not only to electrons and subatomic particles, but
to normal-sized objects as well. The reason that the wave-
like properties of normal objects are not seen in the normal
course of events lies in the extremely small value that
Planck's constant has. For example, if we talked about a
one-kilogram mass moving at one meter per second (this
corresponds to a velocity of about two miles per hour—a
slow walk), the momentum of the particle would be one
kilogram meter per second. From the de Broglie relation,
this would correspond to a wavelength of

$$\lambda = \frac{h}{p} = 6.6 \times 10^{-34} \text{ meters,}$$

which is smaller even than the size of the atomic nucleus!
A wavelength that small, or the diffraction patterns resulting
from it, would be impossible to detect. Consequently, we

301

do not usually worry about the wave-like properties of large objects, even though, in principle, they exist.

SUGGESTED READING

In addition to the standard texts, all of which contain descriptions of the Bohr atom, Planck's constant, and the de Broglie relation, the following may be of interest:

Andrade, E. M. daC. *Rutherford and the Nature of the Atom.* New York: Anchor Books, 1964. A life of Rutherford together with some reviews of his work and a number of anecdotes (there are many about Rutherford).

Gamow, George. *Thirty Years that Shook Physics.* New York: Anchor Books, Doubleday, 1966. An anecdotal and partially autobiographical account of the development of quantum mechanics that is easy to read.

Oliphant, Mark. *Rutherford—Recollections of the Cambridge Days.* New York: Elsevier Press, 1972. Recollections of Rutherford and his work by a former student.

QUESTIONS AND DISCUSSION IDEAS

1. Prove that
    a. for a wave of a given velocity, as the frequency of the wave increases, the wavelength decreases. (You should be able to do this from the definitions of wave quantities given in the previous chapter.)
    b. a quantum of blue light therefore has more energy than a quantum of red light.
2. Consider an ordinary particle moving at a velocity v and having a mass m.
    a. Give an argument that shows that if v is increased, the energy of the particle is increased, as is the momentum.
    b. From Planck's relation, show that if we think of the particle as a wave, increasing its velocity will correspond to an increase in frequency.
    c. Hence, from de Broglie's relation, show that increasing the velocity of a particle corresponds to decreasing the wavelength of the de Broglie wave associated with it.

What you have shown in this problem is that the faster
a particle moves, and the more energy it has, the shorter
is the wavelength associated with it.

3. Suppose that we lived in a world where all "electrons"
had a positive charge and the "nucleus" of an atom had
a negative charge. Call this new atom an "anti-atom."

    a. On the basis of the discussion of circular motion
of electrons given in the text, show that the Bohr
orbits in the new universe would be exactly the
same as they are in ours.

    b. Could you tell by looking at the light emitted from
the "anti-atom" we are discussing whether it had
come from an "anti-atom" or an ordinary atom?
Why or why not?

    c. Hence discuss the possibility of finding out whether
distant galaxies are made up of ordinary atoms or
"anti-atoms."

4. In the discussion of the Bohr orbits, there was nothing
that said that we had to restrict our attention to atoms.
In principle, the quantization principle should apply to
any orbital system. Let us apply it to the earth-moon
system. The following facts should be useful:

    radius of moon's orbit: $385,000$ km $= 3.85 \times 10^8$
      meters

    time to complete one orbit: 29.5 days $= 2,600,000$
      sec

    mass of moon: 7.4×10^{22} Kg

    a. What is the circumference of the moon's orbit?

    b. What is the average velocity of the moon in orbit?

    c. What is the angular momentum of the moon?
(Hint: The answer will be in units of kg meters^2 /
sec $=$ joule sec.)

    d. Using Bohr's quantization rule, calculate the value
of "n" in the Bohr relation. (It should be a very
large number.)

5. Continuing with the example of the moon, let us work
out the distance between Bohr orbits.

    a. Suppose that we have two possible orbits, at radii
r_1 and r_2, and that these orbits differ by one unit
in "n" in the Bohr relation (that is, if r_1 corresponds to a value n_1, r_2 corresponds to a value
$n_1 + 1$). Suppose further that the difference in

velocities between these orbits is so small that it can be neglected. Show that

$$r_2 - r_1 = \frac{h}{mv}.$$

b. Find the distance between quantum orbits for the moon.

c. Hence suggest a reason why the Bohr principle was not discovered by Isaac Newton or the astronomers who followed him.

6. Suppose that instead of weighing what it does, the moon weighed only one kilogram, but still moved with the same speed.

a. What would the distance between Bohr orbits be?

b. What would it be if the moon weighed one microgram $(10^{-9}$ kg)? (Drug doses are sometimes measured in this unit.)

c. If the moon weighed as much as an electron $(9.1 \times 10^{-31}$ kg) what would the distance between orbits be?

d. From your answers to this problem, make some comment on the use of quantum mechanics to deal with ordinary-sized objects and with atomic-sized objects.

7. In the text, we worked out the de Broglie wavelength associated with a normal object moving at a normal speed. Let's work out some more examples:

a. What is the de Broglie wavelength associated with a car of mass 2000 kg moving at 80 meters/sec (this would correspond to a large car going 60 mph)? Compare this to the size of the car.

b. What is the wavelength associated with a bullet, mass 10 grams, moving at 1000 meters/sec (about 3000 feet/sec)? Compare this to the bullet's size.

c. What is the wavelength of an electron (see Problem 6) moving at 10^8 meters/sec (1/3 the speed of light). Compare this to the size of an atom (about 10^{-8} cm).

d. Hence comment on the relation between an object's size and the de Broglie wavelength as a way of determining when the laws of quantum mechanics should be used.

8. In order for Rutherford to "see" inside of the atom, he must have been using a "wave" whose wavelength was at least as small as the atom itself. Given that the mass of the alpha particle is about 6.7×10^{-27} kg and that the size of the atom is roughly 10^{-10} meters, calculate the minimum velocity the alpha particle could have had and still been able to provide information about the structure of the atom (HINT: Use de Broglie's relation.)

9. In Fig. 11.12 we saw a sketch of a wave that can be fit on a Bohr orbit. Draw the next two shortest wavelengths fitted on to an orbit. Draw both the wave on the string and on the orbit.

10. The reader is probably wondering why we have not drawn the lowest wavelength wave (pictured below) on an orbit. To investigate this point, do the following:

   a. If the above diagram shows the "string" at one moment in time, draw the string as it would appear one-half period later (i.e., when the highest points on the string are the lowest, and vice-versa).
   b. Now convert both "string" drawings to orbital drawings—i.e., fit the above waves onto an orbit.
   c. Hence show that this wavelength corresponds to a steady back and forth motion of the orbit.

11. Let us contrast the result in Problem 10 to what we would see for a higher frequency wave of the type shown below:

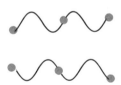

   a. Make a series of sketches showing the string moving from the first position shown above to the second as it vibrates.
   b. Convert each of your sketches in (a) to an orbit diagram
   c. Comment on the differences between the wave in Problem 10 and this one.

12. In our discussion so far, we have always talked about fitting standing waves on orbits. For the wave in Problem 11, discuss (with appropriate sketches) what would happen to the wave if we allowed the positions of the nodes to move around the orbit in such a way that we could always fit the wave on the orbit. Make sketches which convince you that such a wave would appear to be moving around the orbit.

CHAPTER
XII

QUANTUM MECHANICS

"If you want to know the taste of a pear, you must change the pear by eating it."

Mao Tse-tung, "On Practice"

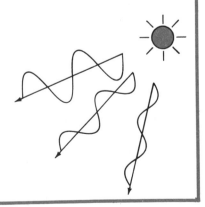

Werner Heisenberg (1901-1976).

A. The Heisenberg Uncertainty Principle

We have already seen in our study of special relativity that physical laws and physical concepts that are appropriate for one region (e.g., for the region of normal objects moving at normal speeds) are not always applicable in other regions (e.g., for objects moving at the speed of light). It should not surprise us, therefore, that when we go to the world of the very small—to the world of the atom—some rather fundamental changes must take place in our physical ideas. The change that takes place in this case has to do with the concept of what we mean by a measurement.

When I say that I see a table, or I say that a table is in the corner of the room, I am reporting the results of a long chain of circumstances. Light originated somewhere (perhaps in a lightbulb, perhaps in the sun), traveled through space, hit the table, was reflected from the table, and finally entered my eye. In other words, when I say that I "see" something, what I really mean is that light was bounced off that something and came to me. When I talk about measuring the position of the table, what I am really talking about is causing light to bounce off the table and enter some measuring apparatus. The measuring apparatus might be my eye, it might be a telescope, or it might be some very sensitive and precise photocell. In principle, however, the measurement is always the same. Something is bounced off the object to be measured and then detected by some sort of receiving apparatus.

In the case of normal objects in our everyday world—the objects that are the subject matter of classical physics—we assume that this process of bouncing light off the object does not affect the object in any way. Clearly, if we are talking about measuring the position of the moon, the fact that we are bouncing a few photons off of it will not affect either the position or the velocity that the moon has. The same is true of a large object, such as a table or chair. In other words, in classical physics it is assumed that it is possible to make the measurement without affecting the object being measured. A billiard ball rolling along the table is assumed to roll along and not be affected by the fact that light is bouncing off it and coming to our eye as we watch it and measure its progress.

Niels Bohr (1885-1962) as a young man with his fiancée, Margarethe Norlund.

Three of the founders of modern physics (left to right) Enrico Fermi (1901-1954), Werner Heisenberg (1901-1977), and Wolfgang Pauli (1900-1958).

When we come to dealing with objects on the atomic scale, however, it is obvious that we are going to have to re-think this basic assumption of classical physics. The only things we have at our disposal to bounce off the object that we are trying to measure are things like photons, electrons, and atoms. In other words, it is no longer possible to make the classical assumption that measurement does not affect the object being measured.

The reason that the assumption works in classical physics is that the energy of the photons that are being bounced off any object are infinitesimally small compared to the energy of the object being measured. The situation in quantum physics, however, is much more analogous to what we would have if the only way we could "see" one billiard ball was to bounce another billiard ball off it. If the billiard ball to be measured were rolling along a table, and in order to measure its position we had to use another billiard ball, it is obvious that the act of measurement would deflect the first billiard ball. Its motion after the measurement would not be the same as its motion before the measurement. In just the same way, when we bounce a photon or an electron off an atom, the motion of the atom must be affected. We are, after all, hitting the object being measured with another object of approximately the same energy. Thus, the funda-mental unspoken assumption of classical physics—that it is possible to measure an object without affecting it—has to be replaced in quantum physics by the following statement:

It is impossible to make a measurement of a quan-tum object without affecting that object.

This is the basic fact behind one of the most puzzling new ideas in quantum mechanics—the Heisenberg uncertainty principle. In what follows, we shall first try to understand what the principle is, and then we shall try to understand how the principle is related to the simple, physical fact quoted above.

We have discussed in a previous section the idea that every experiment has associated with it a certain experimental error. This error may be built into the apparatus itself. It may consist of human error (e.g., errors in judgment) or it may be an error that is inherent in the design of the experiment.

Nevertheless, in classical physics, we would imagine a limit in which the experimental error associated with any measurement becomes as small as we like and approaches zero. We could imagine this limit as being the one that we would reach by using ever finer and finer measuring techniques. For example, in measuring position we could go from using a yardstick to using a very finely graded vernier ruler to using fancy optical techniques. Each improvement in instrumentation would involve less uncertainty in knowing the position of the object in question. There is no limit *in principle* to how fine these measurements could be made. The only limitation in the case of classical physics is our ability to devise instruments in measurement techniques that have sufficient accuracy. Since the actual act of measurement does not affect the object in the classical scheme of things, the object will "sit still" as these finer and finer measurements are being made. In this way, we could say that the uncertainty in position and the uncertainty in velocity (and hence momentum) could be made as small as we like. If we call the experimental uncertainty in position Δx and the uncertainty in momentum Δp, we would then say that in the classical scheme of things we could, in principle, have the situation in which both

$$\Delta x = 0$$

and

$$\Delta p = 0.$$

In the classical case, it is also obvious that there is nothing to prevent us from making the measurement of position and momentum simultaneously. In other words, I can measure where a moving car is and how fast it's going at the same time. There is no reason why I cannot have both Δx and Δp equal to 0 at the same time.

The case in quantum physics is different. We have already seen that it is impossible to measure an object without affecting it. The precise mathematical statement of the Heisenberg uncertainty principle is that if an attempt is made to measure both the position and momentum of a subatomic object at the same time, both cannot be measured simultaneously with infinite precision, and the relation between the uncertainty in position, Δx, and the uncertainty in momentum, Δp, is given by

$$\Delta x \cdot \Delta p > h \,,$$

where h is Planck's constant.

Max Planck (1858-1947).

The meaning of this equation is quite clear. Even if there is no error in a measurement due to the instruments themselves, it is still impossible to have precise knowledge of position and momentum at the same time. This does not mean that we cannot know the position exactly or the momentum exactly. What it means is that if we know the position exactly, we do not know the momentum at all, and vice versa.

How does this follow from the above equation? If we know the position of an object exactly, this means that Δx must be 0. The above equation tells us that the product of multiplying Δx, the uncertainty in position, times Δp, the uncertainty in momentum, must be greater than or equal to a particular number; Planck's constant. If Δx is 0, we have a situation in which we have an equation that tells us that 0 multiplied by something (that "something" being the uncertainty in momentum) must be equal to a number that is not 0. The only number that can be multiplied by 0 and give us a number that is not 0 is infinity. Thus, if the position of an object is known precisely, the uncertainty in our knowledge of the momentum must be infinite. This is just another way of saying that we do not know what the momentum is at all. The same argument could be reversed to show that if we knew the momentum (or velocity) of an object precisely, we would have no knowledge at all of where it was (i.e., of its position).

The philosophical implications of the Heisenberg uncertainty principle are profound. It tells us that the classical idea of a "divine calculator," able to predict the entire future course of the universe once given the position and velocity of every particle in the universe, is pretty much of a dead letter. It is not so much that such a calculation would be impossible, but the uncertainty principle tells us that such a calculator could never get the information needed in order to carry out the calculations. Although the idea of the "divine calculator" has always been something of theoretical abstraction rather than an actual possibility, what the uncertainty principle tells

311

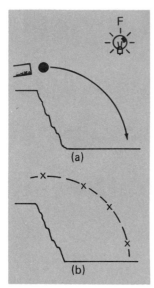

(a)

(b)

Figure 12.1.

us is that is it impossible *even in theory* to predict precisely the future course of the universe in the classical sense. This impossibility arises from the fact that it is impossible in principle to know the position and velocity of even one particle at a given time. Thus, in a sense, the existence of the Heinsenberg uncertainty principle marks a step backward in the aspirations of science. It tells us that when we are dealing with the world of the subatomic particle, there are certain things that we cannot know even in principle, simply because of the nature of the process of measurement itself.

We now turn to a discussion of the meaning of the uncertainty principle. We shall look at this in two separate ways. First, we shall consider the "particle" aspects of the problem, and then we shall consider the "wave" aspects of the problem.

Let us consider an example of a classical measurement illustrated in Fig. 12.1. Let us suppose that we want to trace the trajectory of a cannonball that is shot off the top of a cliff. Now we know from our above discussion that when we talk about "measuring" the position of a cannonball, we are really talking about bouncing light beams off the cannonball and letting them come to our eye. Let us make this interpretation very clear by supposing that it is dark out, and that we have a flashbulb (labeled F in the figure) which will flash at predetermined intervals during the flight of the cannonball. Thus, there will be no photons except when the flashbulb goes off. What shall we see?

Obviously, whenever the flashbulb goes off, photons will strike the cannonball and will come to our eye. Therefore, we shall "see" the cannonball at those positions where it happens to be when the flashbulb goes off. For example, in Fig. 12.1b we show the positions of the cannonball as they would appear in a typical experiment. If we were given the location of the position of the cannonball corresponding to each of these measurements and asked to determine where the cannonball was when no measurements were being taken, it would be a relatively simple thing to do. We would simply draw a line (like the dotted line between the points in the figure) and say that even though there was no measurement being taken at a particular time, the cannonball would have been on the dotted line. We say this because we believe that the measurement of the position of the cannonball made by bouncing photons off it does not effect its flight, and so it is possible to interpolate smoothly between measured points. In other words, we believe that the cannonball moves along in whatever path it happens to be following, unaffected by the measurements. If this path happens to lie along the smooth line (as it does in Fig. 12.1), it is a simple matter for us to guess where the cannonball would be between any two measurements. If we wished to check this, we would also know how to proceed. We would simply fire the flashbulb more often. This would then give us the position of the cannonball at many more spots along the curve; the interpolation would be over smaller distances, and we would have even more faith in the results of our interpolation than we had in the first case. The process of "watching" the cannonball fall, of course, is simply the limit of this experiment in the case when the flashbulb is going off continuously, and we are constantly receiving photons that bounce off the cannonball.

Suppose now that instead of a cannonball we wish to trace the path of an electron that was shot off of a cliff. The situation would then be quite different. We trace the "path" of the electron in Fig. 12.2. Suppose the electron is at point A when the flashbulb goes off for the first time. When the photon that eventually comes to our eye hits the electron at this point, the electron will recoil. However, we do not know precisely how it will recoil. Thus, we do not know in which direction the electron is moving after the collision. The only way to find out where it is going is to make another measurement. Thus, we flash the flashbulb again, and this time the photon hits the electron at point B. Now we know where it was going when it left point A, but we do not know where it was going when it left point B. To find this out, we have to make yet another measurement. The sequence of measurements that we make results in a sequence of points such as that shown in the figure. No longer do they fall in a smooth curve. No longer is it possible to interpolate smoothly between the points. The reason for this change of events, is simply the fact that when a photon hits an electron the electron is affected by the collision, whereas when the photon hits a cannonball, the assumptions of classical physics tell us that the cannonball is not affected by the collision.

How can we find out the actual trajectory of the electron? One way that might occur to us would be to argue as follows: Since it is the collisions between the electron

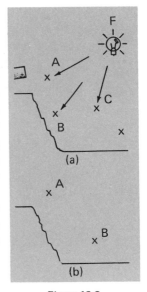

Figure 12.2.

and the photons that cause the "jitteriness" in the path of the electron as we have observed it, perhaps the thing to do is simply to have fewer collisions. In other words, maybe what we ought to do is flash the flashbulb fewer times. This would mean that the electron would be deflected fewer times in its fall.

While it is true that if we flash the bulb fewer times we would perturb the motion of the electron fewer times, it is also true that the fewer measurements we have the more ignorant we are of the position of the electron at any point along its path. For example, in Fig. 12.2b we show a sample result that would occur if we flashed the flashbulb only twice. We would then know that the electron was at point A at one point in its fall and at point B at a later time. However, we would have even less information on where it was at other points in a path than we had in the original situation. Therefore, decreasing the number of collisions will not increase the accuracy of our knowledge about the position of the electron.

A second way that one might think of getting around the difficulty is to argue as follows: Since the cause of our uncertainty in the path that the electron follows is really related to the energy of the photon that is bouncing off it, why don't we just decrease the energy of the photon to the point where its effect on the electron becomes as small as we like?

In principle, this sounds like an extremely good argument. After all, the basic physical principle operating here has to do with the effect that the photon has on the electron it is measuring. Obviously, if the photon had less energy, the electron would recoil less after the collision and there would be less perturbation of the electron path.

Unfortunately, we know too much about photons to allow this particular argument to pass muster. We know from Planck's relation that the energy of the photon is related to its frequency by the formula $E = hf$. Thus, in order to decrease the energy of a photon, it is also necessary to decrease its frequency. If you think about this for a minute, you will realize that decreasing the frequency of a photon corresponds to increasing the wavelength of that photon.

Why?

The frequency of a wave is the number of times that a crest will pass a given point in a given time interval. For example, we might have a wave in which three crests passed a given observer in one minute. This would then be a wave whose frequency was three cycles per minute. Obviously, a wave whose frequency is three cycles per minute will have a shorter wavelength than a wave whose frequency is one cycle per minute. The one-cycle-per-minute wave will have one crest going by that observer every minute, while in the same minute the three-cycle-per-minute wave will have three crests going by that observer. Therefore, the wavelength (which is the distance between crests) will be three times as long in a one-cycle-per-minute wave as it is in a three-cycle-per-minute wave (assuming, of course, that both waves more at the same velocity).

Thus, when we decrease the energy of the photon that is making the measurement in the above example, we increase its wavelength. As this process of lowering the energy of the photons so as not to deflect the electron goes on, eventually we shall come to an energy that corresponds to a photon wavelength that is the same size as the cliff. At this point we can no longer measure the position of the electron. Thus, in our attempt to minimize the perturbation in the path of the electron due to the energy of the photon, we lose the ability to determine its position in a single measurement.

So in the simple example of trying to trace the trajectory of an object, we see how the fact that in quantum physics it is no longer possible to measure anything about an object without perturbing it leads us to conclusions that would be very alien to the mind of the classical physicist. In this case, we are forced to the conclusion that it is impossible to trace the path of an electron with infinite precision. This differs greatly from the equivalent classical statement about a cannonball.

Another way of looking at the uncertainty principle as it is stated in its mathematical form is to discuss the idea of the position and momentum of a subatomic particle in terms of its wave properties, rather than its particle properties. From the de Broglie relation, we know that the wavelength of an object is related to its momentum. It is obvious that if we wish to know the momentum of an object precisely, it will do to measure the wavelength with which it is associated. Given an infinitely precise measurement of the wavelength, we could then use the de Broglie relation to get an infinitely precise determination of the momentum.

315

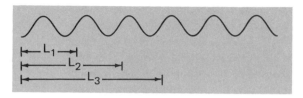

Figure 12.3.

How would we go about getting an infinitely precise measurement of a wavelength? If we had a wave such as that shown in Fig. 12.3, we would know how to go about measuring the wavelength. We would, for example, measure the length L_1 shown in the figure. This would be one wavelength of the wave.

However, we know that we could not measure the distance L_1 with infinite precision. There would always be some error, ΔL, associated with this measurement. The error in determining the wavelength would then be $\Delta L/L_1$. Depending on the kind of ruler we used, we could make this wavelength error very small, but we could never make it vanish completely.

Suppose that instead of measuring length L_1, we measured length L_2: the length corresponding to two wavelengths. The measurement uncertainty would still be ΔL, since we are still using the same ruler and the uncertainty in measuring any distance with the same ruler is the same. However, we are now dealing with an error that is given by $\Delta L/L_2$. Since L_2 is twice as long as L_1, the error associated with determining the wavelength in the case where we measure L_2 is only half as big as the error associated with measuring the wavelength when we measure L_1. Similarly, if we measured L_3, the distance associated with three waves, the error would be only one-third as big.

It is simple to extrapolate this procedure to the case of a large number of waves as we measure longer and longer distances. Although the error ΔL associated with each measurement is the same, as we make the distance being measured longer and longer we make the uncertainty in the wavelength smaller and smaller. In this way, we could imagine going to the limit of measuring an infinite length (or, equivalently, an infinite number of wavelengths) and coming up with an uncertainty in the determination of the wavelength that was zero. This would be the case even though the uncertainty in measuring the length was ΔL. Thus, it is possible in principle to determine a wavelength exactly, but to do so it is necessary to measure an infinite number of wavelengths.

If we are dealing with a subatomic particle for which we have measured the wavelength precisely, and hence for which we know the momentum exactly, we can ask the question "Where is the subatomic object?" Since the wave must be of infinite length in order for us to determine the wavelength exactly, it is obvious that we have no idea at all where the subatomic object is if we want to locate it the way we would locate a particle. In other words, when we use the wave nature of a subatomic particle to determine its momentum exactly, the position of that "particle" is totally unknown. It is impossible to determine both the momentum of a subatomic object and its position precisely if we think of that object in terms of its wave-like properties. A similar argument could be advanced in which we thought of the particle as being localized (i.e., in which we thought of the particle as being located at a particular point) and then tried to determine its wavelength. We would find that if we knew exactly where the particle was, we would then require that we have no knowledge at all of its wavelength. Thus, we would also have no knowledge at all of its momentum.

We see that whether we look at subatomic objects as being particles or waves, we come to the same conclusion. There is something new in the subatomic world that we have not seen before. This "something new" is contained in a statement that when we measure subatomic objects we must use another subatomic object to make that measurement. Consequently, the measurement itself will perturb the system being measured. For physicists, this was a very shocking and very surprising turn of events. This state of affairs has been well known in other sciences for a long time, however. It is well known in the biological sciences that it is very difficult to make measurements of biological systems without disturbing them. For example, if we wished to know how a heart valve functions during the normal beating of the heart, it is almost impossible to devise an experiment to get this information without at the same time disturbing the system. If we cut into the heart in open-heart surgery to watch the valve, it is obvious that the heart which has been surgically opened will not be the same as the heart beating normally. Any other technique (such as injecting dyes into the veins and taking

X-ray movies) also affects the person being measured, and hence affects his heart. What the uncertainty principle seems to be doing is telling us that when we deal with subatomic particles we have finally reached the stage in physics where we can no longer ignore the effect of measurement in an experiment. In the next section, we shall go on to explore the consequences of the uncertainty principle and the shape of the kind of scientific theories that have been developed to deal with this new state of affairs.

B. Quantum Mechanics: The Science of the Subatomic Particle

In 1926 and 1927, two German scientists working quite independently of each other, Werner Heisenberg and Erwin Schrödinger, developed a new way of looking at the problem of subatomic particles. The mathematical techniques that resulted from this work led eventually to what is now called "quantum mechanics." Just as Newtonian mechanics concerns itself with explaining the motion of normal-sized objects, quantum mechanics concerns itself with explaining the motion of subatomic objects. Since these objects exist in quanta, the name is self-explanatory.

The basic idea of quantum mechanics is not difficult to understand, although some of the logical consequences of the ideas are. We can break any quantum mechanical calculation into two parts. One part deals with the wave nature of a subatomic object and the other with the particle nature. A general procedure that one would follow in doing a quantum mechanical calculation would be as follows:

1. The *calculation* is done assuming that the object is a wave.
2. The *interpretation* is made assuming that the square of the amplitude of the wave at any point in space is related to the probability that the particle would be found at that point in space.

This sounds very mysterious and complicated, but actually it is quite simple. Let us take an example in which we have already carried out step 1 in the above procedure. Suppose we had a situation, such as that shown in Fig. 12.4, in which a stream of subatomic objects was being directed against a screen in which there were two slits. The objects

Erwin Schrödinger (1887-1961).

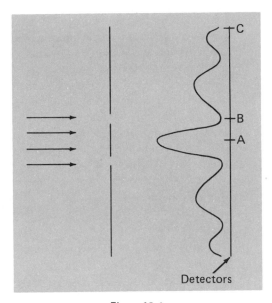

Figure 12.4.

are then detected far away from that screen with a bank of detectors. We have already discussed what would happen if a wave were incident on a screen with two slits in it. For example, we know that at point A, midway between the two slits, there will be positive interference between the waves coming from each of the two slits, and hence there will be a large wave. In fact, we know from our previous discussion that the amplitude of the wave at point A will be twice as large as the amplitude of the wave from one slit alone. We also know that if we move up the bank of detectors to point B (which is taken to the point at which the difference in path lengths from the upper and lower slit is exactly one-half of a wavelength) there will be negative interference and hence a null. By definition, the amplitude of a null wave is 0. If we kept going up the bank of detectors, we would come to the point at which there was again positive interference and hence a large wave. If we continued, we would find another null, another point of positive interference, and so on. This result (that there will be "light" and "dark" bands along the plane of detection) is an example of a calculation that is done assuming that the subatomic objects are waves. It is, in other words, an example of step 1 in the development of quantum mechanics.

319

The carrying out of step 1, then, is a relatively simple matter of working out the interference pattern that would occur for any particular experimental apparatus that was set up. For example, in our above "experiment," the apparatus is the beam of particles incident on the two slits, and the calculation involved finding the amplitude of the resultant wave at the position of the detectors. The important point, and a point that cannot be emphasized enough, is that we have made this calculation assuming that the subatomic objects exhibit interference. This means that we have made the calculation assuming the subatomic objects exhibit wave-like properties. This assumption is justified in light of the many experiments we have discussed which tell us that these objects do, in fact, exhibit such properties.

But suppose that the subatomic objects that we are directing toward the slit are electrons. We think of electrons as particles, and we would like to interpret the results of our calculation as results having to do with particles rather than waves. This is where the second step in the above discussion enters.

We know from our calculation that the amplitude of the wave at the position of the detectors is very large at point A and is 0 at point B. The precise statement of quantum mechanics is that the *square* of the amplitude of the wave is proportional to the probability of finding a particle at the point in question. Let us consider point B as an example. At point B, the amplitude of the wave is 0. Zero times 0 is still 0, so that the square of the amplitude, and hence the probability of finding a particle at point B, is 0. This means that if we shoot a million particles at the screen, none of them will wind up at point B. At point A, on the other hand, we have seen from our calculation that the amplitude of the wave will be large. Suppose that we denote by A' the amplitude of the wave at point A. A' is the amplitude of the actual wave that appears, and hence is twice as large as the amplitude of the wave that is arriving from either slit at point A. The probability of finding a particle at point A, then, would be proportional to $(A')^2$, which is most definitely not 0. We would interpret this result to mean that if we shot a million particles at this double-slit apparatus, many of the particles would wind up at point A. In fact, if the amplitude at point A is given by A', and the amplitude at point C on the diagram (a point at which the interference is neither totally negative nor totally positive) is given by C', then the number of particles that arrive at point A will be precisely $(A'/C')^2$ times the number of particles that arrive at point C. In other words, from quantum mechanics as it is discussed here, we could predict the relative probability of finding a particle at point A or at point B or at point C.

If, instead of a double-slit apparatus, we were dealing with an electron that was in an atom, we would follow the same procedure as the one we have outlined here for the double slit. That is, we would first make a calculation of the electron orbit assuming that the electron was a wave, and then we would interpret that result on the assumption that the square of the amplitude of the wave that resulted from the calculation was proportional to the probability of finding the particle at any point around its orbit. Any calculation that involves the behavior of subatomic particles could be made in the same way.

It is important to realize that this formulation of quantum mechanics is by nature *statistical.* What this means is the following: We can always predict the probability of finding a particular particle at a particular point in space, but we cannot predict what a single particle will do.

An example of a statistical type of theory is the science of demography, especially as it is applied to voting patterns. It is quite possible that a demographer could predict that in a particular election a particular ward in a particular city will vote 80% Democratic and 20% Republican. He could tell us beyond a shadow of a doubt that if 1,000 people voted in that ward, 800 would vote Democratic and 200 would vote Republican. He could also tell us that if one person votes in that ward, the chances are 80% that he will vote Democratic and 20% that he will vote Republican. What he *cannot* predict for us is the vote of any given individual. He can only predict the result of a large number of votes or the probability of a particular vote in an individual event.

The situation in quantum mechanics is much the same. If a million electrons are used in a particular experiment, the results of quantum mechanics will tell us the distribution of those electrons in the final state. For example, if we shot a million electrons at the screen with the two slits in it, quantum mechanics would tell us how many electrons would wind up at each point in the detectors. Quantum mechanics could also tell us the probability that any given electron would wind up any place in the collectors. What it cannot do is tell us with certainty where a single electron will wind up. And yet, just as a single person casts one vote, which is either Democratic or Republican, a single electron will indubitably wind up at only one point in the detectors. This statistical aspect of quantum mechanics is very different from what we saw in classical mechanics. In classical mechanics, it was assumed that a particle could be observed without affecting it and, from the measurements made on the particle, we could predict the future course of that particle *as an individual particle*. In quantum mechanics, on the other hand, we cannot measure the particle without disturbing it, and we cannot trace its development *as an individual particle*. We can only give probabilities as to where it will be and what it will be doing in any given time after the measurement, or, equivalently, we can predict what will happen to large numbers of particles. In a sense, quantum mechanics represents a "step backward" in our ideas of what science is capable of doing.

Before we despair of ever getting significant results from quantum mechanics, however, it should be realized that quantum mechanics is fully capable of predicting the result of *any* experiment that can actually be carried out. It is impossible in principle to design an experiment in which a single electron with a known position and velocity starts out on the left-hand side of our double slit and winds up on the right-hand side. The reason that it is impossible to imagine such an experiment is that in order to know the final state of the electron exactly, we would have to know the initial state of the electron. In other words, to do the classical type of calculation for a single electron going through the double slits, we would have to know both the position *and* momentum of the electron when it started out. But the uncertainty principle tells us that we cannot know both of these quantities simultaneously. In other words, it is impossible to imagine an experiment in which we could define exactly what the initial classical state of the electron was. It should come as no surprise, then, that we are unable to predict exactly what the final state will be.

What *can* quantum mechanics predict, then? In the example of the two slits that we have been discussing, if we knew either (1) the number of electrons at each point in space, and the number of electrons moving with each particular momentum, or (2) the probability that a given electron is at each point in space, and the probability that that electron has a particular momentum, then we could predict either (1) the number of electrons arriving at each point at the screen, or (2) the probability that the given electron will arrive at any point on the screen. In other words, if we know the statistical distribution of electrons on the left-hand side of the screen, or if we know the probability of a single electron being at any point moving with a particular momentum on the left-hand side of the screen, we can then predict the statistical distribution of electrons on the right-hand side of the screen or the probability of a single electron arriving anywhere on the right-hand side of the screen. Since all we can measure about the electron is either the number of electrons at a particular point moving with a particular velocity, or the probability that a particular electron is at a particular point and moving with a particular velocity, it doesn't make sense to demand of our theory that it predict more than we can measure. In the case of quantum mechanics, this means that that theory will predict probability distributions rather than actual precise classical deterministic results.

Thus, what appeared first as a step backward in science, is actually seen to be nothing more than a statement that since we are dealing with subatomic particles, and since there are certain things we cannot do with subatomic particles (i.e., we cannot measure them without disturbing them), the resulting theory will necessarily be statistical in nature. You may not approve of this kind of statistical theory, but if you don't you ought at least to have the intellectual honesty never to use a transistor radio or solid-state stereo system, because both of these rely on the use of the statistical theory of quantum mechanics as it is applied to the transistor and other solid-state electronic devices.

One final point should be made about quantum mechanics. We saw when we discussed the principle of relativity that when we took the relativistic results of time dilatation and length contraction and applied them in the region where velocities were very small compared to the velocity of light, we recovered the classical Newtonian mechanics. In the same way, if we applied the laws of quantum mechanics to objects that were very large on the atomic scale or to normal everyday objects around us, we would find that the quantum mechanics would reduce to the normal Newtonian mechanics as well. This was discovered by Niels Bohr, and is sometimes referred to as the "principle of correspondence." In other words, just as the laws of relativity can be thought of as "containing" the laws of classical mechanics in a particular limit, so too the laws of quantum mechanics can be thought of as "containing" classical physics in the limit that we deal with very large objects.

C. Quantum Mechanics in Action

Just as the development of Newtonian mechanics led to a tremendous increase in our understanding of the motion of material bodies, and just as the development of Maxwell's equations led to a tremendous increase in our understanding of electromagnetic phenomena, the development of the science of quantum mechanics led to a tremendous increase in our understanding of phenomena at the atomic and subatomic level. In this section we shall consider two examples of this new understanding. First, we shall consider the chemistry of the atom and particularly the role of quantum mechanics in explaining the periodic table of the elements; second, we shall consider a modern device—the laser—and see how it works at the atomic level.

1. Periodic table of the elements

In 1871, the Russian scientist Dmitri Ivanovich Mendeleev noticed that if the chemical elements were arranged in a certain way, a great deal of regularity in their chemical properties could be observed. In Fig. 12.5, we show a rough sketch of the grouping of elements that he proposed, which we now call "the Periodic table of the elements." Mendeleev noted that if the elements were arranged in this way, with the weight of each element increasing as we move from left to right in the table, then if we read down the rows of the table we find that the elements in each row have similar chemical properties. For example, the elements hydrogen, lithium, and sodium have similar chemical properties even though they are rather different in appearance. Hydrogen is a gas, and sodium is a metal. Nevertheless, both of them are very reactive chemically and are quite

Dmitri Mendeleev (1834-1907).

324

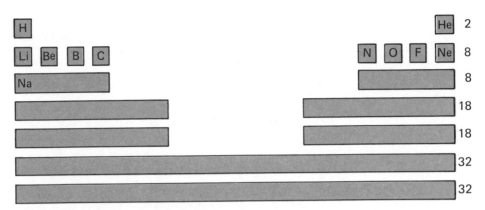

Figure 12.5. The Periodic Table

dangerous to handle. Mendeleev also found that when he arranged the chemical elements in this way, there were certain gaps in the table—elements that were missing at the time he put the table together. The discovery of Germanium and Scandium were both instigated by the gaps in Mendeleev's periodic table. The discovery of the "missing" elements was the first strong experimental observation that tended to say that there was some fundamental law of nature in operation that led to the periodic table. However, at the time of Mendeleev, the periodic table remained an unexplained regularity in nature, much as the law of definite proportions had been an unexplained regularity before the time of Dalton.

One of the rather puzzling things about the periodic table is that if we arrange things according to chemical properties, we find that there are only two elements (hydrogen and helium) in the first row of the table, eight elements in the second row, eight in the third row, 18 each in the fourth and fifth rows, and then 32 elements each in the sixth and seventh rows. Although it was well established on chemical grounds that this is the way the elements were arranged, there was no understanding of why these particular numbers should play an important role in the chemistry of the elements.

With the new understanding of the atom and the development of quantum mechanics, it became possible for the first time to answer the following question. "Why is the periodic table the way it is?" We shall see that the chemistry of an atom is determined primarily by the arrangement of the outermost electrons in the atom. Therefore, we shall

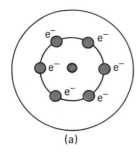

(a)

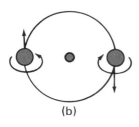

(b)

Figure 12.6.

begin our quest for understanding of the periodic table by examining the way in which electrons exist in the Bohr orbits in a given atom.

Up to this point, we have been talking primarily about only one atom—hydrogen. This atom is very simple, having only one electron in its outer orbits. However, it is obvious that if we think of a heavy element that has a large, positive charge on its nucleus, it will also have many electrons spread out among the Bohr orbits. The question we want to ask is: How are electrons arranged among the Bohr orbits when there is more than one electron in an atom?"

In Fig. 12.6a, we show one possible arrangement that electrons can have in an atom. We saw when we were discussing the process of the "quantum jump" that electrons in an atom will move to the lowest possible energy state, which corresponds to the innermost orbit available to them. Thus, if there were nothing to stop them, we would expect that all of the electrons in an atom would eventually wind up in the first Bohr orbit. If this happened, then every atom would be pretty much the same as every other atom, with its entire complement of electrons packed into the lowest orbit around that atom.

However, from the fact that the chemical properties of atoms are so different, we know that this doesn't happen. Atoms that exhibit different chemical properties must have very different arrangements of electrons. Therefore, there must be some new law of physics that prevents all the electrons from falling down into the lowest possible orbit. This new law of physics was stated by Wolfgang Pauli and is called the "Pauli principle." A precise statement of the Pauli principle is:

> No two electrons can be in the same quantum state.

What does this law mean? What is a quantum state? We have seen that in the case of the Bohr atom, an electron could be in any one of a number of allowed orbits. This means that if we want to specify exactly where an electron is, it is necessary for us to specify in which orbit it finds itself. For example, we could say that an electron is in the first orbit, or the third orbit, or the fifth orbit. Another way

326

of stating the same fact is to assign each orbit a number. For example, the first orbit would be assigned the number 1; the second orbit would be assigned the number 2; the third, the number 3, etc. The number that is assigned to an orbit is given the name "orbital quantum number." Thus, instead of saying "the electron is in the fifth orbit," we could say "the orbital quantum number of the electron is 5." These statements are completely equivalent, and are just ways of stating the same fact in two different languages.

It was also discovered shortly before Pauli's work that the electron had another property. It was found that the electron had spin. In other words, we could think of the electron as a particle that goes around the Bohr orbit and at the same time spins around its own axis. This is similar to the motion of the earth, which goes around the sun once a year but spins on its own axis once a day. In Fig. 12.6b, we show two electrons in the lowest Bohr orbit. Obviously, there are two possible ways that an electron can spin. It can spin clockwise, and it can spin counterclockwise. Rather than talking about clockwise or counterclockwise spin, however, physicists customarily talk about spin in a slightly different language. If we wrap the fingers of our right hand in the direction of rotation of the electron, then the direction of the thumb is taken to indicate the direction of spin. For example, an electron that is spinning counterclockwise is said to have spin up; an electron that is spinning clockwise is said to have spin down. The up and down refers to direction of the thumb.

Thus, in Fig. 12.6b, the electron on the left-hand side of the figure is quickly described in the following way: this electron has orbital quantum number equal to 1, and has spin up. In a similar way, we could say that the electron on the right-hand side has orbital quantum number equal to 1, and spin down. A "quantum state" is defined as that collection of numbers or descriptive phrases needed to specify exactly which electron in a given atom or in a given collection of electrons we are talking about. For example, when we are discussing electrons in a Bohr orbit like the one in Fig. 12.6b, it is necessary to specify two things about the electron. First we have to say that its orbital quantum number is equal to 1. This tells us that the electron is in the lowest Bohr orbit. Then we have to specify whether the spin is up or down.

This distinguishes between the two electrons that are in the orbit.

Now that we understand what is meant by a quantum state, we can go back and look at the consequences of the Pauli principle for the structure of electrons in an atom. This structure, in turn, will lead us to a more fundamental understanding of the periodic table of the elements.

The Pauli principle tells us that no two electrons can be in the same quantum state or, equivalently, no two electrons can have the same collection of quantum numbers describing their state. In the case of electrons in an atom, this means that there are a limited number of electrons that can be fit into each Bohr orbit. There are, in other words, a limited number of ways in which electrons can be put into orbits before we begin to duplicate electrons that are already there. For example, in Fig. 12.6b, if there is no quantum number describing the state of the electron other than the orbital number and the spin, there are only two possible ways that an electron can be put into the first Bohr orbit. It can be put in with spin up, or it can be put in with spin down. If we tried to put in a third electron, it must have spin either up or down and hence it will have to have the same set of quantum numbers (orbital quantum number 1, spin up or down) as one of the electrons that is already there. Therefore, according to the Pauli Principle, there is no room for a third electron in the first Bohr orbit.

Actually, however, the situation is not as simple as it would seem from this first example. Although there is, indeed, room for only two electrons in the first Bohr orbit, in subsequent Bohr orbits another quantum number enters the picture. To understand what this new quantum number is, we have to realize that an electron going around the orbit can traverse its orbit either in a clockwise or a counterclockwise manner. Thus, the direction of the electron in its orbit becomes another quantity that must be specified in describing a quantum mechanical state. This means that in higher Bohr orbits we can fit in more electrons than we would be able to fit into the first Bohr orbit.

With this understanding, let us go back and examine the Bohr orbits one by one, beginning with the lowest orbit allowed. When the motion of the electron in the lowest orbit is calculated according to the principles of quantum mechanics, a rather simple picture emerges. The wave that

328

is associated with the electron in the lowest orbit can be thought of as something like the following: Imagine a piece or membrane like a drumhead stretched across a circle whose circumference represents the outer orbit. Imagine pulling this membrane will move up and down describing the motion of a three-dimensional standing wave. The electron in the lowest orbit is described by this type of standing wave. That is, the motion of the wave is up and down, but there is no orbital motion at all. Electrons in the lowest Bohr orbit do not "go around the nucleus" in the sense that the earth goes around the sun. Thus, there is no possibility in the case of the lowest orbit that a third electron could be inserted whose state was different from the state of the two electrons that were already there.

When we get to the second Bohr orbit, however, the situation changes. In this orbit, there *is* orbital motion, and the electron can be thought of as going around the nucleus in a way very similar to the earth going around the sun. As we have hinted above, this introduces a new degree of freedom and allows room for more electrons than could be fit in the first orbit. In Fig. 12.7a, we show four different ways in which electrons could move in an orbit around the nucleus. The electrons could move either clockwise or counterclockwise in their orbit in a horizontal plane, and they could move either clockwise or counterclockwise in their orbit in a vertical plane. Any other motion of the electron (i.e., orbits in a plane at an angle to the ones already shown) can be thought of as a combination of the motion in the two planes that we have shown. To each direction of motion around the orbit, there correspond two possible ways for an electron to exist—spin up or spin down. This means that there are eight possible slots into which electrons can be put in the second orbit. There are four orbital states and into each orbital state there are two spin states; 4 x 2 gives 8. In Fig. 12.7b, we draw the eight electrons that could be fit into the second Bohr orbit.

The third Bohr orbit is similar to the second Bohr orbit in that there are precisely eight spaces for electrons. The higher orbits get more complicated in shape, and it is somewhat more difficult to visualize the motion of the electrons. However, the principles involved in determining the number of spaces for electrons in higher orbits are precisely the same as they are for the lower orbits, which we have discussed. It turns out that there is room for precisely 18 electrons in the

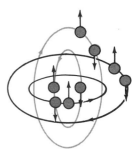

Figure 12.7.

next two Bohr orbits, and 32 electrons in the Bohr orbits following that.

The similarity between the number of electrons that can be fit into each allowed orbit in an atom and the number of elements that appear in the periodic table is too striking to be a coincidence. Let us see if the nature of chemical reactions can be used to give us some insight into the connection between the structure of the atom and the periodic table.

Consider as an example of a chemical reaction what happens when an atom of sodium and an atom of chlorine are brought near each other. These two atoms are sketched in Fig. 12.8. In Fig. 12.8a, the atoms are shown as they exist normally. The sodium has both electron slots in its lowest orbit filled, the eight electron slots in the next orbit are also filled, and there is one extra electron that is sitting in the outer orbit. In technical jargon, we say that sodium has one electron outside of a closed shell. Chlorine, on the other hand, has both of the two inner orbits filled, and seven of the eight slots in the third orbit are filled. Consequently, there is room for one more electron in the outer orbit of the chlorine atom. We represent this situation by showing a box in the outer chlorine orbit. It turns out that the energy of the total system will be less if the one lone electron in the sodium atom jumps across and fills the one vacancy in the chlorine atom. The result of this transition is shown in Fig. 12.8b. If we look at what is left after the electron has moved from the sodium to the chlorine atom, we see that the sodium atom has been left as an ion. It has one less electron that it would have in its normal, electrically uncharged state, and consequently the sodium atom has a positive charge of one unit.

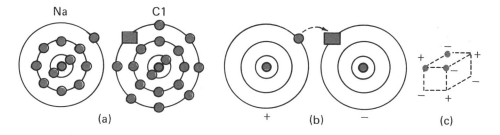

Figure 12.8.

Similarly, the chlorine atom has one more electron than it would have in its normal state, and it therefore has a net electrical charge of -1 unit. The sodium and the chlorine ions, then, are now of different electrical charges, and consequently, from Coulomb's law, we know that there will be an electrostatic force of attraction between them. Thus, the sodium and the chlorine ions will tend to stay near each other. We say that there is a bond between them and this type of bond is called an "ionic bond." In Fig. 12.8c, we show an example of the ionic bond in action—the crystal structure of sodium chloride (NaCl), or common table salt. The sodium and chlorine atom are arranged in a regular array, and the crystal is held together by the electrostatic, or ionic, forces between the atoms composing it.

Another type of chemical bonding that can occur is called the "covalent bond." Suppose two atoms of carbon were brought near each other. In Fig. 12.9, we show such a situation. In its normal state, carbon finds itself with the first electron orbit completely filled, and with four electrons (of the eight possible spaces) in the second orbit. There is no possibility of two carbon atoms forming an ionic crystal. However, another phenomenon can take place which can bind two carbon atoms together. They can "share" an electron. In Fig. 12.9b, we show such a phenomenon, in which one electron from one carbon atom and one electron from another carbon atom jump back and forth between the two atoms. This sharing of the electron creates an attractive force between the two carbon atoms, and it is this force that holds together crystals of carbon (such as graphite or diamond or coal). Covalent bonding is also the primary chemical attraction that exists in organic and living matter.

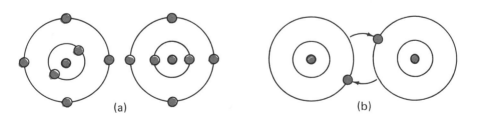
(a)   (b)

Figure 12.9.

Without going any deeper into the nature of the chemical bond, it is obvious that the way in which a particular element will react chemically is determined by the outermost electrons. In other words, the chemical properties of sodium are determined by the fact that there is one electron outside of a closed shell. Similarly, lithium, which has three electrons (two in the first Bohr orbit and one outside the first closed shell), will have very similar properties to sodium. The reason for this is that the outer electron structure of lithium is precisely the same as the outer electron structure of sodium. Both of them have one electron outside of a closed shell. Hydrogen also has one electron available for chemical reaction.

If you read down the first column of the periodic table, you will find the elements hydrogen, lithium, and sodium. The reason they appear in the first column of the periodic table is that they have similar chemical properties. The reason that they have similar chemical properties has to do with the fact that they happen to be those elements that have one electron available for chemical reactions. This fact follows from the Pauli principle and from the calculations of electron orbits that are made according to the principles of quantum mechanics. Similar arguments can be made for any other column in the periodic table.

This column of the periodic table that contains the noble gas elements is interesting in that it contains those elements having no electrons outside of a closed shell. As you may know, these elements are very reluctant to undergo chemical reactions, and are usually classified as "inert" substances. The reason that they are reluctant to undergo chemical reactions is now quite obvious. All of their electrons are bound tightly into the closed electron orbits. Consequently, when other atoms come near them, these other atoms find no place to either deposit or pick up electrons in the way that they would have to if a chemical reaction were to take place.

Thus, with the advent of quantum mechanics, it became possible for the first time to understand the chemical nature of matter and to understand the regularity that had been discovered in the periodic table of the elements.

2. How does a laser work?

Once the behavior of the atom is understood, a multitude of applications of atomic physics can be developed.

332

Many of these applications of the new ideas are already familiar to us in our everyday life—transistors, miniaturized computers, hand calculators, etc. In this section, we are going to discuss one of the important new applications of quantum mechanics—the laser.

The term "laser" stands for "light amplification by stimulated emission of radiation." If we understand what this phrase means, we shall also understand how a laser works and what it does. Let us begin by trying to understand what stimulated emission of radiation is.

We know that if an atom finds itself with an electron in an excited state, that electron will eventually make a quantum jump back to its lowest state and emit radiation in the form of light. We could call this process "natural" emission of radiation. It is a consequence of quantum mechanics, however, that if an atom having an electron in an excited state happens to be in the presence of a light wave from some other source, and if that light is of precisely the frequency associated with the quantum jump, the atom will make the quantum jump more quickly than it would ordinarily. When it makes this quantum jump, of course, it will emit its normal frequency of light. This process (by which the atom in the presence of an external light wave is induced to give off its own light very quickly) is called stimulated emission of radiation. In this process, the initial photon is not affected, but continues on its way.

Suppose, then, by some technique or another we had a situation such as that pictured in Fig. 12.10 where we had a population of atoms, each of which had electrons in an excited state. We shall discuss below how such a population of atoms might be produced, but for the sake of discussion let us simply assume at this time that such a population of atoms exists. Suppose that from the right-hand side of the figure a photon enters the system. This photon might be emitted by natural processes by one of the atoms that we haven't shown, or it might be light coming in from the outside. When the photon passes the atom on the right-hand side of the figure, stimulated emission will occur, and this atom will give off radiation. In fact, it follows from quantum mechanics that the radiation given off by the atoms will be in phase with the radiation that stimulated the emission. This means that if we move toward the left in the figure, there will no longer be just one photon going through the population of atoms, but two. The first photon will still be there, but so will the photon that was emitted by the atom on the right-hand side. Thus, as the wave moves to the left across the figure, it becomes progressively stronger and stronger. Each atom that it meets adds its bit to the strength of the wave. The process in which we start out with the small wave and wind up with a very large one is called "amplification." Thus, if we wanted to amplify a light signal, one way we could do it would be to make an arrangement such as that shown in the figure. A very weak light signal coming in from the right would, by the time it reached the left, be a strong light signal. This would be an example of amplifying light rays by stimulated emission of radiation; it is an example, albeit oversimplified, of a laser.

Actually, a laser is not quite as simple as what we have shown in the diagram. As a light wave moves through a collection of atoms, the chances of it being close enough to any given atom to cause stimulated emission is pretty small. In terms of Fig. 12.10, the light wave coming in from the left might hit only one or two atoms in our entire population as it went through, thereby giving an amplification of only two or three times its original amplitude. This would not be a very useful device if it were operated in the way that we have shown. Some way must be found to make the light collide with more than just a few atoms as it goes through the material.

One way to accomplish this end is to have an arrangement such as that shown in Fig. 12.11, where there are carefully aligned mirrors on both ends of our collection of excited atoms. The light entering from the left will then move through the system, striking one or two atoms as it goes, until it comes to the mirror on the right. When it reaches the mirror on the right, it will be reflected and will go back through the system toward the left, again striking one or two atoms. When it

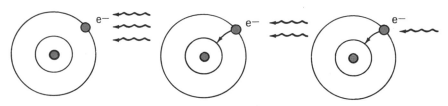

Figure 12.10.

reaches the left-hand side, it will be reflected and move back toward the right again, etc. Just as the original light wave is moving back and forth (perhaps several million times) between the mirrors, any light wave that is emitted by the stimulated emission process and that is *moving in the correct direction* will also pass several million times through the collection of atoms.

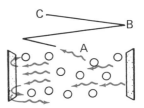

Figure 12.11.

What do we mean by "the correct direction?" Obviously, if a light wave is emitted by the stimulated emission process and it is traveling in an angle to the original wave (such as the photon labeled A in Fig. 21.11) it will be bounced when it hits the mirror on the left. When it hits the mirror on the right, it will bounce further and will follow the path labeled A,B,C in the diagram. Eventually, that particular photon and any photon traveling in the same direction will move out of the system of atoms. On the other hand, if a photon is emitted in precisely the same direction as the first wave that we drew (i.e., in a direction such that its motion is exactly perpendicular to the two mirrors), it will move back and forth through the population of atoms on the same line. In this way, it will not leave the system, but instead will continue to move back and forth and will continue to cause stimulated emission in other atoms that are in the system. Thus, if we have a population of excited atoms, the stimulated emission process will eventually build up a very large number of photons moving back and forth perpendicular to the mirrors.

When this wave becomes strong enough, it will "leak" through the mirrors and will be seen as an external beam of light. This beam of light will have several very important properties. First of all, all of the photons in the beam will be in phase. Secondly, this beam will not spread out very much, since only light that is moving precisely in the direction perpendicular to the mirrors will leak out of the laser. This means that laser light does not "fan out" the way light from a flashlight does. It is precise, pencil-thin, straight line. In fact, when lasers were shown on the moon in a series of experiments by scientists at NASA, the laser beam had a lateral extent of only a yard or so by the time it reached the moon.

Thus, the laser is a device that will produce a very fine, very energetic beam of light. The only question about the laser that we have not answered is the question of how one goes about getting a population of atoms in an excited state in the first place.

You will remember that when we discussed the Bohr atom we pointed out that electrons got into excited states by a number of processes. The most common one is simply by collisions with other atoms. We know that when the temperature of an object is raised, the atoms in it move faster, collide more frequently, and have more energy available in each collision. Thus, by heating a material we produce more atoms in the excited state than would be there normally, so we could "pump" electrons into an excited state in a sample of atoms by heating it. If we could find atoms in which the electrons normally stayed in the excited state for a fairly long time (these states, which last for a long time, are sometimes called "metastable states"), then heating the system would produce a large number of atoms in an excited state, and this particular configuration of atoms would last for a long time. Although this method of "pumping" a laser is not used in most laboratory and commercial devices, it does illustrate the general method of making a laser "lase." Some outside agency adds energy to the system, and some mechanism is exploited which allows this energy to be transferred to the individual atoms, raising electrons to higher Bohr orbits. The photons are then admitted to the system and the amplification process proceeds as described above.

The applications of lasers are far too numerous to be studied exhaustively here. We shall mention a few just to give some idea of the tremendous impact this device has had on modern society. The fact that we are capable of producing a very intense, very finely focused beam of light means that we have a new device for delivering energy to very small systems. For example, a common accidental condition in human beings is called the "detached retina." This happens when the retina (a thin film of cells in the back of the eye) comes loose from its normal moorings. Before the invention of the laser, this particular condition had to be corrected by going in and actually cutting the retina, pasting it back onto the wall of the eye socket, and allowing the scar tissue from the cut to seal the retina in its original position. Today, the same operation is performed with a laser that is shown through the pupil of the eye. Small burns are made on the retina, and then the scar tissue from these burns reattaches the retina to the eye. It is a much simpler and much less traumatic operation than the original one.

336

Another application of the laser comes from the fact that the laser is a tool, or device, which produces a beam of light that does not disperse when it travels over great distances. This means that it is very useful in doing long-distance surveying. The beam of light that comes from a laser is a straight line. If we were dealing with something like the Bart tunnel (a tunnel built under San Francisco Bay for the Bay Area Rapid Transit Corp.), we would have a situation in which we would want to be able to get a level, straight line over a distance of several miles underground where there are no external landmarks to guide us. The surveying for this particular job was done with a laser. Similarly, once the surveying was done, putting a laser on one end of the tunnel and shining it through to the other end allows the engineers to detect very precisely the presence of any settling or movement of one end of the tunnel with respect to the other. If the laser is located on a piece of material that is settling slightly into the ground, the beam at the other end of the tunnel will move down along with the laser. This small motion of light is easy to detect. Thus, the laser is used continuously to monitor settling in large structures such as the tunnel under San Francisco Bay.

There are many other applications of modern quantum mechanics that could be discussed, but the laser is a good example, and one that has all the features of the other examples. It is a device that could not be built without a precise knowledge of quantum mechanics. It is a device that has tremendous impact in fields as widely separated as medicine and structural engineering. It is a device whose impact could not have been foreseen either by the men who developed the laser or by the men who developed quantum mechanics. It is an extremely good example of the way that basic knowledge in one area (in this case, in the area of the structure of the atom) gets translated into practical benefits for society at large.

SUGGESTED READING

All of the standard texts have chapters on quantum mechanics and the uncertainty principle. In addition, you may want to look at the following:

Gamow, George. *Mr. Tompkins in Wonderland.* New York: Macmillan, 1940. An amusing and whimsical account of a world where Planck's constant is so large that the effects of the uncertainty principle are seen in everyday life.

Gamow, George. *Thirty Years that Shook Physics.* New York: Anchor Books, Doubleday, 1966. Contains anecdotal and biographical descriptions of the work of Pauli and Heisenberg, and very readable explanations of the material presented in this chapter.

Schalow, Arthur L. "Optical Masers." *Scientific American, 1961, 204* (52). An account of the development of the laser by one of its originators. From the title, it is obvious that the paper was written before the current name became popular.

QUESTIONS AND DISCUSSION IDEAS

1. One way of estimating the value of Δx in the Heisenberg uncertainty relation is to say that Δx is about the size of the object being discussed. Given this method, use the uncertainty relation to calculate the minimum error in velocity for each of the following:
   a. a car with mass 2000 Kg (about the size of a Volkswagen) and measuring 3 meters in length
   b. a baseball 10 cm across and mass 100 grams
   c. a proton which is 10^{-13} cm across and mass 1.67×10^{-24} g.

2. A typical velocity for a car or baseball might be 30 meters/sec (around 60 mph), while for a proton, it might be something like the speed of light. Calculate the percentage error introduced by the Heisenberg principle in each of the examples in Problem 1, and use these percentages to comment on the effect that the principle has in classical versus quantum physics.

3. Oxygen is an atom whose nucleus has a charge of 8, while sulphur has a nucleus of charge 16.
   a. How many electrons are there in the first three shells of these atoms?
   b. Would you expect these atoms to have similar chemical properties? Why or why not?

4. Repeat Problem 3 for boron (charge 5) and magnesium (charge 12).

5. It is observed that when light of a given wavelength is shown on a screen with holes in it, a diffraction pattern like that pictured on p. 339 is observed with the intensity at C being twice the intensity at A and three times the intensity at E. Now electrons whose de

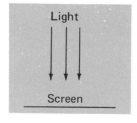

Light

Screen

338

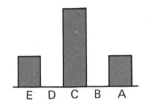

Broglie wavelength is the same as that of the light are shown on the screen.

    a. Where will most of the electrons hit?

    b. Will any electrons hit at B and D? Why or why not?

    c. Will any electrons hit at A and E? Why or why not?

    d. What will be the relative number to hit at A, C, and E?

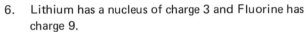

6. Lithium has a nucleus of charge 3 and Fluorine has charge 9.

    a. How many electrons are in the last unfilled shell of each of these atoms?

    b. What kind of chemical bond will these two atoms form?

7. Some molecules are arranged in such a way that it is possible through collisions to rearrange the atoms in the molecule. Over a relatively long time period, the rearranged atoms will make a quantum leap back to their original state, emitting microwave radiation when they do so.

    a. Describe what would happen if a microwave of the emitted frequency started to move through a collection of such molecules?

    b. Could a laser-like device be made from these molecules? Describe how this would have to be done.

    c. What does MASER stand for?

8. Consider the molecule of sodium chloride discussed in the text. The charge of the electron in 1.6×10^{-20} units, the average separation between the atoms in the molecule is about 10^{-8}cm, and the constant in Coulomb's law is the square of the speed of light.

    a. Using Coulomb's law, calculate the force between the two atoms in a molecule of sodium chloride

    b. In order to get some idea of the strength of this force, we note that in these units, the gravitational force on one gram of matter is 980 units. Given the mass of the proton from Problem 1, and the fact that the masses of sodium and chlorine are about 23 and 35 times that mass, respectively, calculate the gravitational force on the molecule.

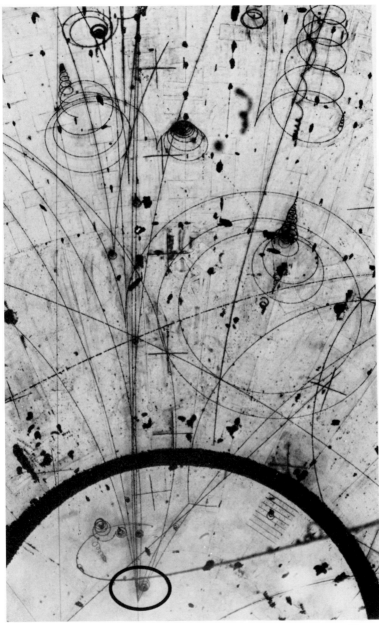

A photograph of the tracks of elementary particles in a bubble chamber. When a lot of particles come from a single point, as in the event circled above, we know that a collision has taken place and that the energy of the incoming particle (seen here as a single line entering the circle) has been converted into a spray of other particles. The tracks of the particles curve because there is a strong magnetic field in the chamber, and the "corkscrew" tracks are electrons or positrons.

Testing 10-watt argon laser for use in earth-to-space tracking.

CHAPTER XIII

THE NUCLEUS

"Anyone who thinks that anything useful will ever come out of nuclear physics is talking pure moonshine."
Attributed to Ernest Rutherford

Wilhelm C. Roentgen (1845-1923).

A. The Discovery of Radiation and What It Meant about the Nucleus

We have already discussed the experiment done by Rutherford which gave direct proof that inside the atom there was a tiny, dense, and very heavy nucleus which contained most of the matter in the atom. Actually, this discovery was not an isolated, unique proof of the existence of such a nucleus. In the years preceding Rutherford's experiment a great deal of evidence had been accumulated that there was something new—something never before seen in nature—going on inside of the atom. This "something" was the discovery that the atom was capable of giving off radiation other than light, and that this radiation interacted with matter in a very different way from the way light interacted with matter.

The first discovery of what we now call "ionizing radiation" (the reason for this name will become obvious later) was made by Wilhelm Roentgen at the University of Wurzburg in 1895. It actually happened by accident. Like many scientists at that time, Roentgen was studying what were called "cathode rays." This was the name given to electrons that were moving through a tube filled with gas (see Fig. 13.1). These "cathode ray tubes" were the forerunners of the modern television tube and of the vacuum tubes, which used to be the backbone of the electronics industry. In 1895, they were still a very new phenomenon, and the idea that electrons could be studied by the way in which they behaved in such a tube was considered to be very important. This device is essentially the same as the apparatus for investigating discrete spectra discussed in the previous chapter. During the course of his experiment, Roentgen wanted to check that no light was escaping from his apparatus. Consequently, he surrounded the apparatus with black cardboard and then turned on the tube. By accident, he had left a sheet of fluorescent metal lying on a table near his apparatus. When he turned the electron current on, he noticed that the sheet of metal began to glow, even though no light was coming out of the tube. In other words, something had left the cathode ray tube, gone through a sheet of cardboard, and reached the metal plate which had been inadvertently left on the nearby table. This unknown kind of ray, which was capable of penetrating cardboard, was called an X-ray by Roentgen. The

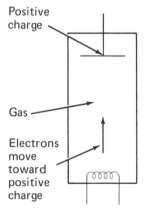

Positive charge

Gas

Electrons move toward positive charge

Figure 13.1.

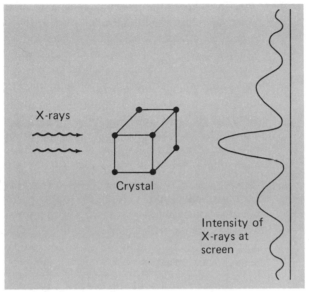

X-rays

Crystal

Intensity of
X-rays at
screen

Figure 13.2.

name was chosen to emphasize the fact that its nature was very unexpected and also very mysterious.

It did not take very long to discover that the source of the X-ray was stray electrons in the cathode ray tube hitting the glass in the side of the tube. It did not take very long, either, for scientists to begin asking the question "What is an X-ray?" On the basis of what we already know about quantum mechanics, it is obvious that the first question that would have been asked is "Is the X-ray a wave or a particle?" This question was answered by M.T.F. von Laue in 1912. He created a beam of X-rays and directed it at a crystal, as shown in Fig. 13.2. The crystal can be thought of as a series of slits through which the X-rays are allowed to propagate. In other words, what von Laue did for X-rays is precisely the kind of experiment we have already discussed in the context of determining whether electrons or photons are particles or waves. He did an experiment to see whether they would exhibit interference patterns. What he found was sketched in the figure. X-rays did indeed exhibit interference patterns, and were in every way similar to light except that they were much more energetic. In other words, von Laue discovered that the mysterious X-rays were simply a very energetic form of electromagnetic radiation.

343

In modern terminology, energetic photons such as X-rays are often called "gamma" rays. Although it was not realized at the time, we now can understand what an X-ray is in terms of our quantum picture of the photon. We remember that according to Planck's relation, the energy of an electromagnetic wave is proportional to its frequency. Therefore, if the X-ray is simply an ordinary photon with much higher energy, it must be a photon whose frequency is much higher than that of ordinary light. But where would a photon of this energy come from in an atom? In the old raisin bun atom, there was no way in which such a photon could be emitted. Even in the Bohr atom, there is no way that a photon of this type could be emitted from the electrons in the atom, since the largest electron jump possible (the jump from an infinite distance away from the atom to the lowest possible orbit in the atom) corresponds to light in the visible range. Thus, whether we look at the discovery of X-rays in terms of the picture of the atom that was available in 1905, or whether we look at the discovery of X-rays in terms of the picture of the atom that we now have, we come to the conclusion that the existence of X-rays implies a totally new kind of process from those with which we are familiar.

In 1896, in France A.H. Becquerel provided another important insight into the nature of radiation. He reasoned that since electrons striking a fluorescent glass tube in Roentgen's experiment produced X-rays, then materials that were naturally fluorescent (i.e., materials that normally glow in the dark when there is no light shining on them) might also give off radiation. To test this hypothesis, he performed what has become a classic experiment. He took a piece of fluorescent material containing the element uranium and other heavy minerals and brought it near an apparatus such as that shown in Fig. 13.3, in which a metal key was laid in front of a photographic plate. The entire apparatus was wrapped in black material to shield it from light.

It had already been discovered that X-rays had the property of exposing (or "fogging") photographic films. The reason for this will become obvious when we talk about the effects of radiation on materials. What Becquerel hoped to do was to show that the radiations coming from the uranium compounds would also fog a photographic plate, and hence to provide experimental evidence for his suggestion that fluorescent materials ought to give off this mysterious new kind

Fluorescent
material

Metal
key

Photographic
plate

Figure 13.3.

of radiation. What he found was that the photographic plate was exposed in the areas that were not in back of the key, but that the key had, by some process which was not understood at the time, prevented the radiation that fell on it from getting to the plate.

This discovery—that the radiation was capable of moving through materials such as cardboard and black paper— caused an immediate sensation in Europe. There were large spreads in major European papers dealing with the possibility that X-rays could be used to see through clothing. On the more serious side, medical teams quickly realized that the X-ray might give them a way of looking inside the human body without actually having to perform surgery. The applications of this particular line of investigation are well known to all of us.

The reader will have realized by this point that the X-rays studied by Roentgen and the X-rays studied by Becquerel came from entirely different sources, even though they were both simply high energy photons or gamma rays. We shall discuss the sources of X-radiation in some detail later, but we simply emphasize here the fact that Roentgen worked with X-rays that were created in his apparatus by electrons being stopped in the glass walls of his tube. Becquerel, on the other hand, was working with X-rays generated in nuclei found in natural materials. This distinction is one of the important points about his discovery, because it showed that the existence of radiation, far from being a freakish sort of thing only evident in specialized laboratories, was actually a phenomenon that was rather widespread in nature. Common materials, particularly those compounds of the heavy elements which were well known and had many commercial uses at the end of the 19th century, were found to give off radiation of some kind.

It was quickly discovered that in experiments of the type done by Becquerel the amount of fogging of the photographic plate was directly proportional to the amount of uranium used as a source of radiation. In other words, doubling the amount of uranium would double the amount of exposure, tripling the amount of uranium would triple the amount of exposure, etc. It was also known that if natural uranium ore (as opposed to chemically refined uranium compounds) were used in Becquerel's experiment, there was

345

roughly five times too much radiation. In other words, there was more radiation coming out of natural uranium ores than one would expect on the basis of the uranium content alone.

At this point, a rather remarkable person comes on the scientific stage. In 1898, Madame Marie Sklodowska Curie began looking at the problem of the excess radiation coming from natural uranium salts. She postulated that the extra radiation must be coming from some as yet undiscovered element which, like uranium, had the property of giving off radiation. Elements that had this property were called "radio-active." To test this hypothesis, she obtained about one ton of pitchblende from Czechoslovakia and began a careful chemical analysis aimed at identifying these new elements. The end result of this search was the discovery of polonium and radium, two elements whose existence had not been suspected before that time.

Madame Curie was the first person ever to win two Nobel prizes during a lifetime. She could be said to be the founder of the science of radiochemistry. She was also the first modern woman to make important contributions in the physical sciences, and her example has been followed by many other women. As a trail-blazer in the area of sexual equality, she often ran into difficulties which seem hard to understand today. For example, she was never elected to the French Academy of Science, even though many men whose accomplishments were vastly inferior to hers did achieve this recognition. English newspapers at the time took great delight in pointing out that this form of discrimination by the French Academy was based solely on her sex and not on her scientific achievements.

By 1899, then, it had been firmly established that a new type of radiation existed, and that this radiation did occur in nature. Furthermore, the new science of radiochemistry was becoming more and more competent in providing experimenters with sources of these new radioactive elements. In 1899, Ernest Rutherford (who at the time was working at McGill University in Montreal) did a series of extremely important experiments on this new radiation. Rather than concentrating on the source of the radiation, Rutherford decided to try to be more precise about the nature of the radiation. He performed a series of experiments in which he took a radioactive source (which in these experiments was radium) and shielded the source by placing different thicknesses of foils between the source and the detector. He found that some kinds of radiation were stopped by the thin foils and others were not. Both of these kinds of radiation were different from X-rays. The radiation that was easily stopped by a thin foil he called "alpha radiation," and the radiation that was not easily stopped by thin foils he called "beta radiation."

In a series of experiments culminating in 1907 in Manchester (and leading to the awarding of the Nobel Prize in physics in 1908), Rutherford proceeded to analyze these new kinds of radiation. He showed first that the beta radiation, if allowed to enter a cathode ray tube of the type we described in relation to Roentgen's experiment, had all the properties of an ordinary electron. Consequently, he identified the beta radiation (the radiation that had experimentally been determined to be more penetrating) with the electron. Just as the X-ray had turned out to be nothing more than an energetic photon, the beta ray turned out to be nothing more than an energetic electron.

The identity of the alpha radiation was somewhat harder to establish. It was obviously much heavier than the beta ray and was not capable of penetrating nearly as far through matter. In order to learn the identity of the alpha radiation, Rutherford set up an experiment such as that shown in Fig. 13.4. A material which gave off alpha rays, was placed near a tube in which known quantities of gases were present. In particular, Rutherford set the experiment up so that there was no helium gas in the tube when the experiment started. The tube was allowed to sit for a period of time during which alpha rays from the source would enter the tube. After this period of time was over, careful chemical analysis showed that there were now traces of helium inside the tube. Therefore, Rutherford identified the alpha particle with the helium atom. More precisely, since the alpha particle was known to have an electrical charge of +2, he identified the alpha particle with the helium atom minus its two electrons. In technical terms, this is called "doubly ionized" helium. In terms of the Bohr atom, or in terms of Rutherford's later experiments, we would say that the alpha particle was simply the nucleus of the helium atom, and was composed of two protons and two neutrons. However, at the time that Rutherford did the experiment, neither the existence of the nucleus, the proton, nor the neutron was even suspected.

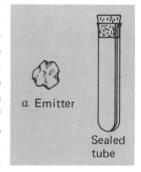

α Emitter

Sealed tube

Figure 13.4.

Thus, when Rutherford did his famous experiments proving the existence of the nucleus in 1911, there was already a long history of discoveries about the atom which needed to be explained and which could not be explained in terms of the simple raisin bun atom that had been previously suggested. If we ask the question "Where does all this radiation come from?" It is obvious that it must come from the nucleus. There is simply not enough energy in the electron circling the nucleus to give rise to either X-rays, beta rays, or alpha rays. In the case of alpha particles, it is obvious that they must come from the nucleus since there are no protons or neutrons anywhere in the atom except in the nucleus. Students are often confused, however, by the fact that beta rays (which are, after all, simply energetic electrons) do not come from the electron orbits in the atom but come from the nucleus instead. We shall discuss the process by which a nucleus can emit an electron in detail later on, but at this point we simply have to stress the fact that the beta ray that we see coming out of an atom does not originate in the electron orbits. It is *not* a normal Bohr electron that has somehow acquired a lot of energy and been kicked out of its orbit. After a nucleus has emitted a beta ray, the normal orbital electrons are still all present, at least briefly.

The same thing can be said for the X-rays that come from an atom. They do not result from electron transitions of the type that give rise to visible light. Their energies are so high that they could only come from inside the nucleus. There is not enough energy in the electron orbits to produce an X-ray.

By the time that Rutherford's nuclear theory had been accepted by scientists, then, there were already many questions that needed to be answered. There were questions having to do with the existence of radiation, and other questions having to do with the very possibility of the existence of a nucleus. It is to these questions that we now turn.

B. The Enigma of the Nucleus: How Can It Possibly Exist?

Once we accept the picture of the atom as a structure in which the negative electrical charge is carried by electrons

that circle in orbits around a tiny, dense nucleus, we immediately run into problems. Let us ask ourselves what the nucleus could possibly be like in such a picture of the atom.

In the first place, the nucleus (according to Rutherford's experimental data) is incredibly small compared to the
atom. In fact, the atom is almost entirely empty space. We
can think of the size of the nucleus in relation to the atom as
being analogous to the size of the sun in relation to the solar
system. In metric units, a typical nucleus might be 5×10^{-13}
cm. across.

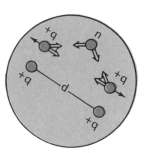

Figure 13.5.

We remember that according to Coulomb's law the
force between two electrical charges increases as the distance
between the two charges decreases, just as the gravitational
force between two masses increases as the distance between
them decreases. Since all the negative electrical charge in the
atom resides in the electrons, and since these electrons are, to
all intents and purposes, a large distance away from anything
inside the nucleus, we are faced with the question of how the
nucleus could possibly exist. After all, we are taking all of
the positive charge of the atom and cramming it into a small
space where Coulomb's law tells us that incredibly strong
repulsive forces must exist. In other words, if we think of
two little bits of charge inside a nucleus (see Fig. 13.5) there
must be a repulsive force between these two little bits of
charge given by

$$F = k \frac{q^2}{d^2}.$$

We know that the nucleus does not come apart in the
normal course of affairs. This means that even though there
is a strong electrostatic repulsion between the two bits of
charge in the above diagram which, if it were acting by itself
would cause them to be repelled from each other, in the
normal state of affairs inside the nucleus these bits of charge
do not move away from each other. In other words, we have
a situation in which a large electrostatic force is operating to
push two pieces of charge away from each other, but they do
not move apart. The only way that we could understand a
situation such as this would be to postulate that there must
be another kind of force—a force that we have not yet encountered in nature—which tends to hold the two pieces of
charge together and to bind them into the structure that we

349

call the nucleus. This conclusion is an inescapable consequence of the Rutherford experiment, just as the existence of stable electron orbits was an inescapable consequence of the same experiment.

The new kind of force must be very powerful indeed in order to overcome the large electrostatic repulsions that must exist inside the tiny positively charged nucleus. This new kind of force is called the "strong force." The study of the strong force (or the "strong interaction," as it is more usually called) has been one of the major concerns of physics since the 1920s, and remains in the forefront of modern physics research today. It is a force that we do not really understand very well at the present time, although we know enough about it to understand the structure of the nucleus. In a later chapter we shall deal more fully with the question of what the force is and how it works. For the purposes of our discussion right now, however, we shall simply note that such a force must exist in order for the nucleus to remain stable.

In discussing the atom, the primary problem we have faced was finding out what the structure of the atom was. The situation with the nucleus is no different. We know that the nucleus is a tiny blob of positive charge located at the center of the atom. If we want to go beyond this rather crude description of the nucleus, we have to ask ourselves two questions. First, we have to ask what it is that makes up the nucleus (i.e., what kind of particles or material exists inside the nucleus) and second, we want to know how the material inside the nucleus is arranged.

The first question—the one concerning what actually is inside the nucleus—was answered by Rutherford in a series of experiments in 1919 in which the nuclei of different atoms were bombarded with alpha particles and the debris resulting from the collisions was analyzed. It was found that many of the particles that were knocked out of the nuclei were actually identical with the nucleus of the hydrogen atom, having a positive charge of one unit and having the correct weight. Rutherford proposed the name "proton" for these particles (from the Greek meaning "the first one"). He argued that a nucleus having a positive charge of Z must have within it Z protons, and hence the atom must also have precisely Z electrons.

350

He also noted that many atomic nuclei weighed more than Z times the weight of the proton. For example, carbon has six protons, but the nucleus weighs about as much as 12 protons. Rutherford suggested that there must be another particle in the nucleus that weighs as much as the proton but has no electrical charge. He called this particle a "neutron." Most physicists accepted the idea that the neutron actually did exist, even though it was not until 1932 that Sir James Chadwick discovered it. In this picture, the carbon atom would have a nucleus containing six protons and six neutrons.

The existence of neutrons gives us a little insight into how the nucleus can be held together by the strong force. The electrical repulsion that tends to blow the nucleus apart exists only between protons. Neutrons, being uncharged, are not affected by this force. On the other hand, both neutrons and protons are equally affected by the strong force. Thus the addition of neutrons to the nucleus provides us with a particle that contributes to the forces binding the system together, but is unaffected by the forces tending to make it fly apart.

The answer to our first question, then, is that the nucleus is composed of two kinds of particles, the proton and the neutron. Let us turn to the second question we posed, which had to do with the way in which protons and neutrons are arranged inside the nucleus.

In order to understand how we came to a knowledge of the architecture of protons and neutrons inside of a nucleus, let us go back for a moment and consider a way in which we might have discovered the existence of Bohr orbits experimentally. Suppose that instead of the historical procedure that followed from the Rutherford experiment, a different series of experiments had been done on atoms. Suppose that instead of bombarding atoms with heavy particles like the alpha, Rutherford had instead bombarded atoms with light particles like the electrons. Suppose further that Rutherford had conducted an experiment such as that sketched in Fig. 13.6. Suppose that electrons of a known energy E were incident on an atom from the left, and that these electrons interacted in some way with the atom and were then detected at the detector at the right. Suppose that the energy of the electron after it comes out of the atom was given by E'.

Electron

E

Electron

E$'$

Detector

Figure 13.6.

Then the quantity E-E$'$ (the energy of the electron after the interaction subtracted from the energy of the electron before the interaction) would be the energy that was lost by the electron in the interaction. Since the only two objects that are taking part in this interaction are the electron and the atom, if the electron loses a certain amount of energy, the atom must gain that same amount of energy. This follows from the conservation of energy.

Suppose that Rutherford had decided to plot his data in a particular way. Suppose that he had decided to count the number of electrons that lost a certain amount of energy, and to present the results of his experiment in a graph such as that shown in Fig. 13.7 in which the vertical axis is the number of electrons involving the particular amount of energy loss E-E$'$ and the horizontal axis is the measure of E-E$'$. According to the simplest version of the raisin bun atom theory, there is no reason why an electron could not lose any amount of energy whatsoever. For example, the electron could lose 10% of its energy, 11% of its energy, 12% of its energy, etc. This follows from the way in which physicists thought of the raisin bun atom. If you have a bunch of electrons floating around inside of a diffuse positively charged background, then when another electron comes in from the outside and collides with this arrangement, the springs in electrons can absorb any amount of energy at all. Thus, if the simple raisin bun atom had existed, Rutherford would have expected to get a curve such as that shown in Fig. 13.7, in which there is a smooth, continuous line resulting from the plot in the experimental data. Such a line would be called a "continuous energy loss spectrum," and would correspond to the statement that there was no restriction on the amount of energy that an incoming electron could lose to the atom. This is completely equivalent to the statement that there is no restriction on the amount of energy that the atom can gain from the incoming electron.

Number of electrons

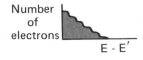

E - E$'$

Figure 13.7.

352

If we had a Bohr atom, on the other hand, the situation would be quite different. In order for the atom to gain energy in the Bohr theory, it is necessary for an electron in the atom to make a quantum jump to a higher orbit. Thus, the incoming electron can lose energy only in discrete units. It can lose no energy (which would correspond to a collision in which no electrons were shifted around between orbits in the atom), or it can lose the amount of energy corresponding to moving the electron up one orbit, or it can lose the amount of energy corresponding to moving the electron up two orbits, etc. It cannot lose an amount of energy corresponding to moving an electron up half of an orbit or a third of an orbit. Thus, if we made the same graph for scatterings from a Bohr atom as we did for scatterings from the raisin bun atom, we would get something like that pictured in Fig. 13.8. There would be a large number of electrons for which the energy loss was zero. These electrons would have interacted with the atom in such a way that they did not change the orbital structure of the electrons inside the atom. Then, at a particular energy loss (which we have labeled by the numeral 1 in Fig. 13.8, we would again see electrons coming out of the interaction. These are electrons that would have lifted one of the orbital electrons in the atom to the first higher orbit. No electron could come out of the interaction with an energy loss between zero and this energy. Such an energy loss would correspond to moving an atomic electron to an orbit that is not allowed. Thus, the shell structure of the orbital electrons in an atom would lead to an energy loss spectrum that was not continuous, as in the first example which we discussed, but that was discrete, as shown in Fig. 13.8. Had Rutherford performed this experiment instead of his famous alpha particle experiment, he probably would have come to the conclusion that there must exist shells inside of the atom, and that electrons could exist in those shells and nowhere else.

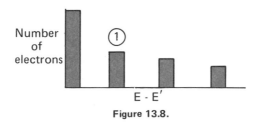

Figure 13.8.

Number
of
electrons

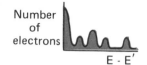

E - E'

Figure 13.9.

Thus, we see that looking at the energy that is lost by a particle interacting with a system is one way in which we can deduce things about the structure of the system. In particular, if we find upon doing an experiment that the energy loss spectrum is discrete, such as that shown in Fig. 13.8, we would be justified in concluding that there must exist something like Bohr orbits or shell structure inside of the object we are probing.

It is possible to make studies of nuclear structure by using this technique. A large number of experiments have been done in which some probe (usually an electron) is directed against the nucleus and the energy loss of that electron when it emerges from the nucleus is measured. When such experiments are done, results such as that pictured in Fig. 13.9 are observed. It is found that energy loss experiments performed on a nucleus give rise to a discrete energy loss spectrum. From our discussion of the Bohr atom, we know that this must imply the existence of some kind of orbits inside of the nucleus itself. In other words, rather than being a chaotic, random collection of protons and neutrons, the nucleus is instead a rather highly ordered system, just as the atom itself is a highly ordered system.

Our modern picture of the nucleus, which was derived in part from experiments similar to that discussed above, is called the "shell model" of the nucleus. The model was proposed by Maria Goeppert Mayer and J.H.D. Jensen in 1949. It is basically similar to the Bohr picture of the atom. In the shell model of the nucleus, we picture the protons and neutrons as existing in orbits around the center of the nucleus, much as electrons exist in orbits around the nucleus in the atom. The force holding the protons and neutrons in their orbit is, of course, the strong interaction. The strong interaction plays the same role inside the nucleus that the electromagnetic attraction between the nucleus and the electrons plays in the Bohr atom. It is the force holding the system together and providing the cohesiveness needed to maintain a stable system.

There is an important difference between the strong force in the nucleus and the electromagnetic force in the atom, however. In the atom, the nucleus sits at the center. This means that there is a strong *central* attraction between the electrons and the nucleus. Each electron interacts primarily with the nucleus and is attracted primarily toward it.

The repulsion between the electrons in the Bohr orbits, while it exists, is very small because of the large distances between electrons. Hence, it does not play a major role in determining atomic structure.

When we look inside of the nucleus, however, there is no "center" as there is in an atom. Although the nucleons themselves may travel in orbits, if we look at the very center of the nucleus, we would not find one central body holding the whole thing together. The only forces that exist inside the nucleus are the strong and the electromagnetic forces. The strong forces act equally between protons and neutrons. The electromagnetic forces are repulsions between protons. However, there is no central body that attracts all of the protons and neutrons. This situation is illustrated in Fig. 13.5.

By some complex mechanism (which was not understood until the late 1950s) the forces between individual protons and neutrons inside the nucleus arrange themselves in such a way that the protons and neutrons find themselves moving in orbits. Actually, the orbits of the protons are slightly different from the orbits of the neutrons. This follows from the fact that the protons have an extra force acting on them (the electromagnetic repulsion), while the neutrons do not.

Our picture of the nucleus, then, is something like this: There are a series of orbits, similar in type to the electron orbits in an atom, for the protons. Because the Pauli principle operates between protons, each orbit can sustain only a finite number of protons. As each orbit is filled, the next protons to be added come in the next highest orbit, etc. The neutron orbits are also filled according to the Pauli principle, but they form an orbital system that is separate from and independent of the proton system. The nucleus, itself, then, is made up of a series of proton orbits and a series of neutron orbits which are overlaid one on top of the other. For example, in Fig. 13.10 we show the orbital system for the element carbon, which has six protons and six neutrons.

With this picture of a nucleus, we can understand the results of electron energy loss experiments. Electrons that collide with the nucleus can either lose no energy or they can lose enough energy to lift either a proton or a neutron to the next higher shell. They cannot lose the amount of energy equivalent to lifting a proton halfway to the next shell or a neutron halfway to its next shell. Thus, a discrete energy loss

355

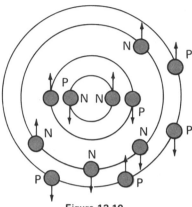

Figure 13.10.

spectrum will result from such an experiment. By the present time, a large amount of other data besides electron loss experiments has been accumulated by nuclear physicists, and the shell model (with some minor corrections) is now accepted as the correct picture of the structure of the inside of a nucleus.

Although we now understand the structure of the nucleus, it is obvious that there is an important factor lacking in our understanding, and this is an understanding of the force that causes that structure. We find that in order to understand the nucleus completely, it is necessary that we understand the strong interaction. Before turning to a discussion of the strong interaction, however, let us first discuss some of the important consequences of the existence of the nucleus, of the existence of nuclear shells, and of the phenomenon of radioactivity.

C. Radioactivity: The Philosopher's Stone

Up to this point, we have discussed radiation and radioactivity only in its historical context. We now turn to a discussion of what the existence of radioactivity means in terms of nuclear and atomic structure. We shall find that the phenomenon of radioactivity is intimately tied to processes whereby one chemical element is changed into another. In a sense, then, radioactivity can be thought of as the key to the transmutation of the elements that the medieval alchemists sought but were never able to find.

356

Let us begin by discussing X-rays, since they are the simplest form of radiation to comprehend in terms of the nuclear shell model. We know that protons and neutrons inside the nucleus exist in orbital shells, much as electrons and atoms exist in electronic orbital shells. The primary difference between the electronic shells in the atom and the nuclear shells in the nucleus are that they are of such different sizes, and hence they involve very different energies. Roughly speaking, the size of a nuclear orbit is about 1/100,000 times the size of an electronic orbit. This is just another way of stating what we already know about the atom—that it is mostly empty space. The energies involved in the two different kinds of orbit are, therefore, very different.

In the case of the electronic orbits in an atom, the energies involved in moving from one shell to another, or in moving an electron from the orbit to a place outside of the atom, is typically on the order of a few electron volts. The electron volt is the amount of energy that it would take to move something that had the charge of one electron through one volt of potential energy. (To move one electron from the positive side of a car battery to the negative side of a car battery, for example, would take 12 electron volts.)

The forces binding the protons and neutrons inside of the nucleus, however, are much stronger than the forces holding the electron into the atom. Consequently, the energies involved in moving protons and neutrons between orbits are much bigger. Whereas energies involved in electron orbits are typically on the order of a few electron volts, energies involved in nuclear orbits are typically on the order of thousands and even millions of electron volts. Thus, when we have a situation such as that pictured in Fig. 13.11, in which one of the protons or neutrons inside of the nucleus finds itself in an excited state and subsequently makes a quantum jump to go back to its ground state, the energy that will be emitted (which is simply the difference in energy between the energy of the initial orbit and the energy of the final orbit) will be much greater than the energy involved in a quantum leap between electrons in the outer fringes of the atom.

This process, which allows the nucleus to emit electromagnetic waves in a manner very similar to the way in which atoms emit light in the Bohr model, gives rise to nuclear gamma or X-rays. The fact that gamma rays are so much

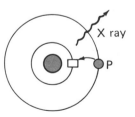

Figure 13.11.

357

more energetic than ordinary light arises from the fact that they originate in transitions between orbits involving a much higher energy than the transitions between the electronic orbits in an atom. Thus, when a sample of material is found to emit X-rays in a spontaneous and natural way, we know that what is happening is that protons and neutrons inside the nuclei are making quantum jumps to lower levels and emitting electromagnetic radiation whose energies are characteristic of the difference in energy between the levels of the jump.

As we have pointed out, however, this is not the only way that X-rays can be generated. It is obviously not the way that X-rays were generated in Roentgen's original experiment, since he did not have at his disposal a lump of radioactive material that emitted gamma rays. To understand how Roentgen generated X-rays, we have to remember an important fact about classical physics. You will recall that one of the arguments about the impossibility of the existence of the nucleus rested on the fact that in classical physics whenever a charged object is accelerated it must emit electromagnetic radiation. If you think of the electrons in Roentgen's apparatus, you know that they were moving with a particular velocity. When they hit the glass walls of the tube, they were stopped. This means that the electrons were decelerated very quickly when they collided with the glass walls. A deceleration can be thought of as a negative acceleration. In other words, according to classical physics, an electron that is being decelerated must also give off electromagnetic radiation. It was this radiation, associated with the stopping of the electron in the glass walls, that Roentgen detected as the X-ray in his apparatus.

Both of these ways of generating X-rays are found in nature and have found uses in everyday life. For example, the X-rays that pass through your body when you have a clinical diagnosis done in a hospital or in the dentist's office are the result of an apparatus that accelerates electrons and then stops them suddenly. On the other hand, many clinical treatments of cancer involve the use of radiation to destroy cancerous tissue. Often, this radiation comes from lumps of naturally radioactive material. In this case, the X-rays that are being used to treat the disease arise from internal transitions within a nucleus.

There are many other uses of naturally occurring X-rays, and they are discussed in the books listed at the end of this chapter. The use of trace elements both in medicine and in biological research is well known. There are many industrial uses for X-rays in which flaws in metal castings are detected on assembly lines by the use of naturally occurring X-rays as well. Finding out more about uses for X-rays in nature is left as an exercise to the student.

We have seen, then, that X-rays arise from internal transitions inside of the nucleus. However, these transitions do not change the nature of the nucleus in question. If we started out with a nucleus like Carbon 12, which has six protons and six neutrons, and if for some reason a transition were made that gave rise to X-radiation, we would still have six protons and six neutrons left in the nucleus at the end of the transition. We would have started out with an excited carbon nucleus and wound up with a carbon nucleus in its ground state, but we would have had a carbon nucleus throughout the entire process. As far as the electrons in the atom were concerned, they would see a nucleus of a given mass and of charge +6 throughout the entire transition, and the nuclear process would take place without the electrons in the outer orbits being affected very much. Consequently, the chemical properties of the atom, which have to do only with the electrons, would not change following the emission of nuclear X-rays. As far as an external observer is concerned, the emission of X-rays has no effect on anything other than the nucleus itself.

The situation with alpha and beta radiation is quite different. Let us consider alpha radiation first. We know that Rutherford proved that the alpha particle that is emitted by the nucleus consists of two protons and two neutrons. Let us consider what happens to an atom when its nucleus emits an alpha particle.

Let us take, for example, the carbon atom that we discussed in the above paragraph. (As a matter of fact, the carbon atom does not normally emit an alpha particle, but this exercise will serve to illustrate what happens to atoms that do emit alpha particles.) In Fig. 13.12a, we see the carbon atom with its full complement of six protons and six neutrons, and six electrons in orbit around the nucleus. Suppose that something happens to cause the carbon atom to emit an alpha

359

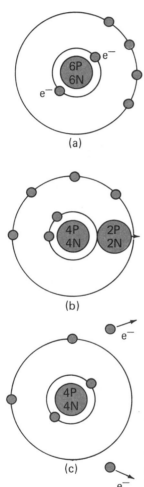

(a)

(b)

(c)

Figure 13.12.

particle. An alpha particle, as we know, is composed of two protons and two neutrons. In Fig. 13.12b, we show the atom just after the emission of the alpha particle. There are still six electrons in the outer orbits, but the nucleus has changed drastically. The original six protons and six neutrons have been divided into two sections. One section, the alpha particle, consists of two protons and two neutrons, and the remaining nucleus (the part that is left behind) now has four protons and four neutrons. There is not nearly enough energy inside of the nucleus to create an alpha particle from scratch, so when the alpha particle comes out, it must be made from constituents of the original nucleus.

In Fig. 13.12c, we show the atom as it exists long after the alpha particle has gone. After the emission of the alpha particle, the charge of the nucleus has changed from +6 to +4. This means that the electrons in the orbits around the atom suddenly find that they are less strongly attracted to the nucleus than they were before. In fact, this particular nucleus (which corresponds to the element beryllium) is capable of holding only four electrons in orbit, rather than the original six. Consequently, the two extra electrons will wander off—perhaps they will be scraped off in collisions with other atoms, or they will just drift away from the atom in one way or another. The net result is that whereas before the collision we had an atom with six electrons in its outer electronic shells, after the reaction we have an atom with four electrons in the outer shells. This new atom will have markedly different chemical properties than the original atom. (Why?) In fact, it will be an entirely new element. We started the reaction with an atom of carbon, and we ended the reaction with an atom of beryllium.

In other words, when a charged particle is emitted from the nucleus, the character of the nucleus and of the entire atom changes. The nuclear reaction takes place very quickly, but the electrons in the atom eventually rearrange themselves when they find that there is a new nucleus at the center of the atom. The process by which one chemical is changed into another is called "transmutation." From the physical point of view, a transmutation occurs when the nucleus of the atom changes. Changes in the electronic structure that do not affect the nucleus do not transmute an element. For example, removing one electron from the carbon atom in the

360

above example would not change its identity. We would simply have ionized carbon. However, changing the identity of the nucleus itself causes the electrons (at a later time) to rearrange themselves in such a way that we are dealing with an entirely new chemical element.

The situation with beta radiation is very similar to that of alpha radiation. Although the process by which a nucleus gives off an electron is rather different from the process by which it gives off an alpha particle, the net result is the same— the transmutation of elements. In fact, the process of beta decay is an example of what is called a "weak interaction." This will be discussed when we deal with elementary particles. For the time being, however, let us concentrate our attention on what happens to a nucleus when an electron is emitted.

The first question to ask is where in a nucleus an electron could come from. We know that the primary building blocks of the nucleus are the protons and the neutrons. Neither of these particles could be mistaken for an electron. Thus, there must exist a reaction in nature by which one of these two particles can give off an electron. This follows from the undeniable fact that beta decay is observed in nature.

Obviously, the proton cannot give off an electron. If the proton did give off an electron, the conservation of charge would require that whatever was in the final state of the interaction besides the electron have charge +2, since the total charge of the system before and after the reaction must be +1. We know of no particles in the nucleus with charge +2. This means that the origin of the beta rays must be the neutrons.

In fact, it is very easy to see how a neutron could give rise to beta radiation. The charge of the neutron is zero. If there were some reaction by which a neutron produced an electron, which has negative charge, then the object that must exist in the final state along with the electron would have to be something that weighed about as much as the neutron and had a positive charge. In this way, the negative charge of the electron and the positive charge of the "something else" would balance and give the same net charge—zero— that the neutron had at the beginning of the interaction. The identity of the "something else" is obvious. It must be the proton. In other words, in order for a nucleus to give off beta radiation, there must be a reaction of the type where the

$$n \rightarrow p + e^- + Q ,$$

neutron in the initial state is converted to an electron (which is what we call beta radiation) and a proton along with any collection of particles (which we call Q) that are allowed by conservation of energy and momentum and that have no charge. We could infer the existence of such a reaction from the simple fact that electrons are observed to come out of some nuclei. Actually, once the neutron was discovered in 1932, it was seen directly that a neutron by itself is not stable. It decays in a matter of seconds into a proton plus an electron, plus a third kind of particle called the neutrino (which we shall discuss later), which has zero charge. When this process occurs in a free neutron, we call it the beta decay of the free neutron. When it happens to a neutron inside of a nucleus, we call it the nuclear beta decay.

Let us go back to the hypothetical carbon atom, which we used to illustrate alpha decay, to understand what happens when beta decay occurs. In Fig. 13.13a, we show the unperturbed carbon atom. It has six protons, six neutrons in the nucleus, and six electrons in the standard Bohr orbits around the outside. If a beta decay occurs, one of the neutrons in the original carbon nucleus emits an electron and a neutrino and changes itself into a proton. This is shown in Fig. 13.13b.

The electron that is emitted is moving very quickly; it rapidly leaves the atom and disappears from the scene. What we have left, then, is what we show in Fig. 13.13c. We have a nucleus which, instead of six protons and six neutrons, now has seven protons and five neutrons. As in the case of alpha decay, the atomic electrons at first do not change their configuration or number immediately because of the nuclear reaction, but over a period of time the six electrons that were there originally will rearrange themselves, and a seventh electron will be picked up from the surroundings to balance the seven positive charges which now exist inside the nucleus. Thus, although we started out with an atom of Carbon 12, we now wind up with an atom that has moved over one place to the right in the periodic table. This atom has seven electrons rather than six, and in fact is simply a species of nitrogen. Once again, we have seen that the existence of radiation implies the transmutation, or changing around, of chemical elements.

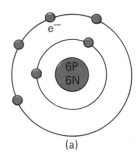

(a)

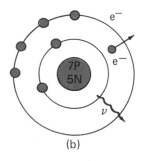

(b)

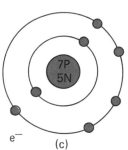

(c)

Figure 13.13.

362

One point that commonly causes confusion in dealing with nuclear beta decay is the question of why the electron emitted from the nucleus doesn't simply join up with the orbital electrons to balance charges. The answer to this question, again, lies in the enormous energies available in the nuclear process. The emitted electron is moving much too fast to be captured into an orbit. Consequently, it escapes completely from the nucleus, and a slower moving electron from the surroundings is eventually captured to fill the newly formed vacancy created by the beta emission.

D. Radioactive Systematics: Isotopes, Half-life, and Decay Chains

Now that we understand what the radioactive process does to an individual nucleus and to the atom in which that nucleus is situated, we can go on to talk about real radioactive processes in nature. Before we do so, however, there is one term that we need to understand.

We know that the number of electrons in the outer regions of the atom is what determines its chemical properties. The number of electrons, in turn, is determined by the charge on the nucleus. In every nonionized atom the number of electrons is precisely equal to the number of protons. In the examples of transmutation that we discussed in the previous section, the reason that chemical elements changed identity was because the number of protons in the nucleus changed during the reaction. For example, in the emission of an alpha particle, the nucleus loses two protons, while in the emission of a beta particle, the nucleus gains a proton. In both cases, the atomic electrons rearrange themselves to accommodate the new charge on the nucleus.

However, it is obvious that changing the number of neutrons in a nucleus will not change the number of electrons in the outer shell. For example, if we had a nucleus like carbon, which normally has six protons and six neutrons in it, the chemical properties would be essentially unchanged if we had a nucleus with six protons and eight neutrons in it. There would still be only six electrons in the outer shells, and they would still be in essentially the same orbits as they would be for the nucleus having only six neutrons. The only difference between the two atoms would be in weight.

The atom whose nucleus has six protons and eight neutrons is normally called Carbon 14, which is written ^{14}C. This is to be distinguished from the nucleus that has six protons and six neutrons, called Carbon 12 and written ^{12}C. These two nuclei are said to be *isotopes* of each other. The atoms with which they are associated have identical chemical properties, but have different atomic weights.

Every element in nature has several isotopes. In other words, every element from hydrogen to uranium comes in several different forms. All of these forms have the same chemical properties because the electron structure of the atoms are the same, but they have different weights because there are more neutrons in some than there are in others. In many cases, such as in carbon, one particular form of the element is much more abundant than all of the others, and some isotopes may be radioactive while others are not. ^{12}C is by far the most abundant isotope of carbon and is not radioactive; ^{14}C is a relatively rare isotope of carbon but it is radioactive, and, as you may know, is very useful in processes involving radioactive dating. We shall encounter the presence of isotopes throughout the rest of our discussion of nuclear physics.

Let us now turn to a discussion of radioactivity as it actually exists in nature. We know that there are a large number of naturally occurring radioactive elements. We have already discussed the discovery of uranium, polonium, and radium as examples of this. We also know that these materials taken in bulk give off the various kinds of particles that we call radiation. We know, however, that these chemical elements were formed many billions of years ago. How is it that they are still giving off their characteristic radiation? Why hasn't all the radiation in the universe died down long before the present time?

The answer to this question is really twofold. On the one hand, we can understand part of the answer to the question when we understand the time scale involved in the emission of radiation; on the other hand, we will understand another piece of the answer when we understand what is meant by the existence of the radioactive decay chain. Let us start by considering the element uranium. The most common isotope of uranium is ^{238}U, which has 92 protons and 146 neutrons. We all know that uranium is a radioactive element. This means that the nucleus of the uranium atom will give off some sort of particle, and the uranium atom itself will change its identity and become another element following the emission of the particle. This process is generally called "radioactive decay."

When ^{238}U decays, we find experimentally that it does so by the emission of an alpha particle. The resulting nucleus, then, must have 90 protons and 144 neutrons. This is the nucleus of the element thorium, isotope 234, which we usually write ^{234}Th. Thus, even if we started with a sample of pure uranium, the nuclear decay process would eventually produce a new element—thorium—in our sample.

How long would it take for the original uranium to disappear? This is an easy quantity to measure, since all that is necessary is to let a sample of pure uranium sit for awhile, and then see how much is left. The number most commonly used to describe the rate of decay of a collection of nuclei is the half-life, which is defined as the time it takes for the nuclei to be reduced to half of their original number by the decay process.

There is no general rule for how long the half-lives of nuclei should be. They can range from fractions of a second to billions of years. At present, we do not have the ability to predict what the half-life of a given nucleus will be, but, as we mentioned above, it is a relatively easy quantity to measure. The half-lives of all elements that exist in large enough samples to be measured are known.

One interesting example is ^{238}U. The half-life for the decay of ^{238}U which we discussed above is 4.5 billion years. This is a rather long half-life, being on the order of the lifetime of the earth. This is part of the explanation as to the continuing persistence of radioactivity. Since the half-life for the decay of ^{238}U is about the same as the known geological lifetime of the earth, about half of the uranium that was here at the creation of the earth is still here in its undecayed state. It will continue to give off radioactivity, but eventually (many billions of years in the future) it will be all gone.

Other elements, however, have much shorter half-lives. For example, once the uranium has decayed into ^{234}Th in our above example, the product of the decay of uranium, the atom of ^{234}Th, will itself decay. The most likely decay for this element is the emission of a beta ray, and the half-life for this second decay is only 24 days. Thus, once the original uranium atom has decayed, the product of the original decay will itself decay in a much shorter time.

This process continues. The product of the decay of the ^{234}Th, Protactinium 234, is also unstable, and will decay into ^{234}U by beta emission with a half-life of two minutes. The successive decays from one element to another to another by the emission of different kinds of radiation is what we call a *radioactive decay chain.* A number of examples of decay chains are given in the problems at the end of the chapter. The existence of decay chains explains why Madame Curie was able to discover so many different radioactive elements in her sample of pitchblende. Each element was a part of a radioactive decay chain, and each one was capable of producing still more radioactive elements.

Eventually of course, the radioactive decay chain will lead to an element that is stable. One example is the isotope of lead, which we call ^{208}Pb. Once uranium has started its decay, the ultimate end product will be lead. Lead is stable, and does not decay by any radioactive process; therefore, at that point, the radioactive decay chain stops. One would think, then, that the earth would eventually become non-radioactive. While this would be true if there were no way that radioactive elements could be produced in nature, the constant bombardment of the earth by cosmic rays (which we shall discuss later) is continuously producing new radioactive elements from the stable ones. For example, if a cosmic ray proton were to strike the ^{208}Pb, it might form a new element, which would begin its own radioactive decay chain and lead to its own stable end product. Thus, there is a constant flux in nature between elements that are decaying into stable end products and the creation of new radioactive elements, and hence new decay chains, leading to new stable end products.

E. National Radioactivity: Everything is Radioactive

This statement is true today, it was true 10 years ago, and it has been true ever since man first evolved from lower animals. The presence of radioactivity in our environment has nothing to do with the development of nuclear science since World War II. Our environment has always been radioactive, and there is no great difference between the amount of radioactivity today and the amount that would have been measured by a Roman had he been able to measure it. In this section, we shall discuss this naturally occurring radioactivity and try to understand where it comes from, how much there is, and how much it varies from one place on the face of the earth to another.

First we have to find some way of measuring radio-activity. Actually, there are two ways that we could imagine doing this. We could imagine taking a source of materials that was giving off radiation and simply counting the total number of particles coming from the material per second. In other words, we could take a measure of radioactivity that had to do with the number of nuclear decays in a sample. The unit of radioactivity measured in this way is called the Curie, which is commonly abbreviated C. The definition of the Curie is that one Curie corresponds to 370 billion nuclear decays per second in a sample.

Obviously, there is no necessary connection between the amount of radioactivity in a sample of material, in Curies, and the amount of material there is. A sample of material that decays very quickly will give a very high number of Curies for a short period of time. A sample of material of the same weight, but which decays slowly (i.e., has a long half-life) will give a reading that is very low when measured in Curies, but it will not die out quickly with time. Thus, the Curie is simply a measure of how many nuclei are decaying in a given sample, and does not measure either the length of time that the sample will remain radioactive or the kinds of particles or the energy of the particles that are being emitted.

Another way in which one could imagine measuring radioactivity is to ignore the source and to concentrate on the effects that the radiation has on materials. For example, we might want to know how much energy is deposited in a given material by a given source of radiation. Such a measure would obviously depend on two things: first, it would de-pend on the strength of the radioactive source and, second, it would depend on the interaction between the specific kind of particle that was being emitted and the specific material that was absorbing it. There are, for example, materials that are much more efficient at absorbing radiation than others. Lead is an example of a material that absorbs radiation readily. A one-Curie source surrounded by a lead sheet will deposit almost all the energy of radiation in the lead. On the other hand, a one-Curie source surrounded by paper may not. Thus, the effect of the radiation on the material around the source depends on both the amount of energy coming from the source and on the type of material itself. In this discussion we shall be primarily concerned with the effects of radiation on biological tissue.

The unit that is used to measure the absorption of energy from radiation in biological materials is the rem, usually abbreviated R. It stands for "Radiation Equivalent in Man." The definition of a rem is a little complicated. There is a unit called the rad, which corresponds to an amount of radiation that deposits 100 ergs of energy in one gram of material. The rem is defined as the radiation dose to biological tissue that will cause the same amount of energy to be deposited as would one rad of X-rays.

A millirem (abbreviated mR) is one one-thousandth of a rem. To give some idea of the size of things we are discussing, you get about 20 mR of radiation from a dental X-ray, and about 150 mR of radiation from a chest X-ray.

Obviously, the two units of radiation that we have defined are complementary. The Curie tells us something about the source of radiation, and the rem tells us something about the effects of radiation on whatever it is that absorbs it. It is important to realize that the measurement of radiation effects in rem is totally independent of where the radiation came from. That is, an alpha particle of a given energy will deposit a certain fraction of that energy in each gram of biological tissue encountered by it. It does not make any difference whether that alpha particle came from the sun, from uranium in the rocks around us, or from a nuclear reactor. Once a given alpha particle is emitted somewhere, and once it is absorbed by a given piece of tissue, the radiation dose measured in rem is the same.

Before going on, it would probably be useful to discuss briefly what happens when a particle enters any kind of material. Exactly how does damage occur as a result of radiation?

We know that all material is made of atoms, and we know that in biological systems these atoms are arranged in a highly structured way in cells. When a fast moving particle comes near an atom, several things can happen. If the particle collides with the atom, the atom can be set into faster motion.

368

We record this faster motion of the atom as a raising in temperature of the material that is being bombarded by the radiation. In addition, if the particle is charged (i.e., in the case of alpha or beta radiation), there will be an electrical force between the charged particle going by and the electrons in the atom. If this force is large enough, the electron may actually be knocked off its atom, leaving an atom that is ionized behind it. Thus, radiation may cause ionization of atoms in its path. Gamma rays can also cause ionization by direct collision with electrons in the atomic orbits.

Thus, when radiation enters a sample of material, it can cause heating of the material and it can cause ionization of the atoms in the material. If that material happens to be a living cell, and if the atoms in question happen to be atoms in the DNA chain or in some other spot in the cell that is vital to the continued existence of the cell, the radiation can either destroy the cell completely or cause defects in the DNA which cause the cell to replicate incorrectly or to cease replicating entirely. In any case, massive doses of radiation deposited in biological tissue will obviously have the effect of destroying the delicate balance in the cell which allows it to keep going.

How is it, then, that human life has been able to evolve at all in a radioactive environment? To answer this question, we have to find out how much radiation there is in our natural environment and how much radiation we are exposed to in the normal course of affairs.

We have already hinted at one important source of radioactivity in our environment. In the sun and other stars, heat is generated by nuclear reactions. These reactions have as their by-products the emission of high energy particles, which we call radiation. These particles travel through space, and some of them eventually strike the earth. When they do so, they are called cosmic rays. When they enter the earth's atmosphere, these cosmic rays suffer collisions with molecules of oxygen and nitrogen and cause showers of fast-moving particles, which we also perceive as radiation. The net result of this direct process by which energy from nuclear reactions in the sun and other stars is converted to radiation in the earth's atmosphere is a dose at sea level of about 40 mR per year for any individual. In order to understand what this means, let us examine a couple of examples.

369

Every minute about four or five cosmic rays are going through your body. This has been going on ever since you were born, and will continue until the day you die. If you ever have access to a Geiger counter, it is an instructive experiment to take it outside, far away from any "radioactive sources," and see what happens. The counter will click away, triggered by cosmic rays. At sea level, it is obvious that the earth's atmosphere is absorbing a lot of the radiation that would normally strike the ground. Actually, the annual dose goes up by about 10 mR per year for every thousand feet of elevation. The average dose from cosmic rays to an individual in the United States works out to about 50 mR per year when this elevation effect is taken into account.

Another major source of natural radioactivity has to do with radiation that is emitted by minerals in the ground. This should not be too surprising. We know that the ground is radioactive. After all, where does the uranium that fuels our reactors come from? As it turns out, uranium is actually a common element, being much more abundant by weight in the earth's crust than materials such as mercury and silver, with which we are more familiar. In the table below, we give the amount of uranium that would be found on the average in a million tons of the earth's crust (corresponding to a cube less than 100 yards on a side).

Element	Tons of Element in a Million Tons of Earth's Surface
Copper	70
Uranium	4
Mercury	0.5
Silver	0.1
Gold	0.005

(adapted from *Handbook of Chemistry & Physics,* Sec. F–81)

This uranium, of course, has been around since the creation of the earth. Its extremely long half-life prevents its quick disappearance. However, much of the other radiation that we receive from the ground comes from the decay products of uranium, particularly radium. Finally, the bombarding of materials in the ground by cosmic rays that reach the earth's surface also generates new radioactive elements constantly, and each of these has its own decay chain, which adds its own particular radioactive isotopes to the ground.

The dose of an average individual in the United States from radiation in the ground is about 50 mR per year.

During the course of our lifetime, we absorb many of these radioactive trace elements when we breathe and when we eat or drink. Some of them (particularly the isotope of potassium known as ^{40}K) are absorbed into our body and remain there throughout our life. During our lifetime, they decay and deposit radiation in the tissues around them. The average individual in the United States gets about 24 mR per year from all the radiation in his body (including potassium, which accounts for about 20 mR per year).

There are other miscellaneous sources of natural radiation, the primary one being radioactive trace elements in the air. They contribute about 5 mR per year to our annual dose. In the table below, we summarize the discussion of the last few paragraphs on the sources of natural radioactivity in our environment.

Cosmic Rays	50 mR
Ground Radiation	50 mR
Body Radiation	24 mR
Air Radioactivity and Misc.	5 mR
	129 mR

We see that the average dose to an individual in the United States is about 130 mR per year. This does not, however, mean that *every* individual in the United States receives this dose. How much is received depends greatly on where the person lives. For example, in Texas the average dose is about 100 mR per year, while in Colorado and Wyoming it is as high as 250 mR per year. It is obviously higher in the mountains because there is less air to shield individuals from cosmic rays and also because there are more radioactive elements in the soil. Actually, there are areas in Brazil where the dose can be as high as 2000 mR per year because of the large amounts of thorium in the soil.

In addition to these sources of radiation (which we cannot control), the average American is exposed every year to controllable radiation in the form of medical and dental X-rays. Each dental X-ray delivers about 20 mR of radiation if the patient is not shielded by a lead apron. A chest X-ray may give as much as 150 mR and X-rays for internal organs typically may give that much or more. It is not hard to see, then, that it would be very easy to absorb in medical and dental X-rays more radiation than one absorbs from natural causes. Actually, the average American receives about 70 mR per year from these controllable sources of radiation. This would bring his total radiation to about 200 mR per year. It is very interesting that the "international safe standard" for radiation is 170 mR per year. We shall discuss the question of radiation safety in the next chapter.

We see, then, that there are sources of radioactivity in nature that have nothing to do with any man made activity, and that have always been present at roughly their current levels since the beginning of the human race. Why are people so worried about radiation now?

Although this radiation has always been present, the human race has not been aware of it until very recently. We know that the discovery of radioactivity took place only at the end of the 19th century. Large-scale uses of radioactivity in medicine and in the generation of power are even more recent than that. Consequently, radiation remains a very mysterious and threatening thing to most people. It obviously has potential for doing great harm, as we shall see in our subsequent discussion. However, it could also be argued that the evolution of the human race itself could not have taken place without our normal background radiation. After all, something had to cause the mutations which, through the processes of natural selection, resulted in the human race as we know it today.

The lack of awareness of radiation in our environment causes many people to feel that *any* increase in radiation levels must be harmful. Let us, therefore, discuss some differences in radiation levels that we willingly accept in order to get some idea as to what might be a permissible dose of radiation to add to our environment.

We have already discussed the fact that there is as much as 150 mR per year difference between different areas of the United States in terms of the normal radiation dose received by people in those areas. It is interesting to note that the cancer rates are actually much higher in New Orleans than

Madame Marie Curie (1867- 1934).

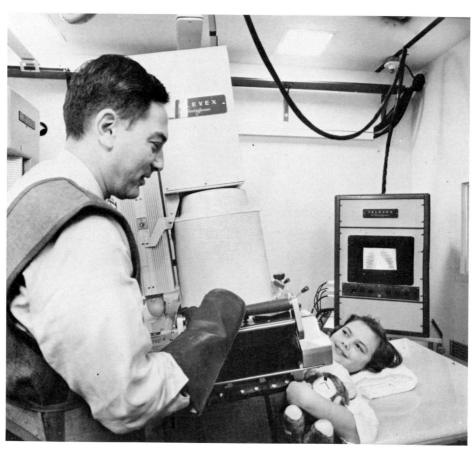

A modern diagnostic X-ray unit at work in a hospital.

they are in Colorado and Wyoming, even though the radiation dose is higher in the latter places. This emphasizes one very important fact. Radiation is not the only source of cancer, and probably is not even the leading source of cancer in the United States. We shall come back to this point later.

If you live in a house built of brick, the materials in your house were dug out of the earth and fired in a brick factory. We already know that there are certain trace elements of radioactive materials in the ground everywhere. These trace elements, of course, will be found in the clay from which the brick is made, and will follow the clay into its formation as a brick, and will eventually wind up in your house. Thus, if a house is made out of brick, the walls will be radioactive. In fact, the average dose that you would receive from a brick house is about 40 mR per year greater than the dose you would receive if you lived in a wood house. In the same way, a person living in Brooklyn, which is on sandy soil, will receive less radiation than a person living in Manhattan, which is on rock.

Flights in jet aircraft commonly carry their passengers at very high altitudes. At these high altitudes, most of the shielding effect of the atmosphere is gone, and we are exposed to rather high doses of cosmic rays. In fact, six transcontinental air trips can deliver a dose equivalent to the normal yearly dose from cosmic rays.

We see, then, that there are sources of radiation all around us, and that the kinds of doses that we normally receive measure in the hundreds of millirem. More important, depending on the area of the country and the type of activity in which one is engaged, one can increase this radiation dosage by 100 or 200 millirem without any difficulty at all. It will be very important to keep these numbers in mind when we discuss the question of radiation safety.

The most important point that should be gained from this section is the knowledge that everything in the world around us contains some radioactive elements, so that everything in our environment is radioactive. The important question that we must ask, then, is not "Is it radioactive?" because the answer to this question will always be yes. The question that we must always ask is "*How* radioactive is it?" It is to questions of this nature that we now turn.

SUGGESTED READING

All of the standard physics texts have sections on nuclear physics. In addition, the following may be useful.

Bethe, Hans A. "What Holds the Nucleus Together?" *Scientific American,* 1953, *189* (58). A discussion of meson exchange as the mechanism of the strong attraction.

Cohn, Bernard L. *Nuclear Science and Society.* New York: Anchor Books, Doubleday, 1974. An extremely readable text with a clear discussion of what the nucleus is like and a large number of examples of practical uses of nuclear physics.

Heofstadter, Robert. "The Atomic Nucleus." *Scientific American,* 1956, *195* (55). An account of how we measure the size and shape of the nucleus by the man who received the Nobel Prize for this work.

Inglis, David R. *Nuclear Energy: Its Physics and Social Challenge.* Reading Mass.: Addison-Wesley, 1973. Although mainly devoted to discussions of nuclear reactors, this book has several useful (though somewhat mathematical) appendices on nuclear physics.

Mayer, Maria G. "The Structure of the Nucleus." *Scientific American,* 1951, *184* (22). A good discussion of nuclear shell structure.

Wahl, Werner H. and Kramer, Henry H. "Neutron Activation Analysis." *Scientific American,* 1967, *216* (68). A discussion of still another practical use of nuclear physics and radioactivity.

QUESTIONS AND DISCUSSION IDEAS

1. Let's try to get some idea of the strength of the "strong" interaction. Consider two protons that are a distance 10^{-8} cm apart. This is about the size of an atom.
   a. Calculate the repulsive force between the two protons due to their electrical charges.
   b. Calculate the attractive gravitational force between the two protons.
   c. What would happen to these two protons if they were allowed to move?

2. Work Problem 1 for the case of two protons located 10^{-13} cm apart. This is a typical size of a nucleus. How much stronger is the electrical force for this case than it was for the case of Problem 1. This is the force that must be overcome in order to hold the nucleus together.

3. The half-life of ^{14}Carbon is about 5600 years. Suppose that a piece of wood is taken from a tree. The processes by which carbon is taken from the environment and added to the wood will then stop, and the ^{14}C will decay.
   a. How long will it be before only one-half of the original amount of ^{14}C is left?
   b. How long before one-fourth is left?
   c. Suppose you found a sample of such wood that had only about one-eighth of its normal concentration of ^{14}C. How long ago would you say the the wood had been cut?

   NOTE: This problem illustrates the method of radio-carbon dating that is so important in archeology.

The following information about some unstable nuclei may be helpful in working the next few problems.

name	number of nucleons	protons	half-life	particle emitted
Uranium	238	92	4.5 billion yrs	alpha
Uranium	234	92	250 thousand yrs	alpha
Protactinium	234	91	1.2 minutes	beta
Thorium	234	90	24 days	beta
Thorium	230	90	80,000 yrs	alpha
Radium	226	88	1620 yrs	alpha
Radon	218	84	37.8 days	alpha

4. How many neutrons in each of the above nuclei?
5. Suppose we started with 1,000,000 Uranium 238 nuclei at the beginning of the world (about 4.5 billion years ago). How many nuclei would have decayed by the present time?
6. Consider those U 238 nuclei which did decay.
   a. What is the first new nucleus that would have appeared in our sample?
   b. If there were 100 such nuclei at any given time, how long would it be before only 50 were left?
   c. During the time it took to reduce the number of new nuclei by one-half, would any new nuclei be added to the pool by the decay of the U 238?

   (HINT: Figure out how many U 238 nuclei of the original million would decay in the time necessary for the second nucleus to be depleted by half.)

375

7. Continue the process started in Problem 6, listing each new nucleus that appears in the decay chain until you reach Radon. Will there be other nuclei in the chain beyond Radon?

8. From your answers to the previous questions, discuss the question of whether or not you would expect to find Radium and Radon (two rather dangerous substances, the latter being a gas) in the environment, given the fact that uranium is a fairly abundant natural element.

9. Suppose that we have two radioactive elements, A and B. Suppose further that A decays into B with a half-life of 100 days, while B decays into something else with a half-life of one day. Suppose that we start with 1,000,000 atoms of A. In what follows, you can neglect the decline in the number of nuclei of type A caused by the decay (i.e., you can proceed as if each time a nucleus of type A were to decay, it would be replaced by a fresh one).

    a. How many nuclei of type B are created in the first day?

    b. Suppose that we neglect the fact that some of the B atoms created will decay during the first day, assume the B atoms created in the first day do not start to decay until the start of the second day, and that B atoms created in the second day will not start to decay until the third day, etc. How many B atoms created in the first day will decay during the second day?

    c. How many new B atoms will be created during the second day? Hence, say how many B atoms there will be in the sample at the end of the second day.

    d. Repeat part (c) for each day up to the end of day 8.

    e. If I wait a long time, how many atoms of type B will there be in the sample? Can you give a simple explanation of what is happening in this case?

10. How would the answer to part (e) in the previous question be modified if I did not make the assumption that each atom of type A was replaced after it decayed?

CHAPTER
XIV

NUCLEAR REACTORS AND RADIATION SAFETY

*There is no gathering
the rose without being
pricked by the thorns*

Fables of Pilpay

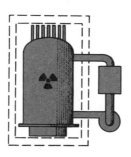

Enrico Fermi (1901-1954) who built the first nuclear reactor in Chicago.

A. Introduction

The aspect of nuclear physics that most closely touches modern society has to do with the use of nuclear reactors to generate electricity in a world where oil and gas are running out. In the United States, a furious debate over all aspects of nuclear-generated power has been building up over the past few years, and any text on physics would be incomplete without some mention of the problem.

The first step, obviously, is to understand what a reactor is and how it works. That is the purpose of this section. The next question to be discussed has to do with the risks that accompany the operation of reactors. However, it would probably be a good idea to think a little about the general philosophy of making choices before going into details.

When choosing a course of action, we have to balance two things: the benefits to be received from following that course against the risks of damage that it will cause. The choice about whether to build nuclear reactors is no different. There are obvious benefits to be derived—relatively clean power, no strip mining, no dependence on Arab oil, etc. But there are also dangers—large amounts of radioactive material, waste disposal, the chance of accident, etc. If we are to make an intelligent choice, we have to understand what these risks are and how serious they are. Only then can we decide whether the benefits of nuclear power outweigh the disadvantages. Before talking about risks, however, we have to understand what a nuclear reactor is, and it is to that question that we now turn.

Stripped to its essentials, a nuclear reactor is a rather simple device. It depends for its operation on a nuclear process known as *fission*. Fission means the splitting of the nucleus.

Any nucleus can be "split" or made to undergo fission if enough energy is applied to it. Even carbon and oxygen, atoms that make up a good deal of our surroundings and that we normally think of as stable, can be broken up if a particle with sufficient energy is made to collide with them. There are in nature, however, a few elements where the fission process takes on a rather extraordinary character. In these few cases, the weight of the pieces that are left after the fission process has taken place is less than the weight of the

378

nucleus before the fission process. This is not the case when we talk about oxygen and carbon, but it is the case when we talk about one particular isotope of uranium—this is ^{235}U, an isotope of uranium that has 92 protons and 143 neutrons.

Obviously, if there is less mass after a reaction than there was before a reaction, something must have happened to it. If you recall our discussion in the section on special relativity about the equivalence of mass and energy, you will realize that the only thing that could have happened to that mass would be that it had been converted into energy, according to the formula $E = mc^2$. Thus, in these few reactions, the fission process actually releases energy. The energy released is quite large. About 170 million electron volts are released in one fission, and the fissioning of a pound of ^{235}U would release as much energy as 2.8 million pounds of coal.

Among those few nuclei that undergo fission in such a way as to give up energy, there are a few that have another very important property—the property of being able to sustain a chain reaction. How does a chain reaction work? In the case of ^{235}U, the fission is caused by a collision of a neutron with the uranium nucleus. The reason that it is a neutron and not a proton that causes the fission is actually very simple. Remember that the uranium nucleus has positive charge of 92 charge units. A proton trying to approach the uranium nucleus will thus be repelled by the rather large positive charge that resides there. A neutron feels no such repulsion and can penetrate to the nucleus much more easily than a proton.

In the case of ^{235}U, when the neutron strikes the nucleus, the nucleus undergoes fission. This is a rather complicated process that is only now beginning to be understood by nuclear scientists. The end result of the fission, however, is easy to understand. The uranium nucleus breaks up, usually into two large pieces of nuclear matter, along with two or three free neutrons. In other words, in addition to the large fragments and the large release of energy caused by the conversion of mass, the net result of sending one neutron into a ^{235}U-type nucleus is to bring two or three neutrons out. If there are other ^{235}U nuclei in the immediate vicinity, the neutrons from the first fission will strike them, causing secondary fissions, each of which will release two or three more neutrons, which are then free to go on and cause further fissions. This process is called a "chain reaction."

In each fission, an enormous amount of energy is released in the form of gamma rays and fast-moving particles. When these particles collide with atoms in the surroundings of the uranium nucleus, they give up some of their energy. After this collision, the atom is moving faster, and so we would say that the temperature of the material around the uranium had been raised. This is the mechanism by which a chain reaction can heat material in its immediate vicinity.

Actually, the isotope of uranium that has an atomic weight of 235 is not the "normal" isotope of uranium. Over 99% of naturally occurring uranium is the isotope ^{238}U. A very complicated and costly process called "enrichment" is carried out with naturally occurring uranium ore in order to increase the amount of ^{235}U to the point where the material will sustain a chain reaction. In normal uranium, the type you would dig out of the ground, the ^{235}U atoms are spaced so far apart that neutrons from one have a very small chance of finding another. Hence, natural uranium will not maintain a chain reaction. The process of isotope separation, or enrichment, yields as its end product a material that has about three percent ^{235}U, and that will sustain a chain reaction. Uranium at this level of enrichment will not, however, be useful in the construction of weapons, which may require up to 90% ^{235}U.

A nuclear reactor is a device that allows the chain reaction to heat a material, usually water, which is then used to run a standard electrical generating system. A sketch of a reactor (highly schematic) is given in Fig. 14.1. The uranium is contained in fuel rods, which are composed of long slender objects in which a rod of uranium material about the diameter of a piece of chalk is wrapped in a metal coating. These fuel rods are stacked up in a nuclear reactor; typically there will be hundreds and even thousands of fuel rods in a commercial reactor. The space between the rods is filled with a material, usually water, which is called a moderator. The function of the moderator is crucial to the sustaining of the chain reaction. The neutrons emitted from the original fission of the ^{235}U nucleus are moving very fast. If these original neutrons came near another nucleus, they would be moving too fast to cause it to fission. If, however, they are slowed down by one means or another, they will spend a long time in the vicinity of the next nucleus, and will have a much

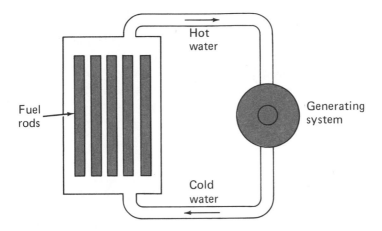

Figure 14.1.

higher probability of interacting with that nucleus and causing fission. The function of the moderator is to slow down neutrons. It accomplishes this by the simple process of collision. A neutron emitted in a fission of one fuel rod will leave that fuel rod and enter the moderator. It will make many collisions before it reaches the next fuel rod. Consequently, it will have slowed down to the point where it now has a very high probability of interacting with the uranium nuclei there. In this way, the chain reaction is maintained at its optimum efficiency, but the maintenance of the chain is critically dependent on the presence of the moderator.

Once the water has been heated to several hundred degrees centigrade (well above its boiling point), it is taken off and used to run a conventional power generator. A nuclear generating plant, then, differs from a conventional oil or coal powered plant only in the way in which the steam that drives the turbines is generated. From the reactor on, a nuclear power plant is essentially identical to a coal or oil power plant. Thus, when we talk about the dangers of nuclear reactors, or the safety features of nuclear reactors, we are really comparing the "burning" of uranium in the reactor to the burning of coal in a normal furnace.

Obviously, the facts that uranium is undergoing fission and that a radioactive decay chain is going on inside of the reactor mean that there is an enormous amount of radioactive material in the reactor. A typical reactor may contain as many as 20 billion Curies. The question that is being debated on the national level today has to do with whether or not the presence of so much artificially created radioactivity may cause pollution of our environment and harm to the health of individuals. In order to discuss this question, we have to understand how to make rational estimates of risk to individuals from the presence of given amounts of radiation. It is to this question that we now turn.

B. The Calculation of Risk

In this section we shall discuss two things. First we shall talk about how one can go about quantifying the idea of risk, and simple procedures for calculating risks will be given. We shall also discuss some examples of risk in our daily lives. This will give us some basis for comparison when we go on to talk about the risks involved with radiation.

There are two categories of risk, and they should be thought about differently. The first category are those risks over which we have little or no control. Most accidents fall into this category, as would natural disasters such as flood or hurricane. These risks are a part of life, and all we can do is calculate what they are.

There are other risks, however, that are voluntarily accepted. People choose to smoke, for example, even though they know that it is harmful to them. It is to this category that our earlier remarks about weighing risks and benefits applies, and it is into this category that nuclear reactors fall. In this section we shall discuss both kinds of risks, and we shall generate some numerical estimates of risk that will be useful in discussing reactors later on.

To get some idea about the risks sustained in our daily lives, let us take a few statistics (which are available in any almanac) and calculate the risks that we sustain every day just from living in the United States.

In the year 1971, there were 115,000 accidental deaths per year in the United States. A little less than half of these involved automobile accidents, most of the rest involved falls, burns, drowning, firearm accidents, etc. Let us ask the question "What is the chance per year that any individual in the United States will die from an accident?" Obviously, if there are 115,000 deaths per year, and if there are 200 million people in the United States (a population figure that we shall assume throughout the rest of our discussion for ease of calculation) then the chance that any individual will meet with a fatal accident in any year is given by

$$\frac{115,000}{200,000,000} = .00057 \text{ chances of accidental death/year.}$$

This means that there are about six chances in 10,000 that any particular individual will die from an accidental death per year. We shall call this risk the *individual risk,* which is simply the chance per given unit of time that an individual will die from the cause we are cosidering (which, in this example, is an accident).

Let's put this number in a little more perspective. Let's ask the question "What is the chance that any individual in the United States will die by an accident within the next hour?" Well, we know that there are 365 days in the year and 24 hours in a day. Thus, there are 8,780 hours in each year. Since there is a chance of .00057 that a person will meet with an accidental death in a year, there is a chance of $.00057/8,780 = 8 \times 10^{-8}$ that any person will die by accidental death in the next hour. In other words, each of us has roughly one chance in 10 million of dying by accident in any hour of our life. This is an unavoidable risk, which we sustain simply from living where we do.

You could try to reduce this risk if you wanted. For example, you could go to some rural environment in which there were not many cars, and hence reduce the risk of death from an automobile accident. However, if you did so, you would introduce new causes of accidental death (for instance, snake bites, accidents aggravated by the absence of medical care, etc.), which would, in themselves, bring some unavoidable risk with them.

The point of this discussion is not to be ghoulish. It is simply to make us realize that we sustain a risk of death every minute that we are alive. When we consider something like nuclear power, we have to find some way of comparing that risk to other risks with which we are familiar.

Let us go on with our calculation. We have calculated the risk to an individual per year of death by accident, and called that the "individual risk." We can also define a "life-time risk," which is the probability that an individual will die accidentally some time during his life. If we assume that a person is equally likely to die by accident at any age, and take 70 years as the average life expectancy in the United States, then the lifetime risk of accidental death is

$$70 \times .00057 = .04.$$

In other words, the chance that any one of us will die by accident is about one in 25. Obviously, the other 24/25 chances are related to others means of death.

There is yet another way of discussing the risk that we sustain from any source of danger such as accident. This is called the *relative risk.* It is calculated as follows: We have already assumed that an individual is probably just as likely to die from an accident in his first year as in his last year of life, and that the risk each year that an individual will die from accident is about the same. The "average" death from an accident, then, will occur at age 35—halfway through one's life. This does not mean that everyone who dies in accidental death will die at age 35. Obviously, some will die at age 1, some at age 2, some at age 40, some at age 65, etc. It is only when we consider large groups of people that we can talk about the "average" age at death.

Nevertheless, if a person, on the average, dies as a result of an accident at age 35, he will have lost 35 years of his life (this number is obtained by taking his expected lifetime, 70 years, and subtracting the actual time he lived, 35 years). The relative risk to an individual from accidental death, then, is defined as the lifetime risk times the amount of life lost in the average case, so that for an accidental death we would have

$$35 \text{ yrs} \times .035 \text{ chances/lifetime} = 1.6 \text{ yrs.}$$

Thus, we can say that the average person in the United States loses 1.6 years of his life because of the fact that by living here he subjects himself to the risk of accident.

There are a couple of points that ought to be made about the calculations of risk that we have presented here.

The calculation of individual risk is relatively straightforward. If a certain group of individuals (in our example this group was the population of the United States) suffers a certain number of deaths from a particular cause per year, then the individual risk is defined as

individual risk =

$$\frac{\text{total number of deaths in the sample population per year}}{\text{total number of people in the sample population}}$$

In our example, the sample population included everyone in the United States. There are populations, however, that do not include everyone in the United States. For example, we shall discuss later the risks attendant on smoking. In this case, the population (the "number of people in the sample" in the above equation) will not be the entire population of the United States but the population of heavy smokers. If we were dealing with the risk due to a disease affecting only children, then the sample population would be only the total number of children in the United States and not the total number of people.

The lifetime risk—the total risk that an individual faces during his entire lifetime—is defined as

lifetime risk = (individual risk) x
(total period of time that the individual is at risk).

In our example, the total time that the individual is at risk is the same as his lifetime, because accidents can happen to anyone at any time. On the other hand, when we deal with other sources of risk, this will not be the case. Using the example in the previous paragraph, if we were dealing with the risk of a childhood disease, we would not put 70 years in the above expression for a lifetime risk, but 10 or 15 years—the time during which an individual is a child, and therefore subject to the risk.

The relative risk is the risk that we shall use to compare different sources of danger in our lives. There are several points that should be made about it. The equation for relative risk is

relative risk = (average loss of life) x (lifetime risk).

385

In other words, the relative risk is the product of the amount that we lose if we encounter a certain source of danger times the probability that we shall encounter that source of danger in our lifetime. In the above example of accidental death, the fact that the average death occurred at age 35 meant that the individual lost $70 - 35$ years of his life, or 35 years. Since we had already calculated the lifetime risk to an individual, the calculation of relative risk was fairly simple. It must be emphasized that not every source of danger in our life results in a loss of 35 years of life. For example, if we were dealing with the danger of childhood disease, and if the average age at death was 10 years, then the life lost would be $70 - 10 = 60$ yrs.

It should also be emphasized that in dealing with relative risk, what we are really doing is sharing the risk from a particular source throughout the entire population. In saying that the *average* American loses 1.6 years of his life due to accidents, we are not saying that *every* American loses 1.6 years. Some Americans die in accidents as children, and lose most of their life, others die in accidents at old age and lose less time. Most do not die in accidents at all. Once again, when we are dealing with relative risks, we are dealing with averages over a large number of people, not with individuals. The point is that if an individual dies from an accident, he loses a great deal. Most individuals, however, do not die from accidents. Therefore, the relative risk is really something like a risk to society at large rather than a risk to any individual person of death from a particular source.

Another source of risk that we can discuss is the risk from smoking. Unlike accidents, this is a source of danger that many Americans accept voluntarily. Consequently, it is useful to compare the risk attendant on smoking with the risk that would result from a nuclear economy in the United States.

According to the World Almanac, about 22 million people in the United States (roughly 10% of the United States population) are heavy smokers. A heavy smoker is defined as a person who smokes more than one pack of cigarettes a day. From this same source, we find that there are about 100,000 cases of lung cancer per year in the United States (this corresponds to 1/6 of all the cases of cancer), and, according to the American Cancer Society, about 75,000 of these cases occur among persons who are heavy smokers. Given these statistics, let us go through the kind of calculation of risk that we did for death by accident.

The individual risk, as we said above, is the total number of deaths from a particular cause divided by the total number of persons in a given population. In our case, the total number of deaths among heavy smokers from lung cancer is 75,000, and the total number of heavy smokers is 22,000,000. Thus, the individual risk to a heavy smoker per year is

$$\frac{75,000}{22,000,000} = .0034 \text{ chances per year.}$$

You will note that this is considerably higher than the risk to the people in the United States in general from accidents that we calculated above.

The lifetime risk as we have defined it is the total risk per year times the period of time during which the individual was at risk. Most people do not start smoking at age 1. Let us assume that people who start smoking do so when they become adults, and let us arbitrarily define this age as 20. Thus, a person who lives 70 years (the expected lifetime) will have been a heavy smoker for about 50 years (the expected lifetime minus the 20 years of childhood when we are assuming that he was not a smoker at all). Thus, the lifetime risk is

$$(50 \text{ yrs}) \times (.0034 \text{ chances per year}) = .17 \text{ chances per lifetime.}$$

You will note that this is also higher than the chance of dying by accident during your lifetime.

We have defined the relative risk as the average amount of time lost times the chance that a particular fatality will occur during one's lifetime. Let us assume that the average smoker gets lung cancer halfway through his life as a heavy smoker. Since we have assumed that a person is a heavy smoker for 50 years of his life, this would mean that the average fatality would occur after 25 years had elapsed from the time that the person started smoking, or at age 45. This, in turn, would mean that the average death would result in a loss of 25 years of life. Thus, the relative risk from smoking due to lung cancer would be given by

$$(.17 \text{ chances per lifetime}) \times (25 \text{ years lost}) = 4.2 \text{ years.}$$

Again, the relative risk of smoking due to lung cancer alone is more than twice as large as the risk from accidental death.

Actually, we have not counted all the risks that an individual undergoes if he smokes. There are, in addition to lung cancer, other diseases such as hypertension and coronary problems that are much more prevalent among smokers than among non-smokers. If we added all of the risks (and not just the risk of lung cancer, which we actually calculated above), we would come up with

relative risk from heavy smoking = 9 years.

It is instructive to convert this number to the relative risk or cost of smoking one cigarette. If a person smokes a pack of cigarettes a day for 50 years, this means that he will smoke a total number of cigarettes during his lifetime given by

(50 years) x (365 days per year) x (20 cigarettes per day) = 365,000 cigarettes smoked in a lifetime.

Thus, the average heavy smoker can be expected to smoke at least a third of a million cigarettes during his life. If this many cigarettes cause a relative loss of nine years, then we can easily calculate the total amount of time lost per cigarette. First let us calculate the number of minutes in nine years. This is given by

(9 years) x (365 days per year) x (24 hours per day) x (60 minutes per hour) = 4,700,000 minutes in 9 years.

To get the total amount of time lost per cigarette, then, what we do is divide the total amount of time lost (in minutes) by the total number of cigarettes smoked. This gives

$$\text{life lost per cigarette} = \frac{4{,}700{,}000}{365{,}000} = 13 \text{ minutes.}$$

In other words, the average heavy smoker loses about 13 minutes of his life every time he smokes a cigarette. Another way of thinking about this statistic is to say that you lose about as much time of your life from smoking as you spend smoking. The same remarks that we made about relative risk above, of course, apply in this calculation of relative risk as well.

With this introduction to the calculation of risks from various sources in our life, we now turn to the question of calculating the risk due to the presence of radiation. The calculation of the effects and costs of other kinds of risks in our life is left to the examples at the end of the section.

C. The Risk from Radiation

The first thing that comes to most people's minds when they think about risks from radiation is something like an atomic bomb blast. Short of war, however, it is highly unlikely that anyone in the United States would ever be exposed to the amount of radiation that would result from something like a nuclear explosion. For comparison, however, we shall begin by discussing the amount of radiation necessary to produce direct death in a human being.

You will recall that the average dose of radiation to people in the United States is on the order of 130 mR per year. The dose required to produce certain death from radiation sickness is about 750 R, which corresponds to 750,000 mR. Thus, the dose that would be required to produce the kinds of death from radiation sickness that were produced in atomic bomb attacks in World War II would be more than 1,000 times greater than the dose that we actually receive on a yearly basis from natural sources of radiation. A dose of 200 R (200,000 mR) will make an individual quite ill, but will not prove fatal. A dose of 25 R (25,000 mR) will produce only very slight symptoms in an individual.

Thus, the amount of radiation needed to cause immediate death or illness is quite large compared to the amounts of radiation that we deal with in our normal everyday existence. Why are people worried about radiation hazards from nuclear reactors then? The answer is that there are, in addition to the short-term effects of large doses of radiation, long-term effects from smaller doses of radiation. We discussed the fact that radiation striking cells can damage or change the DNA content of the cells. It is possible, therefore, to cause mutation in healthy cells which, many years later, may give rise to cancer in the individual who is exposed to the radiation. It is this risk—the long-term risk of cancer due to radiation exposure—that we will discuss in this section.

Many people find a discussion of the type we are about to enter offensive. They feel that a discussion of fatalities

A possible way out of the coal-nuclear dilemma—a 100 kilowatt wind generator being tested at Lewis Research Center, Cleveland, Ohio, by NASA.

among the survivors of the atomic bomb attack on Hiroshima and Nagasaki is in some way almost irreverent. I can sympathize with this point of view. The discussion is not pleasant. On the other hand, we are facing a situation where our national survival and national health may depend on the decisions we make about how much of our power to generate by nuclear reactors, and what kinds of safeguards are needed in reactors to preserve the general health of the population. This is an extremely important decision, and we simply cannot afford to ignore past experience with radiation, unpleasant as that experience has been. We cannot change the fact that nuclear explosives were used in World War II, nor can we change the fate of the survivors of those particular explosions. But with the lives of many millions of people resting on the decision that we make now about nuclear power, we cannot afford to ignore the information that we now have on the victims of past exposure to radiation.

The situation with exposures to radiation is a little bit like early work on brain function which was done in the United States and Europe. It is not possible to experiment on brain function in individuals by taking a group of people, removing different portions of the brain, and then seeing what happens to them. What can be done, however, is to find individuals who have suffered damage to different portions of their brains by accident, and to then see which of the normal functions (such as speech, use of the limbs, etc.) they have lost. In this way, the first primitive maps of functioning of different areas of the human brain were put together.

In the question of long-term effects of radiation we have a similar situation. We cannot go around exposing people to radiation and then seeing how many get cancer. What we can do, however, is to find individuals who have, from one source or another, received large amounts of radiation. We can then study the medical records of these individuals to find what the effects of that radiation were.

There are a large number of people who have been exposed to radiation during the last 30 or 40 years. We have already mentioned the survivors of Hiroshima and Nagasaki. In addition, during the 1930s (before the harmful effects of radiation were well known) doctors often treated various types of illness—from spinal column problems to acne—with massive doses of radiation. This resulted in another group of people who had been exposed to large amounts of radiation. Another group of people of this type were workers in watch factories who painted the luminous dials and numerals on watch faces. They would often lick the tiny brush that they used for painting the numerals to get the point sharp; they therefore ingested large amounts of radium. All of these groups can be studied to give us some idea of how radiation effects the human body in the long term.

Let us consider the Hiroshima and Nagasaki survivors, because there are extensive records available on them.

There were 23,979 people who were cared for in hospitals after the atomic bomb attacks and who received large and well-documented doses of radiation. In this population of people the average dose of radiation was 130 R (130,000 mR). Since the war, there have been 81 cases of leukemia reported in this population. In an equivalent number of people drawn from the Japanese population at large, however, we would have expected only 20 cases of leukemia. This means that in this particular population (which differed from the general population only in having been exposed to a large amount of radiation) there were 61 cases of leukemia in excess of what would have been expected.

These cases were not distributed randomly during the course of time. In fact, what was found was that for a period of five years after the war (i.e., up until 1950) very few extra cases of leukemia were seen in this population. After 1950, the number of excess cases rose rapidly and stayed roughly constant for a period of about 20 years, after which it dropped off. These results are summarized in Fig. 14.2.

In Fig. 14.2, we have sketched both the actual number of excess cases and the model that we shall use to calculate excess risks from radiation. We shall assume that after an exposure to radiation, there is a latent period of about five

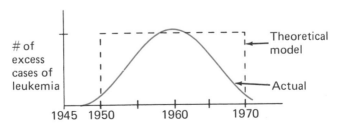

Figure 14.2.

years, during which the individual is not at risk from cancer, but during which time the cancer is developing. Then for a period of about 20 years we shall assume that the risk due to the radiation that was received before the latent period is constant. After this 20-year period, we shall assume that the risk is again zero.

It is important to get clearly in mind, then, how radiation affects the human being. If we received a large dose of radiation today, no immediate effects would be apparent for five years. After this five-year latent period, we would be at risk for about 20 years. In other words, from five years after the radiation until 25 years after the radiation, there is a risk that we will get cancer because of that particular radiation. If we make it through that 20 years without getting cancer, then we are no longer at risk because of the radiation we received at the beginning. In other words, if you are now 20 years old, then you are beginning to be at risk because of the radiation you received at age 15, and in five years you will begin to be at risk because of the radiation you are receiving now. The risk from the radiation you are receiving now will begin in five years when you are 25, and end in 25 years when you are 45. Of course, since you are continuously receiving radiation from natural sources, you are always at risk from some period of radiation in your life.

In the case of the survivors of Hiroshima and Nagasaki (i.e., those persons who survived the initial effects of the bomb), there were 61 excess cases of leukemia distributed over the 20-year risk period, for an average number of cases per year given by

risk per year = 61/20 = 3.05 cancers per year.

Therefore, an individual risk for a person who received 130 R of radiation would be

individual risk for 130 R = 3.05/23,979 = 1.25×10^{-4}.

Given that an individual receiving 130 R of radiation will have an individual risk of 1.25×10^{-4}, what is the risk to an individual who receives 1 R of radiation? This brings up another rather touchy point in the discussion of radiation. It is always assumed that the risk from small amounts of radiation is just scaled down proportionately from large amounts of radiation. This is called the "linear hypothesis." It assumes that the body is incapable of repairing any damage from radiation no matter how small that damage may be. This is probably not true, but on the other hand we do not understand the interactions of radiation with living systems well enough to know how true or untrue it is. Therefore, the linear hypothesis is almost always used in calculations of risk and we can regard it as a very conservative estimate of what the damage from radiation ought to be. It should always be borne in mind, however, that we would think the linear hypothesis was ridiculous in most situations in real life. For example, we know that dropping a thousand-pound weight on a person will kill him. If we dropped a one-pound weight on a thousand individuals, however, it is unlikely it would kill any of them. On the other hand, the linear hypothesis would say that dropping a one-pound weight is one-thousandth as dangerous as dropping a thousand-pound weight, so that if we dropped the one-pound weight on a thousand individuals, one of them should die.

Under the assumption of the linear hypothesis, the individual risk from 1 R of radiation is given by

$$\text{individual risk per R} = .00125/130 = 10^{-6}.$$

We interpret this number to say that if we get 1 R of radiation today, then for each year in the 20-year period beginning five years from now, there will be one chance in a million per year that we shall get cancer.

Actually, we have been talking in this discussion only about one type of cancer—leukemia. If we included all cases of cancer, not just leukemia, the number would increase by 6. In other words, the total individual risk of all cancers from exposure to 1 R or radiation is given by

$$\text{individual risk per R} = 6 \times 10^{-6} \text{ cancers per year/R}.$$

As an example of the relative risk occurring to an individual because of radiation, let us consider a person who is 20 years old. If this person has been living in the United States, he has been receiving, on the average, 130 mR per year of radiation. This means that by age 20 he has received a total of 2.6 R of radiation from the natural background sources that we discussed in the previous chapter. Let us assume that all of this radiation is delivered at once, and that he gets no more radiation in his lifetime. What is the risk of cancer for this individual?

From our analysis of the Hiroshima and Nagasaki survivors, we can make several statements about this 20-year-old individual's risk of cancer. In the first place, he is under no risk at all for the next five years—that is, until he reaches age 25. From age 25 on for a period of 20 years, there will be a certain risk. This means that he will be at risk from age 25 to age 45 because of the radiation he received at age 20. The individual risk is quite easy to calculate. We know that the individual risk per year during the period when the individual is at risk is given by $6 \times 10^{-6}/R$. Thus, for an individual who has received 2.6 R, the individual risk will be

$$\text{individual risk} = 2.6\ R \times 6 \times 10^{-6} \text{ chances per R} =$$
$$1.5 \times 10^{-5} \text{ chances/year.}$$

The period during which the individual was at risk, as we have said, is 20 years. Thus the lifetime risk to the individual will be given by

$$\text{lifetime risk} = 20 \text{ years} \times 1.5 \times 10^{-5} \text{ chances/yr} =$$
$$3.0 \times 10^{-4} \text{ chances/lifetime.}$$

Finally, we can calculate the relative risk to this individual from the radiation he has received in his lifetime. Since the individual is at risk from age 25 to age 45, we can expect that the average cancer, should it occur, will occur halfway through the period, at age 35. With an average life expectancy of 70 years, this would mean a loss of 35 years in the life of this individual. Thus, the relative risk would be given by

$$\text{relative risk} = 1.05 \times 10^{-2} \text{ yrs} \approx .01 \text{ yrs} \approx 88 \text{ hours.}$$

Thus, the risk to an individual from 20 years of exposure to the average background radiation in the United States is a matter of a few days. It is not at all difficult to see that this risk corresponds to the risk that the individual would stand if he had smoked about 18 cigarettes each year during the

394

period of 20 years (you should be able to work this out for yourself). Most people would not consider this a very large risk, and hence most people are not terribly concerned about the fact that background radiation exists in the United States and everywhere else in the world. When you think about it, you realize that a concern about background radiation would do very little good in any case. There is no way that background radiation can be turned off. We cannot stop the sun from sending out particles, nor can we stop the atoms of radium and uranium in the ground from decaying. Thus, the relative risk of a few days to a 20-year-old individual that we have calculated above (and other relative risks are given in the problems at the end of the chapter) represents sort of an irreducible minimum of relative risk due to radiation in our environment.

With this understanding of the way in which relative risks can be calculated and the way in which we can make quantitative estimates of the risk to persons from radiation levels in the environment, we can turn to the question of controlling those radiation sources that are controllable. Although we cannot control cosmic rays, we *can* control medical X-rays, and we can decide whether or not to allow various different types of nuclear reactors and processing plants to be built. The question that we must answer then is "How safe or how dangerous to the national health would the existence of a given number of nuclear reactors or nuclear processing plants be?"

One further point should be made at this stage. Of all the agents that can cause cancer and that exist in our environment, radiation is the best understood. It is possible, by using methods such as those outlined here, to place a good numerical estimate on the risk associated with any level of radiation exposure. This is definitely not the case with many of the chemicals that have been developed over the last half century and whose dangers are only now becoming apparent. We often have no idea of the cancer risk associated with even small exposures to such chemicals, and many experts in the field of carcinogenesis feel that they constitute a much graver threat to the public health than does radiation exposure. Of course, this does not mean that we should ignore the dangers of radiation, but simply that we should view them in perspective.

D. The Risk Involved with Nuclear Reactors

The question of how much risk to the public health is represented by a nuclear reactor can be debated on three different levels. First, we can ask how much risk there would be if the reactor runs the way it is supposed to run. Second, we can ask what the chances are of an accident occurring in a reactor, and ask what the consequences of such an accident would be. Finally, we can ask what will happen to the radioactive material in the fuel rods once the reactor has burned all the ^{235}U. This is known as a problem of waste disposal. We shall consider each of these questions separately.

1. Normal operation

According to the federal requirements now in effect, the radiation dose at the fence of a nuclear plant can be no more than 5 mR per year. One mile from the plant, this radiation dose must be less than 0.1 mR per year and 15 miles from the plant it must be less than .01 mR per year. Most nuclear plants operate well below these maximum safety standards.

We see immediately that the amount of radiation that would be added to an individual's yearly dose if he lived right at the fence of a nuclear reactor plant would be very small compared to the normal dose we receive from natural sources, and would be very small compared to the natural fluctuations in normal doses that would arise from living in different parts of the United States. It is even less than the difference that would be experienced in moving from a wooden house to a brick house. Nevertheless, let us ask how much risk would be involved to an individual who had lived his entire life on the border of a nuclear reactor plant.

Let us take again the individual who is, at present, 20 years old. Let us assume that he has lived his entire life at the fence surrounding a nuclear installation, and hence has received 5 mR of radiation per year because of the presence of the plant. That means that during the 20-year period he has received 100 mR in total radiation. We can estimate the risk to this individual for the rest of his life because of that extra 100 mR by assuming that it is all delivered at age 20.

In the last section, we saw that the risk per R per year from radiation was 6×10^{-6}. For an individual who has received 100 mR (0.1 R), this means that for each year between age 25 and age 45 there will be an individual risk of cancer given by

$$6 \times 10^{-6} \times 0.1 = 6 \times 10^{-7} \text{ chances per year.}$$

The lifetime risk of cancer to such an individual would be the individual risk per year multiplied by the period during which the individual was at risk (which is 20 years in the case we are considering) so that

$$\text{lifetime risk} = 20 \times 6 \times 10^{-7} =$$
$$1.2 \times 10^{-5} \text{ chances per lifetime.}$$

The relative risk is easy to calculate. If the individual contracts a cancer half-way through his period of risk, he will die at age 35. This means that he will lose 35 years of his life, so that the relative risk is given by

$$\text{relative risk} = 1.2 \times 10^{-5} \times 35 = 4.2 \times 10^{-4} \text{ years} =$$
$$0.16 \text{ days} \approx 3.5 \text{ hours.}$$

In other words, the risk to an individual from spending the first 20 years of his life next to a nuclear reactor corresponds roughly to the risk that the same individual would suffer had he smoked one cigarette per year during those 20 years!

It is relatively easy to show that if the individual had spent those years one mile from the reactor, rather than right at the reactor fence, his relative risk would drop to about two minutes, which would correspond to smoking one-fifth of a cigarette during those 20 years.

We see, then, that with present standards for emissions from nuclear reactors it is not at all dangerous to live near or even next to a reactor that is operating normally. The added risks are so small as to be insignificant. (A good discussion of the risks of reactors under normal operation compared to other kinds of risks in our everyday life is given in the text by B. L. Cohen listed in the Suggested Reading section of this chapter.) Consequently, most of the debate on nuclear reactor safety does not center around the question of reactors under normal conditions, but rather centers around the question of what happens when conditions are not normal—i.e., what happens during an accident. It is to this question that we now turn.

397

2. Reactor accidents

The first fact that we have to have firmly in mind when we discuss reactor accidents is that it is *impossible* for a reactor to explode like a nuclear bomb. No one, not even the most vocal critics of nuclear power, has suggested that such a thing could happen. You will recall that the moderator located between the fuel rods played a crucial role in the maintenance of chain reaction. If the moderator were to be removed from the reactor, then the chain reaction would simply shut down. The neutrons emerging from the ^{235}U would be moving much too fast to cause fission in neighboring atoms, and the entire reaction would stop.

You will also recall that the moderator in most reactors is ordinary water which is pumped through the reactor, doubling as a vehicle for carrying heat away from the reactor core. In the event of a serious accident in which a chain reaction went out of control, the reactor core would heat up and the water would be vaporized. This means that there would be nothing between the fuel rods after the vaporization had occurred, and this, in turn, would mean that there would be nothing left to moderate the reaction. In other words, the first thing that would happen if the chain reaction went out of control would be that the moderator in the reactor would disappear. Consequently, a chain reaction running out of control would immediately shut itself off without the intervention of any human agency at all. It is simply a design feature of the nuclear apparatus.

The question of reactor accidents is a very complicated one. Let us consider an analogy. We can all imagine a truly terrible accident that is not impossible, and that could happen any day. For example, we could imagine a fully loaded and fully fueled 747 crashing into the Rose Bowl during a football game. It is not hard to imagine such an accident causing thousands and perhaps even tens of thousands of deaths among the spectators. Nevertheless, even though this accident would be truly horrible, we do not say that because this accident is possible we must shut down all airlines and stop all football games.

Why not? Because even though the accident would have catastrophic consequences, the probability that the accident would happen is so small that most of us are willing to go on flying in airplanes and attending football games.

Thus, the question of risk in accidents necessarily involves a calculation of the probability that the accident will occur.

To see how such calculations could be made, let us take a simple example. Suppose we design a car that has a system of brakes. In principle, the brakes will never fail. Even if they are properly maintained, however, we know that human fallibility being what it is, and materials that go into brakes being what they are, there is some chance that the brakes will fail. Suppose that we find (based on experience or experiment) that a particular set of brakes could be expected to fail once in 50 years of normal use. What are the chances that the brakes will fail on any given day?

In 50 years there are 50 x 365 days; thus, the chance of failure on any given day is

$$\text{chance of failure} = \frac{1}{50 \times 365} = 5.6 \times 10^{-4} \text{ chances per day.}$$

It is left as an exercise to the student to convert this number into an individual risk, a lifetime risk, and a relative risk (making some reasonable estimate about the cost to the individual of each accident).

It could be that we were very concerned about our safety in the car, and we wanted to improve our chances of going through a lifetime of driving without having our brakes fail. One way to do this would be to design the car so that there were two systems of brakes, set up in such a way so that if the first system fails the second system would take over. This would mean that in order for the car to be out of control, *both* systems of brakes would have to fail *at the same time.* What are the chances that both systems of brakes would fail on the same day? Well, the chances that one system of brakes would fail is given above. The chance that both systems of brakes would fail at the same time is the square of this number, or

$$\text{chances of both systems failing} = 5.6 \times 10^{-4} \times$$

$$5.6 \times 10^{-4} \approx 4 \times 10^{-7} \text{ per day.}$$

Thus, by having two independent systems of brakes (a technique that engineers call "redundancy" in design) we can increase our chances from five in 10,000 to about four in 10,000,000. If we built yet a third redundant system, chances would increase to about two in 10 billion.

We see that we could estimate the probability of an accident to our sample car simply by testing a single set of brakes, and then calculating the probabilities for the redundant systems all to fail at the same time. A similar (although much more complicated) calculation can be done for nuclear reactors. After all, a nuclear reactor does not involve exotic new engineering systems. Water is being pumped through pipes, steam is being generated, and all of this involves equipment that is standard in industries all over the world. Consequently, it is possible to carry out studies and produce estimates of the probability of a serious accident in a reactor.

The "serious accident" normally considered in such studies is called the "loss of coolant" accident, and is usually abbreviated LOCA. In Fig. 14.3 we show the parts of a reactor entering into the LOCA. During the normal course of operations of reactors, water is being pumped through the reactor as shown in the figure. This water serves as moderator, and it serves to carry away the heat of the nuclear reactor. A loss of coolant accident would occur if one of the large pipes carrying the water into the reactor were to rupture suddenly, cutting off the flow of water into the reactor. As we have discussed, the loss of the water would correspond to the loss of a moderator in the reactor, and hence it would not lead to a nuclear explosion. However, even if the chain reaction is shut down, we have a huge amount of very hot, very radioactive material left in the

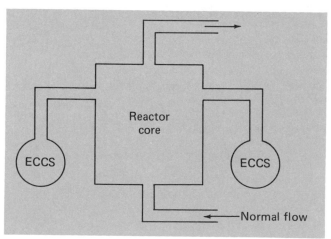

Figure 14.3.

reactor core. This material had been cooled by the water, but when the water disappears it begins to heat up because of its own internal radioactivity. If nothing were done, it would eventually become hot enough to melt its way through the bottom of the reactor vessel and into the ground, thereby releasing large amounts of radioactivity to the environment. In order to prevent such an occurrence, reactors are equiped with something called an "emergency core cooling system," or ECCS. This consists of several reservoirs of water, one stored under very high pressure and others with large pumps. In the event of a major loss of coolant, this water is either pumped or blown into the reactor vessel under high pressure in order to cool down the core.

The question which is being debated at the present time concerns the probability that a number of things will go wrong and that a major loss of radioactive material to the environment will occur. Now let us trace what would have to happen in order for such a thing to occur. First of all, a pipe carrying water would have to break suddenly. Breaks in pipes normally do not occur all at once. Normally, a break in a pipe starts with a small crack or leak and develops into a large one. Reactors are equipped with several systems to detect small leaks or small cracks in the pipe. The pipes are inspected ultrasonically on regular maintenance schedules, there are humidity detecting devices around reactors that would detect a sudden increase in water in the air due to the presence of a leaking pipe, and the water pressure itself is constantly monitored. If all of these detection systems failed, or if the pipe ruptured suddenly, then it would have to break in such a way that all of the water was prevented from reaching the reactor. For example, the kind of break that you normally think of in a pipe in a home would not be of this type. The fact that water was squirting out of a pipe would not prevent a certain fraction of the water in the pipe from getting past the hole.

Given the existence of such a break in the pipe, for a meltdown to occur it would also be necessary for the emergency core cooling system to fail. Thus, in order for a serious accident to occur, we would have to have a failure first of the detection systems that would warn of the weakening of the pipe (and there are at least three of these), then the kind of break in the pipe that would cause a sudden complete loss of

401

flow, and then the complete failure of the emergency core cooling system to operate. This is an example of what engineers call "defense in depth" when they discuss design.

While the LOCA is one way to arrive at a situation where the core would melt down, there are others as well. Major accidents or fires (such as the one which occurred at the Brown's Ferry reactor in 1976) could, in principle, produce a core meltdown and subsequent release of radioactive materials to the environment. Each "route" to a meltdown would have to be examined as we examined the LOCA to determine the probability that a meltdown would occur in that way. The total probability of a meltdown would be the sum of all of these probabilities.

At the present time, most of the scientific debate on the question of reactor safety centers around the Rasmussen Report, which was done under the leadership of Prof. Norman Rasmussen of MIT for the old Atomic Energy Commission (whose regulatory functions are now held by the Nuclear Regulatory Commission). This study came up with estimates of the risk associated with serious reactor accidents. In the original version of the report, separate probabilities were calculated for different degrees of seriousness of the accidents. The most serious (and least likely) accidents involve horrendous consequences—thousands of deaths and billions of dollars of damage. The more likely accidents are less serious. Cohen[1] gives the overall conclusions of the original report as follows:

a) the probability of a core meltdown is one every 17,000 reactor years;

b) the consequences of the accidents were much less severe than had been thought. In the terms of our discussion here, the consequences of a major accident would be

(1) 0.16 deaths per year from direct exposure to radiation,

(2) 1.2 deaths per year from cancers,

(3) $6.5 million dollars in property damage per year.

Of course, these small numbers occur because the most serious accidents occur rarely, so their cost gets spread out over long periods of time.

[1] B.L. Cohen—text listed on page 409.

When the report was published, it became the immediate center of a lively debate. Several independent groups, including the American Physical Society (the professional society of physicists in the United States), carried out their own analyses of some of the accident consequences considered in the Rasmussen Report. In addressing the question of long-term cancer associated with a reactor accident, for example, the Physical Society group found that the Rasmussen study had seriously underestimated the effects of certain radioactive elements, (particularly Cesium 137) which have a half life of tens of years and which would act as long term land contaminants. Mathematicians familiar with statistical analysis of unlikely events from their work in the Apollo program raised technical objections to the methods used in the Rasmussen Report. Other scientists, while agreeing that reactor accidents are unlikely events, have pointed out that the uncertainties involved in calculations of this sort could raise the accident frequency estimates for the most severe accidents in the Rasmussen Report by as much as a factor of 10,000.

The current situation as regards reactor safety was summed up by Dr. Frank von Hippel of Princeton University (and one of the members of the Physical Society study): "The studies following the Rasmussen report have restored the reactor accident as a major cause of concern in the energy picture. This doesn't necessarily mean that we should stop building reactors, but it does mean that very tight controls on factors bearing on safety will have to be enforced by the federal government."

3. *Nuclear waste disposal*

A new controversy which has sprung up very recently has to do with the question of what happens to nuclear wastes once they are generated. The picture that is called to mind by opponents of nuclear power is of the world being strangled by huge amounts of dangerous radioactive wastes that must be stored and kept away from any kind of human contact for millions of years. How did such a picture arise?

The answer has to do with the presence in radioactive wastes of the element plutonium. Plutonium has a half-life of about 24,000 years, and it is a rather dangerous element. You often see the statement "A piece of plutonium the size of a grapefruit would be enough to destroy the entire population of the world." This statement is true, but it is somewhat

403

misleading. I could equally well argue as follows: It is well known that a human being can drown in a cup of water. Therefore, there is enough water in an ordinary swimming pool to drown everybody who lives on the entire eastern seaboard of the United States. This statement is also true, and it is also misleading.

The problem with both statements is that they assume that every bit of plutonium or water finds its way to that area of the human body where it will do the most harm. It is extremely unlikely that such a thing would happen. For example, at least a ton of plutonium was released into the air during the atom bomb tests in the 1950s. There was no immediate upsurge in deaths due to lung cancer from it. The reason for this is that plutonium, in addition to having a radio-active nucleus, also has electrons going around it, and hence has a chemistry. In other words, plutonium will react chemically with its surroundings as well as having nuclear reactions going on in its nucleus. The most common chemical reactions involve having the plutonium being bound into clays in the ground. Once they are in the ground, they represent very little threat to any sort of life. As far as risks to human beings are concerned, the major threat from plutonium has to do with plutonium being released into the air and breathed into the lungs. At the present time, it is not clear how much plutonium is required to cause cancer, although it is surely a rather small amount (something on the order of the size of a grain of dust).

The question of waste disposal, then, boils down to the question of how we are to contain radioactive substances after they have been used in reactors, and how we are to prevent massive releases of this material into the environment. The first question to ask concerns the amount of material there would be. The present plans call for solidifying this waste into something resembling a clinker from a coal furnace, or a glass, and burying it about one-half mile under the ground in geologically stable areas. In solid form, the waste from a large reactor would occupy about 90 cubic feet of space—about the size of a desk.

By "geologically stable" people usually mean burial in salt mines. The presence of salt means that water has not been in the area in a long time. Since the main danger of buried

The first nuclear reactor, 1942, and Fermi team, rendered by artist Gary Sheahan.

The nuclear power station at Indian Point, New York.

Industrial pollution.

Solar Collectors

radioactive waste has to do with the waste being carried by underground water to regions where it can be brought into contact with human beings, the burial site would probably be chosen for its distance from underground water, and will therefore be somewhere in the southwestern United States.

One way of gauging the seriousness of the radioactive waste disposal problem is to inquire as to the amount of radioactivity that is already in the ground from natural sources. We have already seen that the ground is radioactive, and that large amounts of uranium already exist in the ground beneath us. As an example, let us take a piece of land six miles on a side, and about one-half mile deep. In metric units, this would be a piece of land 10 kilometers on a side and one kilometer deep. This is the size of an average moderate-sized town such as Charlottesville, Virginia (where the author lives). How much uranium is there in a block of material 10 kilometers on a side and one kilometer deep? The volume of such a material is about 10^{17} cubic centimeters, which means there are about 3×10^{17} grams of rock present. This corresponds to 3×10^{11} metric tons of rock.

There are four grams of uranium for every ton of material in the earth's crust, which would mean there are $4 \times 3 \times 10^{11} = 1.2 \times 10^{6}$ tons of uranium in the ground. The amazing fact is that there is, under each 100 square kilometers of land area in the United States, about a million tons of uranium on the average.

Uranium itself is not a particularly dangerous element. However, there is a very dangerous side product that comes out of its radioactive decay chain, and this is radium. Radium has a half-life of 1,620 years, and it does stay in the body. In fact, it is one of the major sources of radiation to the body from natural sources. If we ask how much radium there is in the piece of ground that we are talking about, we find that there is about one-third of a ton of radium in a piece of ground six miles on a side and half a mile deep. Some of this radium normally finds its way into our bodies and is part of our normal background radiation.

In order to put things in perspective, let us ask how many people could be killed by the radium that is in the ground below Charlottesville. Normally, we would expect that something like one milligram of radium would be a rather

405

dangerous and perhaps even fatal dose. In other words, we would say that radium and plutonium are roughly equally dangerous to human life. There are 300,000,000 milligrams of uranium in a third of a ton, which means that there is enough radium under Charlottesville to kill every living person in North America.

Obviously, the population of North America is not seriously menaced by the radium that is now lying under Charlottesville. The reason for this is simply that once the radium is underground very little of it actually gets out of the ground and into contact with human beings. There is no reason to expect that plutonium or other radioactive waste would be any different. Consequently, I think it is clear that the danger from the radioactive waste is rather slight provided that it is buried in an appropriate place.

However, like the question of reactor safety, the mere fact that waste disposal *need* not be major health hazard does not guarantee that it will be handled in such a way as to see that the hazard is minimized. Again, stringent federal controls will have to be maintained to ensure that all of the waste winds up underground where it belongs, and that none is accidentally released into the environment.

E. Coal versus Nuclear Energy

Now that we understand the risks that are attendant on the construction of nuclear facilities, we can turn to the question of generation of power in the United States for the remainder of this century. There are many aspects to this problem. For example, by far the easiest way to generate power would be to burn oil. However, the world political situation makes it extremely hazardous for the United States to depend on Middle Eastern sources of oil. Therefore, the use of oil to generate electrical power is probably not feasible for the United States.

The two alternatives that we have for the short term are the use of coal (which the United States possesses in abundance) and the deployment of nuclear fission reactors. Let us consider only the question of safety—i.e., let us consider which method of generating power will cause the least number of fatalities. We have already discussed the sources of danger from the use of nuclear reactors. The sources

of danger from coal are equally obvious. There are accidents that claim the lives of miners who mine the large
amounts of coal that would be needed to generate electrical
power for the United States. There is the unavoidable pollution of the atmosphere by the effluents from coal stacks, and
there are the somewhat less tangible but equally important
problems of destruction of the land from stripmining. None
of these are trivial problems. Therefore, when we begin to
think about how we want to generate electrical power in the
future, it will be necessary to ask the question "Which of the
methods available to us is the least harmful?" rather than the
question "Is nuclear power safe?". *Every* method of generating power could cause injuries to persons and property.
The question that we have to ask is not whether any particular method of generating power is dangerous, but whether it
is less dangerous than the other options that we have open
to us at the present time.

There is, however, one extremely important point that
we have not yet discussed. We know how to estimate the
amount of radiation exposure that people receive from the
routine and accidental emissions from nuclear plants. We do
not know so well how to estimate the exposure to routine
radioactive emissions from coal plants.

What does this mean? You will recall that when we
discussed the occurrence of natural radiation, we made the
statement that everything is radioactive. This statement
meant that every substance—including coal—contains small
amounts of radioactive materials. It is impossible to separate
these materials out in any economical way from the bulk of
the coal. Consequently, when the coal is burned radioactive
materials will be released from the coal and some of them
will enter the atmosphere. As a matter of fact, estimates for
the amount of radioactive material emitted by a coal-burning
plant in the course of a year run very high. The study by
Eisenbud and Petron given in the Suggested Reading Section
of this chapter estimates that a coal-fired plant will release
anything from one-quarter to one Curie per year of radiation
into the atmosphere. This is to be compared with the emissions from the Yankee Nuclear Power Station in Connecticut
of about one *milli*-Curie per year.

The other disadvantages of a coal-powered economy
have to do with the pollution of the atmosphere by sulphur

dioxide and other dangerous chemicals that are subsequently inhaled by human beings and that lead to a large incidence of lung cancer and other forms of disease. The estimates on the damage done to individuals by these agents vary widely but 5,000 to 20,000 deaths per year are typical.

When we draw up the balance sheets on coal and nuclear electrical generation, then, we see that there are plusses and minuses on both sides. Coal is a familiar form of fuel which the United States possesses in abundance. Large scale strip mining, with the attendant destruction of land and social structures, will be necessary to exploit it, and burning coal will introduce chemical and radioactive pollutants into the atmosphere.

Relying on nuclear fission, however, also has drawbacks. While the mining and "burning" of Uranium does not have the environmental impacts that the corresponding processes do for coal, a whole new set of hazards are introduced, mainly associated with accidental release of radioactive materials. Of these, the possibility of a reactor accident is probably the most serious, even though the probability of such accidents is small. In addition, as the nuclear industry expands, there will be more pressure to introduce two new elements into the picture—the breeder reactor and fuel reprocessing. The breeder reactor produces plutonium as it burns uranium, so that more reactor fuel (in the form of plutonium) comes out of the reactor than went in. Fuel reprocessing concerns itself with the extraction of useable fuel (primarily plutonium and U-235) from used fuel elements of commercial reactors. Both of these technologies would result in large scale production of plutonium. It is relatively easy to extract plutonium from fuel rods by chemical means and it can also be made rather easily into bombs. Thus, a whole host of political questions having to do with nuclear proliferation and terrorism have to be considered before this next step can be taken.

The nuclear debate will be with us for a long time, then, and as this is written (May, 1977), the debate is moving away from the technical questions, such as reactor safety, and toward more political questions, such as proliferation, which are beyond the scope of this book.

SUGGESTED READING

Many books and articles on the question of nuclear reactors have been written, and it would be impossible to list them all here. The text by Cohen, cited below contains an extensive bibliography.

Cohen, Bernard L. *Nuclear Science and Society.* New York: Anchor Books, Doubleday, 1974. An extremely readable and thorough account of the uses of nuclear energy, the reactor controversy, other uses of nuclear physics in medicine and industry, and the basic structure of the nucleus.

Ehrlich, Paul. R. *The End of Affluence.* New York: Ballantine Books, 1974. The text contains the environmentalist point of view on nuclear reactors, as well as many references.

Eisenbud, M., and Petron, G. *Science* 1964, *144* (288). A study of radioactive emission from coal-fired generating plants.

Lawrence, William K., "Of Acceptable Risk" William Kaufmann, Inc., Los Altos, Cal. 1976 Discusses the difference between risk (which can be calculated) and safety ("is subjectivity defined) for a wide range of modern problems. An excellent non-technical book for those interested in these questions.

Inglis, David R. *Nuclear Energy: Its Physics and Social Challenge.* Reading Mass., Addison-Wesley, 1973. A careful and detailed discussion of the physics of the nuclear reactor.

QUESTIONS AND DISCUSSION IDEAS

1. It is estimated that coal in the western part of the United States contains about 1.5 pounds of uranium per million pounds of coal. In 1970, the United States burned 330 million tons of coal in electrical generating plants, supplying 260 million kilowatts of power.

    a. About how much coal must be burned on the average to supply one kilowatt of power per year? (NOTE: An electric toaster uses about one kilowatt.)

    b. If the exhaust from the coal burning is not cleaned, how much uranium will be released in one year?

    c. A typical large power plant generates 1000 megawatts. How much uranium is in the coal used by such a plant?

In the problems that follow, a statistic on a particular form of death for the United States is given. In each problem, calculate (a) individual risk, (b) lifetime risk, (c) relative risk, and

409

(d) comment on the assumption you make about the relative probability of dying from that cause at various times during one's life. Assume that the population of the United States is 200 million.

2. There are 55,000 deaths per year in automobile accidents.
3. There are 500 deaths per year from insect bites.
4. There are 1500 deaths per year from being hit by falling objects.
5. There are 400 deaths per year from flood, hurricane, and tornado.
6. From the answers to Problems 2–5, make a comparison of the risks of death from living next to a reactor to other forms of risk.
7. Using the same reasoning that was used in calculating the amount of uranium in Section D3 of this chapter, calculate the amount of copper, silver, and gold in 100 cubic kilometers of the earth's crust (i.e., in a piece 10 kilometers on a side and one kilometer deep).
8. If you live in a brick house, you will receive about 40 additional mR of radiation from trace elements in the brick. Calculate individual, lifetime, and relative risks of cancer to a person who has lived in a brick house for 20 years. State when the risk occurs.
9. Repeat Problem 8 for a person who receives a dental X-ray of 20 mR at age 20.
10. Repeat Problem 8 for a person who receives a series of diagnostic X-rays (such as chest X-rays) totaling 500 mR (the equivalent of three to four chest X-rays) at age 20.

CHAPTER
XV

ELEMENTARY PARTICLES

Big fleas have little fleas
That bother 'em and bite 'em
And little fleas have littler fleas
And so on, ad infinitum.

Anonymous

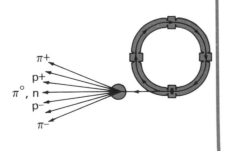

A. The Elementary Particles We Have Encountered So Far

In dealing with the structure of the atom and of the nucleus, we have run across a number of different kinds of entities, which we call "particles." Actually these entities are neither particles nor waves in the classical sense. However, it has become customary to refer to things like electrons as "particles," even though they exhibit the properties of waves in some experiments.

The particles we have discovered so far are the electron, the proton, the neutron, and the photon. In the table below we list these particles, together with their electrical charge and their mass. The measurement of mass in MeV (million electron volts) will be discussed below.

particle	charge	mass (grams)	mass (MeV)
photon	0	0	0
electron	-1	9.1×10^{-28}	.5
proton	+1	1.7×10^{-24}	939
neutron	0	1.7×10^{-24}	939

The profound hope of physicists during the 1930s was that these few particles (with perhaps one or two additions, which we shall discuss later) would describe the entire physical universe. There were three particles (the proton, neutron, and electron) from which everything else could be made. From the proton and neutron, any nucleus could be constructed simply by adding different numbers of each. By adding the appropriate number of electrons, any atom could be built around that nucleus. Finally, by the process of chemical combinations which we have already discussed, any chemical compound could be made from those atoms.

In this picture of the world, a marvelous simplicity is seen to hold. The infinite variety of substances and materials and objects which we see around us is seen to be explicable in terms of three elementary "building blocks." These three building blocks, therefore, were called "elementary particles." The simplicity and beauty of this system led scientists to believe that they were very close to achieving the complete understanding of the world in terms of a few basic entities which had eluded them from the time of the Greeks. The only fly in the ointment was the fact that no one was able to explain how it was that the nucleus stayed together. The presence of the positively charged protons so close together inside the nucleus, as we have already discussed, implied that there must be a stronger force holding them together than any that had been encountered in nature up to this time. What was not appreciated at the time, but became evident during the late 1930s, was the fact that the existence of this stronger force implied that there must be more elementary particles than those in the above table.

When we discussed the covalent bond we saw that it was possible for two atoms sitting next to each other to exchange electrons back and forth. We saw, also, that this exchange of particles led to an attractive force which kept the two atoms together. Virtually all of the molecules in your body are held together by forces of this type. In 1935, the Japanese physicist H. Yukawa suggested that a similar mechanism might be responsible for the strong force which holds the protons and neutrons together inside the nucleus. He suggested that just as atoms can be held together by the exchange of electrons, protons and neutrons can be held together by the exchange of another kind of particle.

According to his calculations, this new kind of particle would weigh more than an electron but less than the proton or neutron. Consequently, he gave it the name "meson," which comes from the Greek word for intermediate, and referred to the fact that these new particles would have to be intermediate in weight between particles that were already known. His original calculations suggested that the mass of the new "meson" would have to be about 200 times the mass of the electron.

Since the new particles were not in evidence in ordinary matter, one could well wonder how one would go about proving or disproving a suggestion like Yukawa's. The answer lies in the relation between mass and energy, which we learned in special relativity. Since the energy required to create a given mass is given by mc^2, it is obvious that if enough energy were available, a new particle could be created from that energy where no particle had existed before. This is the inverse of the process we discussed when we talked about generation of power from nuclear fission. In that process, mass disappeared and energy was created. The other process, in which energy disappears and mass is created, is also possible in nature. The problem is finding enough energy to make the mass that you want.

The way that physicists usually discuss this problem may seem a little strange at first. We know that mass and energy are equivalent. Therefore, what is usually done is to talk about the mass of the particle in terms of the energy that would be available if the mass of the particle were completely converted to energy. Thus, instead of saying that the mass of the electron is 9.1×10^{-28} grams, we say that the mass of the electron is about 500,000 electron volts.

The electron volt, as we have stated earlier, is simply the amount of energy required to move one electron through one volt. Moving one electron from one volt of a car battery to another would correspond to an expenditure of 12 electron volts and moving one electron from one pole of a 500,000 volt "battery" to the other would correspond to 500,000 electron volts. If the mass of one electron were completely converted to energy, this is how much energy there would be. On the same scale, the proton would have a "mass" of 939,000,000 electron volts.

Because the Europeans consider a "billion" to be a million millions, while Americans consider a "billion" to be a thousand millions, some confusion arose in the early days of particle physics about what to call the unit that was a thousand million electron volts. What was eventually settled on was a unit called the GeV, or giga electron volt. The giga electron volt is 10^9 electron volts, and this is the same unit on both sides of the Atlantic. In these units, the mass of the proton is .934 GeV.

In the language of measuring mass in terms of electron volts, then, Yukawa's meson would have a mass of about 140 million electron volts (MeV). If we wished to create this particle, we would have to find somewhere a source for the energy that could be used up to create this particular mass. In the 1930s, no such sources existed in the laboratories. The "high energy" projectiles of the type used by Rutherford in his experiments came from the disintegrations of nuclei and consequently had energies of only a few MeV. None of them had energies of hundreds of MeV, and consequently none of them could be used as a source of energy for the creation of the Yukawa particle.

414

Fortunately, there is a source of naturally available particles whose energy is more than sufficient to create a meson. This source is cosmic rays. We have already discussed cosmic rays in the context of radiation dose. What we did not mention at that time, but what is extremely important to the development of high energy physics, is the fact that these cosmic rays come in a wide spectrum of energies, and that some cosmic rays have energies far in excess of that needed to create new particles.

Consequently, during the late 1930s, a major search was made for Yukawa's particle. It was known that cosmic rays entering the atmosphere would quickly strike atmospheric nuclei. The particles that were created in that collision would go on to strike other nuclei and create still more particles, until a regular cascade had been built up. By setting up instruments at sea level, the parts of this cascade which penetrated all the way through the atmosphere could be analyzed and, in particular, a search could be carried out for particles that were neither protons, neutrons, nor electrons.

In 1938 it was found that there were particles in this debris that had a mass of about 105 MeV. This particle was dubbed the μ meson, following the convention of using Greek letters (in this case "mu") to denote particles.

Unfortunately, it turned out that the experimenters had found the wrong particle. They observed that the μ meson did not interact strongly with the nuclei through which it passed. In the Yukawa picture this would be very difficult to understand, since it was this meson that was supposed to be supplying the strong force that held the nucleus together. It is inconceivable that a meson which got near a nucleus and which had the capability of interacting strongly with the particles inside that nucleus would not do so. The lack of interaction between the μ meson and the nucleus, then, was an indication that the μ meson, interesting as it might be in its own right, was simply not the particle that had been predicted by Yukawa.

It was not until 1948 that experiments with cosmic rays revealed still another meson, which was christened the π (Greek letter "pi") meson and is usually called the "pion." This particle *did* interact strongly with nuclei, and hence was a candidate for the particle that Yukawa had predicted. Although we speak of the π meson as a "particle," there are actually three π mesons. They all have approximately the same mass, but one has a positive electrical charge, another has a negative electrical charge, and a third has no electrical charge whatsoever. They are given the same name because all of their properties other than charge are the same. For example, they all have zero spin (that is, they do not rotate about their own axis as the electron does). Other quantum numbers of the π meson family will be discussed later.

If we think of the proton and neutron as the building blocks of the nucleus, then the π meson plays the role of the glue which holds the proton and neutron together and keeps the protons from flying apart under the influence of the electrical force. They act as a glue by the simple process of being exchanged between protons and neutrons, in a way that is somewhat analogous to the exchange of electrons in the covalent chemical bond. However, as much as the existence of the π mesons may add to our knowledge of the nucleus, it begins to break down the hope of finding a simple way of explaining nature with a small number of particles. With the three π mesons and the μ mesons the proliferation of elementary particles had begun.

B. Technical Digression: The Idea
   of Force as a Particle Exchange

Up to this point we have spoken rather loosely of an analogy between the covalent bond in chemistry and the strong force that holds a nucleus together. In this section, we shall discuss in a little more detail how it is that the π meson provides the glue for nuclear structure.

You will recall that when we studied quantum particles, we found that there was an uncertainty principle which operated and which told us that it was impossible to measure with infinite precision both the momentum and the position of a particle at the same time. In mathematical terms, this uncertainty principle is written

$$\Delta x \cdot \Delta p \geqslant h,$$

where h is Planck's constant. It can also be proven (although it is much more difficult to make plausible on an elementary basis) that there is another uncertainty principle. This second

416

uncertainty principle tells us that it is impossible to measure both the energy and time with infinte precision. In mathematical terms this takes the form

$$\Delta E \cdot \Delta t \geqslant h,$$

where ΔE is the uncertainty in the measurement of the energy of a system, and Δt is the uncertainty in the time during which the system has that energy.

The energy-time uncertainty relation has one rather amusing consequence. Suppose we had a proton sitting by itself. The energy of this proton would be given $E = mc^2$, where m is the proton mass. In classical physics, the energy would be this amount at any time. It could not change unless some outside agency were to act on the proton. But what does it mean to say that the energy cannot change? When we speak of a change in energy, we necessarily are speaking of separate measurements. Speaking of a change in energy implies that we measure the energy at one time, which we can call t_1, and then we measure the energy again at a later time, which we shall call t_2. If the measurement of the energy gives the same results at t_2 as it did at t_1, then we say that the energy was constant during that time.

In the context of classical physics, this is the way we would think about energy conservation. In the context of quantum physics, however, the uncertainty principle introduces a new dimension. Just as we cannot talk about measuring the momentum of a particle in the quantum world without at the same time making some statement about the uncertainty in that measurement and the effect of the measurement of momentum on the measurement of position, we cannot measure the energy of the particle in the quantum world without some discussion of the times involved in that measurement. In particular, the energy-time uncertainty principle stated above tells us that if the time interval during which we measured the energy has the value t_0, then the uncertainty in the energy caused by the disturbance of the system attendant on the same measurement itself will be given by

$$\Delta E \geqslant h / t_0$$

Thus, if we wish to know the energy of a system exactly, we must have an infinite time interval during which to make that measurement. This is exactly analogous to the momentum-position uncertainty principle, in which if we wish to measure the momentum of a particle exactly, the particle cannot be localized anywhere in space but would have some probability of being located anywhere.

Let us go back and consider a single proton sitting by itself. We have already discussed the fact that if the time interval t_0 is large enough, the uncertainty in the energy can be made as small as we wish. However, there is another side to this coin. If the time interval $_0t$ is *short* enough, the energy uncertainty can be as large as we wish. In particular, if the time interval is very short indeed, the energy uncertainty can be so large that we could not tell whether the energy of the proton was simply m_pc^2—the rest mass of the proton—or whether it was $(m_p + m_\pi)c^2$—the rest energy of a proton and a pion combined.

This is a rather revolutionary concept. What it says is that a single proton can change itself into a proton plus a π meson, provided that it gets back to its original state in a short enough time. In high energy physics, we often speak of the proton "fluctuating" into a proton plus a pion and then coming back to its original state in a short time interval. A proton moving along is thought to be something like what we have pictured in Fig. 15.1. Some of the time the proton is simply a proton, but for short time intervals it can fluctuate into a proton and a pion and then recombine, become a proton again, fluctuate again, recombine again, and so forth. The only limitation is the requirement that the π "get home" quickly enough so that the uncertainty principle is satisfied. The actual state of a proton, then, is not a very simple thing. It is basically a simple proton, but at any given time the proton may be in a state of fluctuation, in which there is a π meson present as well.

It is important to realize that the possibility of fluctuations of the type we discussed above is something new that we have not encountered before, It arises through a combination of quantum mechanics (through the energy-time uncertainty principle) and relativity (through the mass-energy relation). In other words, quantum mechanics tells us that there is an uncertainty in the measurements of the energy of the system which depends on the time interval involved in the measurement. Relativity, then, translates the amount of

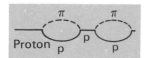

Figure 15.1.

energy in this uncertainty into a mass through the standard formula $E = mc^2$. Thus, by combining these two principles, we come to the conclusion that an elementary particle may undergo fluctuations and in these fluctuations there may appear more particles than we would think should be present. However, these particles can exist for only very short, fleeting moments of time.

Because of their rather ephemeral nature, these particles are usually referred to as "virtual particles." Both the proton and the neutron (and even the π meson) fluctuate into states of virtual particles all of the time. Most of the particles that we shall discuss in the next section do so as well.

The existence of virtual particles has a very important consequence for discussing the strong interactions. We have discussed the fact that a single proton may, at any given time, fluctuate into a state of a proton plus π meson. If the proton is sitting by itself, this will be a rather unremarkable event. The proton will fluctuate into the proton plus π meson and then return to its basic state before we can make a measurement to determine if the fluctuation has occurred. However, suppose that the proton is not sitting by itself. Suppose that the proton is actually sitting inside of a nucleus, where there are other protons and neutrons in close proximity. Then we could have a sequence of events something like that pictured in Fig. 15.2. The first proton could fluctuate into a state of a proton plus π meson. This π can exist for a very short time—the time allowed by the uncertainty principle. In practice, this is about the time that it would take light to travel the distance across a proton, or the distance from one proton in the nucleus to another (these two distances are approximately equal). Suppose that during the short fluctuation of the first proton the π moved away from the first proton a short distance, as shown in Fig. 15.2b. Suppose further that another proton happened to be close enough to the first proton to absorb the π that the first proton had emitted. This is shown in Fig. 15.2c. The net result of this sequence of events could be described as follows—proton number one emits a virtual π meson, which is subsequently absorbed by proton number two. The entire sequence of events takes place in a time interval so short that we could not, because of the energy-time uncertainty principle, make any measurements that would prove that it could not happen. We would then say that the

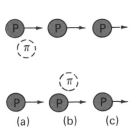

(a)   (b)   (c)

Figure 15.2.

419

two protons (or two neutrons, or one proton and one neutron) had exchanged a π meson. However, this π meson would not be an ordinary π meson, but a virtual π meson.

We believe that the strong interaction originates in this way, by the exchange of vitrual mesons. It is the force generated by the exchange of virtual mesons that we belive holds the nucleus together. This is what is meant when we speak of the π meson as being the nuclear "glue."

Once this idea (that the exchange of a particle could be responsible for force) had been introduced into physics, a number of the forces that we have encountered previously were seen to be associated with the exchange of other kinds of particles. For example, we discussed the electrical force—the force between two charged particles. This force can be explained as the exchange of an elementary particle—the photon—between the two charged bodies. In this picture of things, the Coulomb force between two charged bodies is actually associated with the exchange of a large number of virtual photons between them. Of course, these photons are not associated with light or with radio waves or any other "normal" type of electromagnetic radiation. They are of such long wavelengths that it would be almost impossible to detect their existence. Nevertheless, from a theoretical viewpoint, we can think of the electromagnetic force as being due to the exchange of photons, much in the same way as we think of the nuclear force as being due to the exchange of π mesons.

Even the gravitational force is now thought to be associated with the exchange of a particle called the "graviton." This particle has never been detected, but it is thought to exist by theoretical physicists. The gravitational attraction between the earth and the sun, then, is thought to result from the interchange of many gravitons back and forth between those two bodies. It is left as a problem for the reader to see how this concept—the generation of gravitational forces by the exchange of a graviton—affects the arguments about "action at a distance," which were so prevalent during the 18th century.

We see, then, that the combination of quantum mechanics and relativity leads us to a totally new way of thinking about forces in nature. This new way of thinking is intimately connected with the idea of elementary particles.

Therefore, it becomes essential to understand what elementary particles are and how they affect the nature of the material world.   It is to this question that we now turn our attention.

C. The Elementary Particle Zoo

By the middle of the 1940s the number of elementary particles had grown.   Instead of having three simple building blocks (proton, neutron, and electron) along with the photon, we had the discovery of two different kinds of mesons. One, the π meson, interacts strongly with the proton and neutron and has been identified as the "glue" holding the nucleus together.   The μ meson's interactions seem to be much more like those of the electron than those of the π meson.   In other words, it has an electrical charge, so that it interacts electromagnetically with other particles, but it does not seem to interact strongly with them.   Except for the fact that it is 200 times heavier than the electron, there is very little to distinguish the μ meson from the electron. The fact that nature created the electron, and then created an almost identical particle 200 times heavier remains one of the major puzzles of elementary particle physics.

Nevertheless, the number of elementary particles was still small.   The introduction of the two mesons does not really make us doubt that our ultimate goal—the explanation of matter in terms of a few elementary particles—will ultimately succeed.   However, experiments were already being done in the later 1940s that would lead to discoveries that did cast doubt on the possibility of achieving this goal.

In 1947, investigators at the University of Manchester in England noted some strange events taking place when cosmic rays were allowed to strike sheets of lead in an experimental apparatus.   The apparatus, called a "cloud chamber," is designed so that the path where an elementary particle has passed becomes visible because tiny droplets of water vapor collect on the ionized atom in the wake of the particle.   What they saw is sketched in Fig. 15.3.   A cosmic ray entered the piece of lead from above and created a spray of elementary particles.   Most of these were protons and π mesons.   Because the neutron does not have an electrical charge, it leaves no ionized atoms in its wake, and hence would not leave a track in an apparatus like the one we are discussing.

421

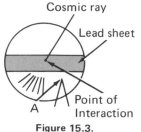

Cosmic ray

Lead sheet

Point of
Interaction

A

Figure 15.3.

Every once in a while, however, a strange type of event would occur. Some distance from the lead sheet (for example, at the point labeled A in Fig. 15.3) two particles would suddenly materialize and go in different directions. This type of event was said to be caused by a "V particle." The only explanation for its occurrence was that an uncharged particle (which would leave no track) had traveled from the point of interaction to point A, where it had spontaneously split into two charged particles (later identified as a proton and a negatively charged π meson), which did leave tracks in the cloud chamber and which were therefore visible.

Two things were immediately obvious about this "V particle." First of all, since one of the visible particles was a proton, the V particle must weigh more than the proton. If it did not weigh more than the proton, it could not have split into a proton plus something else. Simple energy conservation would guarantee this.

The second point that is obvious about the V particle is that it lives for a fairly long time. It lives long enough to go from the point of interaction to a point A of the above figure. Even if it were traveling at the speed of light (the fastest speed at which it could travel), this would imply that it had a lifetime of about 2×10^{-10} seconds measured in its own rest frame. During this time, it could travel several centimeters in the laboratory.

This may seem like a very short time. Indeed, on the human scale, it *is* a very short time. However, on the nuclear scale, it is an extremely long time. A typical time during which a particle like a virtual π meson would have to interact as it goes across a nucleus would be something on the order of 10^{-24} seconds—the time that it would take light to cross a nucleus. The fact that the V particle can be created in a very short time in a nuclear collision but takes a very long time to reach its final state of proton and π meson means that this particle has properties that were totally unexpected when it was discovered.

Actually, particles like the V particle have become more common since 1948, and a number of terms have been introduced to describe them. The first V particle that was discovered is now called the Λ. The process by which one of these heavier particles splits into a proton and a π meson (or any other final states) is called, in analogy with the nuclear

422

processes giving rise to radioactivity, a "decay." Thus, in modern terminology we would say that Fig. 15.3 showed the production of a Λ particle and a subsequent decay into a proton and a π meson.

The Λ particle is not the only particle known in 1948 that had the property of being created in a strong interaction and decaying very slowly. We have already discussed the fact that the neutron decays very slowly. Indeed, this is the process that lies behind the emission of β radiation in the nucleus. A neutron by itself will decay in a matter of minutes into a proton, an electron, and yet another kind of elementary particle, the "neutrino."

Carl Anderson (1905–), the discoverer of the positron.

The existence of the neutrino had been suggested in the 1930s by Enrico Fermi because when the decay of the neutron was examined, it was found that neither energy nor momentum conservation would hold if only the proton and electron were counted. In other words, if we looked at the decay of a neutron and added up the energy and the momentum before the decay and compared it to the energy and the momentum of the proton and electron after the decay, we would find that there was less energy and momentum associated with the proton and electron after the decay than there had been associated with the neutron before the decay. Consequently, Fermi postulated the existence of a particle that he called the neutrino (which is Italian for "little neutral one") which would carry away the extra energy and momentum. Since the particle could not be detected, the laws of conservation of energy and momentum could be preserved. Direct evidence for the existence of a neutrino did not become available until the mid 1950s.

Thus, both the ordinary neutron and the new Λ particle were observed to decay slowly. We have already mentioned that this type of slow decay is called a "weak" interaction. The "weakness" of the interaction can be thought of as associated with the fact that it takes the interaction a very long time to occur when it is compared to a "strong" interaction such as that of the π meson inside of the nucleus. The essential difference between the decay of a neutron and the decay of a Λ particle is that in the decay of the Λ no neutrino is involved. Thus, this type of particle was totally unexpected at the time of its discovery. Later on, because of this unexpected property, the Λ was classed with a group of particles that were given the generic name "strange particles." A more concrete definition of "strangeness" will be given in the next section.

The discovery of other "strange particles" followed quickly on the discovery of a Λ. It was found that there was an entire family of strange particles that decayed ultimately into protons and π mesons. They were called the Σ particles. They were heavier than the Λ and the proton. Like the Λ, they were created in strong interaction and decayed slowly (i.e., through the weak interaction). Unlike the Λ, there were three such Σ particles. One had a positive electrical charge, one had a negative electrical charge, and one was neutral.

Thus, it began to become obvious that the number of elementary particles was not going to be small. Indeed, with so many elementary particles discovered, the use of the term "elementary" could be questioned. What does it mean to talk about "elementary particles" if there are as many of them as we are discovering? Nor does the saga of the discovery end with the Σ particles. We have seen that the existence of particles that were created in strong interactions, and would subsequently decay through weak interactions had been totally unexpected. The particles we expected to exist would be particles that were created in strong interactions and then decayed by the strong interaction. In other words, we would have expected to find particles that were created quickly and then decayed quickly. The fact that such particles were not found in cosmic rays turned out to be the results of the fact that cosmic ray experiments are not suited to search for such particles. Their discovery had to wait until artificial "cosmic rays" could be produced in accelerators.

An accelerator is simply a device for taking a proton or electron and giving it much more energy than it would have had normally. In the early 1950s, accelerators (called cyclotrons) were being built in research laboratories around the United States. We shall not go into a discussion of how they work here, but we shall simply say that the end result of having an accelerator is that one can have a beam of protons of very high energy. This beam can be directed against targets in order to discover new facts about elementary particles. The cyclotron takes the place of the α emitter in the Rutherford experiment, and gives the experimenter control over the probe he is using to learn about nature.

424

In 1952, Enrico Fermi and his co-workers at the University of Chicago used their new cyclotron to begin a systematic search for particles of the type that scientists had expected to find. It turned out that their search was successful. They found that there was indeed a class of particles that could be created in the strong interaction and that decayed by a strong interaction. In fact, they found a family of four such particles (which they called "resonances") to which we now give the collective name of the Δ. These particles weighed more than the proton, decayed quickly into a proton and a π meson once they were created, and came with four different electrical charges. One of the Δ had an electrical charge of +2, one had an electrical charge of +1, one had an electrical charge of 0, and the last had an electrical charge of -1. Although each of these is a separate and distinct particle, it has become the custom in elementary particles to refer to all four collectively as "the Δ." This use of the singular to denote an entire family may be bad English, but it turns out to be good physics.

In the period that had elapsed since 1952, the elementary particle "zoo" has grown beyond the wildest expectations of any of the scientists working at that time. Many more particles of the type that we call "strange particles" have been discovered, as have many more particles of the type that we call "non-strange." In addition, many more particles of the type we refer to as "mesons" have been discovered. At the last count (and most people have stopped counting by this time) there were several hundred particles that had a right to be called "elementary." This proliferation of "elementary particles" is today one of the most puzzling aspects of the nature of matter. The explanation of the elementary particles and their relation to the atomic nucleus that is their natural home is the subject matter of a new field of physics, which is called "elementary particle physics" or "high energy physics." It is the purpose of this field to do the same thing for elementary particle physics that Isaac Newton did for classical mechanics and that Maxwell did for electricity and magnetism. The fact that this task is a long way from completion should not be too surprising, given the complexity that we have already discovered in the subnuclear world.

D. Classifying the Elementary Particles

Obviously, the first task to be faced when confronted

425

with a plethora of particles is to try to impose some sort of order on them. The situation in elementary particle physics is very similar to the situation that existed in chemistry before the introduction of the periodic table of the elements. There were a hundred chemical elements, but in order to make progress it was necessary to realize that these elements came in families. It was not necessary at the time of Mendeleev to know why the elements came in families. The first necessary step was to discover what those families were. In this section, we shall discuss some of the ways of classifying elementary particles, and we shall show that these classifications, although they may appear unusual at first sight, are really nothing more than restatements of experimental facts that we have already discussed.

The first way that we could imagine ordering the elementary particles would be to split them up according to the kind of interactions in which they participate. We already know about four different kinds of interactions in nature. We know that two different particles can interact through the gravitational attraction provided they have a mass. We know that they can interact through the electromagnetic interaction provided they have an electrical charge. We know that other particles can participate in a weak interaction, and still others can participate in the strong interaction. Actually, the hierarchy of forces that are available in nature is very interesting. In the table below, we list the four forces that we know about, together with a relative estimate of their strength, taking the value of the strong interaction as one.

interaction	relative strength
strong	1
electromagnetic	10^{-2}
weak	10^{-5}
gravitational	10^{-38}

Once we see that the forces between elementary particles can be assembled into a hierarchy with the strong forces at the top and the gravitational forces at the bottom, we can begin to classify elementary particles according to which of these interactions they can participate in. For example, a particle like the proton can participate in a gravitational interaction (because it has a mass), the electromagnetic interaction (because it has an electrical charge), the weak interaction (because it is involved in the weak decay of the Λ, and the strong interaction (because it is involved in the binding of the nucleus). Other particles, like the electron, can be involved only with the electromagnetic, the weak, and the gravitational forces. They do not participate in the strong interaction.

Particles that participate in the strong interactions are given the collective name of "hadron" (from the Greek word for strong). Particles like the electron, the μ meson, and the neutrino, that participate in the electromagnetic and weak interactions but not in the strong interactions, are given the name "lepton." Thus, one classification of particles is between the leptons, which do not participate in the strong interactions, and the hadrons, which do.

Among the hadrons, a further set of distinctions can be made. Although all hadrons can be created in strong interactions, some of them decay by strong interaction and some decay only by the weak interaction. The hadrons that decay only by the weak interaction are given the name "strange hadrons," while those that can decay strongly are called "non-strange hadrons."

Another distinction between the hadrons is possible. We have discussed particles like the Λ, the Σ, and the Δ. All of these particles have the property that when they decay, whether it be through the weak or strong interaction, there is one proton in the final state. There are other hadrons we have not discussed, but for which this statement is not true. For these other hadrons, the final state might consist of a number of π mesons, or it might consist of a number of π mesons, electrons, neutrinos, etc., but it does not contain a proton or a neutron. Those particles whose ultimate decay products include a proton and neutron are given the name "baryon" (from the Greek word for heavy), while those whose final decay products do not include a proton and neutron are given the name "meson." The term "meson" was originally used to refer to particles whose weight was intermediate between the electron and the proton. The new definition of "meson," however, leaves open the possibility that there might be particles heavier than the proton that would still be called mesons. All that would be necessary would be to find the particle (and there are many of them in the subnuclear zoo) that was heavier than the proton but decayed into a number of π mesons rather than into a final state containing a proton or a neutron.

427

Thus, we have the possibility of a particle being a hadron or lepton, and if it is a hadron it can be either strange or non-strange, and it can be either a meson or a baryon. All combinations of particles within this hadron scheme are possible, and, indeed, are known.

In the table below we list all of the particles we have discussed, and all the properties we have so far discovered about them.

particle	type	charge	mass	decay
electron	lepton	– 1	.5 MeV	—
mu meson	lepton	±1	100 MeV	slow
pi meson	hadron	±1, 0	140 MeV	slow
proton	"	+1	939 MeV	—
neutron	"	0	939 MeV	slow
Λ	"	0	1115 MeV	slow
Σ	"	±1, 0	1193 MeV	slow
Δ	"	+2,+1,0,–1	1238 MeV	fast

We begin to see a kind of order developing among the elementary particles. They can indeed be classified according to their properties. At the present time, we classify particles by their mass, by their electrical charge, and by whether or not they are "strange." The important point to realize about these criteria of classification is that *all* of them are simply statements of results of experiments. The mass of a particle can be deduced from "weighing" it. The electrical charge of a particle can be deduced by watching its behavior near a known charge. Whether or not a particle is "strange" can be deduced by observing how long it lives.

This way of classifying particles—by their experimental properties—is commonly used by physicists today. However, it is used in a way which, while at first glance may seem rather abstract, actually turns out to be very useful indeed. Instead of talking about the experimentally determined properties of a particle directly, the particle is assigned a quantum number.

Let us take as an example the following statement about the electrical charges of the proton and neutron:

> The proton has a positive electrical charge equal in magnitude to the charge on the electron, and the neutron has no electrical charge whatsoever.

428

What does this statement mean? It means that if we brought a proton near a positive electrical charge, it would be repelled from that charge, and the force of repulsion would be equal in magnitude (although oppositely directed) to the force that would be exerted on an electron if it were brought to the same position near the positive charge. If, on the other hand, we brought a neutron to a position near the positive charge, no force would be exerted on it at all.

The above paragraph is an example of a set of experimental facts about the proton and neutron. These experimental facts could, in principle, be tested at any time to determine whether or not they were true. However, it would become very cumbersome indeed if everytime we wanted to refer to the charge in a proton we had to specify that it was positive and that it was equal in magnitude to the charge on the electron. Consequently, we can develop a shorthand way of stating the facts about the electrical charges of the proton and neutron. We can make the following statement:

The proton has charge quantum number +1, and the neutron has charge quantum number 0.

What does this statement mean? You will recall that a quantum number in atomic physics referred to the state in which an atomic electron found itself. For example, we discussed orbital quantum numbers, which specified which of the Bohr orbits an electron was in. Similarly, a quantum number in the sense in which we are using it here simply refers to the "charge state" or electrical charge that a particle carries. By convention, a particle with charge quantum number +1 is defined to be a particle with a charge that is positive and equal in magnitude to the charge on the electron. A particle with a charge quantum number −1 will be a particle whose charge is negative and equal in value to the charge of the electron. A particle with charge quantum number 0 will be a particle with no electrical charge at all.

We see that the two statements we have given about the charges on the proton and electron are exactly and precisely identical. They are simply two different statements about the same set of experimental facts. The experimental facts have to do with the behavior of the proton and neutron near known electrical charges. The statements that summarize these experiments can be cast either into the mold of discussing the charge of the proton and neutron, or can be cast into the language of charge quantum numbers. However, there is no new information about the particle contained in the statement in terms of the charge quantum numbers. Both statements contain precisely the same amounts of information. There are simply different languages for conveying that information.

Now we know that in experimental situations, we never see electrical charge created or destroyed. We speak of this as a *conservation law* and we say that the total charge in the system is conserved. In other words, if a particle of positive charge is created, a particle of negative charge must be created as well, so that the total charge on the entire system that we are observing remains constant.

The statement of this law of nature in terms of charge quantum numbers is quite simple. It says

the total charge quantum number of a system remains constant.

In order to understand what this statement means, let us consider the process in which a π meson and a proton collide in order to produce a number of particles in the final state. Take as an example the reaction

$$\pi^- + p \rightarrow \Delta\ (++) + \pi^- + \pi^0 + \pi^-.$$

This is a reaction in which a negatively charged π meson collides with a proton and produces a Δ which has positive charge +2, two negatively charged π mesons, and one neutral π meson. Let us examine this reaction from the point of view of the electrical charges of the particles involved. The charge of the negative π is equal to the charge on the electron, the charge on the proton is equal in magnitude to the charge on the electron, but opposite in sign. Therefore the total charge in the beginning of the reaction is 0, since the positive and negative charges cancel each other out. The charge of the Δ, which we produced, is positive and equal to twice the charge on the proton. The charge on the neutral π is 0, and the charge on each of the Π^- is in the final state is equal to the charge on the electron. Therefore, we have a particle with a positive charge of two units, and two particles, each with a

430

negative charge of one unit, so that the net charge in the final state is 0. Thus, the law of conservation of charge is seen to be obeyed in this reaction. There is no net charge in the initial state, and there is no net charge in the final state. However, it must be emphasized that the number of particles and the charges on them both change when we go from the initial to the final state.

Let us make this same analysis in terms of charge quantum numbers. The charge quantum number of the π^- is -1, and the charge quantum number of the proton is +1. Thus, the total charge number in the initial state of the reaction is

initial total charge quantum number = -1+1=0.

The charge quantum number of the Δ (++) is +2, the charge quantum number of the π^0 is 0, and the quantum number of π^- is -1. Consequently, the total charge quantum number in the final state is

final total charge quantum number = +2-1-1+0 = 0.

Consequently, the statement of charge conservation now is given in terms of the charge quantum numbers. The total charge quantum number before the collision in which the new particles are formed is the same as the total charge quantum number after the collision.

The language of quantum numbers, then, is seen to be simply a new way of stating properties of the particles that we are considering and of stating observed experimental laws. For example, in the case of electrical charge that we have just considered, we talk about the charge quantum number of particles rather than their actual charge, with the understanding that the actual charge of the particle is measured in terms of the unit that is equal in magnitude to the charge on the electron. The law of conservation of charge becomes a statement that the total charge quantum number of a system cannot change, no matter what kind of interactions go on. It is obvious that other kinds of laws of nature could be expressed in this same language. We shall discuss two more quantum numbers and their conservation in this section.

431

In discussing the categorization of elementary particles, we mentioned that these particles could be divided into two classes—baryons and mesons. The baryons were defined to be those particles that had a decay chain which ultimately resulted in a final state in which one proton or neutron was present. For example, both the Δ and the Σ were baryons, because each decayed into a proton and a π meson. The fact that the Δ decayed quickly and the Σ decayed slowly does not affect the products of the decay that are ultimately found to exist. Both are baryons.

It is found experimentally that no particle possesses decay chains that could lead to both a state in which a proton was present after the decays and a state in which the proton was not present. In other words, even if a particle has several different ways that it can decay, the final state in which it finds itself will always be characterized by either the presence or absence of one proton or neutron. Consequently, we would say that it is an observed experimental fact that every particle is either a baryon or a meson. No particle shares the properties of both.

This experimental fact can be cast in the language of quantum numbers. Suppose that in discussing the Δ and Σ, we made the statement "the Δ and Σ have baryon number +1," instead of making the statement that the "Δ and Σ decay into products which include one proton or neutron."

Once we have assigned every particle a baryon number (which is +1 for the particles that do decay into a proton or neutron, and 0 for those that do not), we can talk about interactions other than the decays of the particles. In other words, even though we use the decay chain of a particle to define what the baryon number is, the baryon number can be a useful concept in other kinds of reactions, such as collisions in which new particles are created. Take, for example, the production of the Δ and π mesons discussed above. Let us, as an exercise, calculate the total baryon number on both sides before and after that collision.

In the initial state, we have a π meson (baryon number 0) and a proton (baryon number +1). Therefore, we have

initial total baryon number = 0+1=+1.

In the final state, we have three π mesons (each with baryon number 0) and a $\triangle$, baryon number +1. Consequently, we have

final total baryon number = +1+0+0+0=+1.

In other words, in this reaction, the total baryon number, like the total charge, is conserved.

Actually, it turns out to be an experimentally observed law that in every interaction involving elementary particles, whether it be collision, production, or decay, the total number of baryons before the interaction is the same as the number of baryons after the interaction. In terms of our quantum number language, another law of nature, which is observed in the laboratory and can be stated clearly in terms of quantum numbers, can be stated as follows

the total baryon number is the same before and after every interaction.

It is interesting to contrast the behavior of baryons with respect to this new quantum number with the behavior of mesons. If we look again at our initial reaction, we see that there is one meson before the reaction begins and three mesons after the reaction is finished. In other words, the number of mesons is not conserved in the reaction. There are countless examples of reactions in which the number of mesons before is different from the number after the reaction. Consequently, although there is a conservation law for baryons, there is no experimentally observed conservation of the number of mesons. This is why we do not introduce a meson quantum number.

What else can we discuss in terms of quantum numbers? We have seen that there is another important distinction that can be made among the elementary particles, and that this distinction involves the speed with which a decay takes place. The baryon number, as we have seen, simply describes what the final products of the decay will be, but has nothing to do with the speed. On the other hand, we have seen that there are particles that decay into the proton and hence are baryons, but that do so on a time scale that is very long compared to the characteristic strong interaction time. Can we describe this experimentally observed phenomenon in the language of quantum numbers?

The experimental fact that there are two kinds of particles—those that decay quickly and those that decay slowly—can indeed be translated into the language of quantum numbers. Suppose we assign a new kind of quantum number to elementary particles. Suppose that, for the sake of definiteness, we call this new quantum number "strangeness." Suppose that we assign zero strangeness to the "normal" particles—i.e., those that decay quickly. In this case, we would say that the proton, the neutron, and the Δ all had zero-strangeness, or were non-strange particles. Let us assign a non-zero strangeness to the particles that decay slowly. By convention, we say that the Σ and the Λ, for example, have strangeness –1.

Why do we choose –1? The choice of the number –1 for the strangeness of these particles is a historical accident and has no particular significance. What *is* significant, however, is that the strangeness of these particles is not zero. Once we have assigned to every particle that we know a new quantum number corresponding to its strangeness, we can then summarize the observed experimental results of the following statement:

A decay will be fast if strangeness is conserved, and it will be slow if strangeness is not conserved.

In order to illustrate this rule, let us take one of the processes we have already considered:

$$\Lambda \to p + \pi.$$

This was the process by which the Λ particle was first discovered in cosmic ray experiments. Let us calculate the strangeness on both sides of this reaction. The Λ, as we have said, has strangeness –1. Therefore, the total strangeness before the reaction takes place is

total strangeness before reaction $= -1$.

On the other hand, both the proton and the π meson have strangeness zero. Therefore, the total strangeness after the collision is

strangeness of final state $= 0+0=0$.

Therefore, in this decay, we would say that the total strangeness of this system changed by one unit, from -1 to 0. According to our rule, then, we would say that this would be a slow decay.

Let us consider, on the other hand, the reaction

$$\Delta \rightarrow p + \pi.$$

The strangeness of the Δ is zero, and the strangeness of the proton plus π meson is also zero. Therefore, this decay does not involve any change of strangeness. The strangeness is zero before the decay, and is zero after the decay. Therefore, this would be a fast decay (as we know it is).

One important point must be emphasized about introducing a quantum number like strangeness. It does not introduce any new information into the world. It is simply a shorthand way of talking about fast and slow decays. Assigning a particle a strangeness of -1 is totally equivalent to saying that there will be a slow decay at some point in the decay chain of this particle. Similarly, assigning a particle a strangeness of zero means that all of the decay steps will be fast in its decay into proton and π mesons.

In order to make this idea more clear, let us consider one more example. We have already discussed the fact that there are many elementary particles. In particular, elementary particles have been found that are heavier than the Σ and Λ. An example of such a particle would be the Σ^*, which weighs 1385 MeV. This particle decays either into the $\Sigma + \pi$ meson (11% of the time) or into a $\Lambda + \pi$ meson (89% of the time). A decay chain of the Σ^*, then, would look something like this:

$$\Sigma^* \rightarrow \Lambda + \pi$$
$$\quad \rule{0.3cm}{0.4pt}\!\!\longrightarrow p + \pi.$$

The decay chain of the Σ^* is rather similar to the radioactive decay chain of nuclei that we have talked about before. It consists of several steps, each of which has its own characteristics. The strangeness of the Σ^* is -1. In the first reaction, the strangeness before the reaction is therefore -1, and the strangeness after the reaction is -1. Consequently the strangeness does not change. Therefore, we would expect the first decay—the decay of the Σ^* to the $\Lambda + \pi$—to be fast (which indeed it is). On the other hand, the decay of the Λ, as we have discussed, involves the change of strangeness, and therefore is slow. The point of this example is to emphasize the fact that even though a particle may be strange and have a strangeness quantum number which is not zero, this does not prevent it from having fast decays. The only things that our rules tell us is that if the particle is strange and if it has a fast decay, then the end products of that decay must also be strange.

There are many examples of particles like the Σ^*. In fact, there are hierarchies of particles that have been discovered in nature. For example, there is a Σ^{**}, which decays into the Σ^*, which in turn decays into a Λ, which in turn decays into a proton. The complete description of all of these particles is outside the scope of this text, but it is a good thing to keep in mind that the decay schemes of elementary particles are every bit as complicated and as rich in detail as the decay schemes of radioactive nuclei that we discussed earlier.

Finally, we present below another list of the particles that we have discussed which, instead of listing their properties in normal language, lists their quantum numbers. From this point on, we shall use the notation that baryon numbers shall be denoted by B, strangeness by S, and charge by Q.

particle	B	Q	S
proton	+1	+1	0
neutron	+1	0	0
pi meson	0	±1, 0	0
Λ	+1	0	-1
Σ	+1	±1, 0	-1
Δ	+1	+2, +1, 0, -1	0

E. The Periodic Table of the Elementary Particles

By the end of the 1950s, a large number of elementary particles had been discovered. So many, in fact, that the old dream of being able to explain the entire universe in terms of just a few elementary particles was seen to have vanished forever. Consequently, the next task awaiting physicists was to discover some kind of order among the elementary particles, just as chemists had discovered the periodic table of the elements. If such an order could be discovered, then perhaps we would understand the elementary particles in the way that we now understand the periodic table of the elements.

Fermi National Accelerator Laboratory, Batavia, Illinois.

The 15-foot bubble chamber at Fermi National Accelerator Laboratory near Chicago, where collisions between elementary particles are studied.

Professor Samuel C. Ting of MIT with some of the group that discovered the J particle. They are standing above a graph of the final results which convinced him (and the Nobel Committee) that a new particle had been discovered.

The MIT/BNL double armed spectrometer used at BNL to discover the J-Particle (Brookhaven National Laboratory).

Before we can discuss the periodic table of the elementary particles which were discovered in the early 1960s. it is necessary to introduce two more quantum numbers. Unlike the quantum numbers that we have introduced up to this point, however, these new quantum numbers do not refer to any particular kind of experimental results. They are simply combinations of the quantum numbers that we have already discussed.

The first new quantum number that we shall introduce is called the hypercharge, and is customarily denoted by a Y. The definition of the hypercharge is

$$Y = B + S.$$

In other words, the hypercharge is simply the sum of the baryon number and the strangeness. Let us calculate the hypercharge of a few elementary particles so that we can understand what it means and how we go about calculating it.

The baryon number of the proton is +1, and its strangeness is 0. Consequently, the hypercharge of the proton is simply $Y=0+1=+1$. The hypercharge of the neutron is also +1.

The Σ particle, on the other hand, has baryon number +1 and strangeness -1. Consequently, its hypercharge is equal to $Y=+1-1=0$. The π mesons have both baryon number and strangeness equal to 0. Consequently, their hypercharge will be equal to 0. It is easy to see how we could calculate the hypercharge of any particle from the quantum numbers listed in the table in the last section. It is left as an exercise for the reader to calculate the hypercharge for the rest of the particles that are in that table.

The second "new" quantum number that we shall introduce goes by the rather formidable name of "third component of the isotopic spin." Rather than use this cumbersome title throughout the rest of our discussion, we shall call it simply "isospin."* It is defined by the equation

$$I_3 = Q - Y/2.$$

*Actually, the use of the term "isospin" in this context is slightly incorrect. However, a discussion of why it is incorrect and what isotopic spin actually is would simply result in the reader being more confused by the proliferation of quantum numbers than he already is. The interested reader is referred to the text by Pine in the list at the end of the chapter.

In other words, the isospin as we define it is simply the charge quantum number of the particle minus one-half of its hypercharge. All of the remarks that hold for the introduction of hypercharge hold for the introduction of isospin as we have defined it here. It is not a new quantum number, and it does not refer to any new experimental facts that we have not introduced up to this time. It is simply a different combination of other quantum numbers, each of which does refer to some experimentally observed property of the particle in question.

Let us work out a few isospins for the particles we have already discussed. The charge quantum number of the proton is +1. Its hypercharge is +1. Consequently, its isospin will be 1/2. The neutron, on the other hand, has charge quantum number 0 and hypercharge +1. Consequently, the isospin of the neutron will be –1/2. The isospin of the other particles that we have discussed can be easily worked out.

What have we accomplished with this juggling of quantum numbers? If we were to stop at this point, the answer obviously would be that we have accomplished nothing. We have defined a set of quantum numbers—charge, baryon number, and strangeness—which define certain properties of the elementary particles. Each elementary particle has some values for these particular quantum numbers, and each of these numbers can be interpreted in terms of actual physical observable properties of the particle. The purpose of introducing the new quantum numbers—isospin and hypercharge— is not to summarize still more experimental results. It turns out that when we introduce these new quantum numbers, we can find a way of classifying the elementary particles that is very reminiscent of the classification of chemical elements in the periodic table.

As an example, let us consider the baryons that we have discussed so far. In the table below we list these baryons, together with their quantum numbers, and, in addition, include a baryon that we have not yet discussed. This is the Ξ (the Greek letter xi) and is also sometimes called the "cascade particle." It is simply one of the 200 or so particles that have been discovered since the mid-1950s. It decays slowly into a Λ and a π meson, and the Λ in turn decays slowly into a proton and a π meson. Thus, the cascade particle must have a strangeness –2. It is left as an exercise to the reader to figure out why this must be so.

438

particle	Y	I_3
P	+1	$+\dfrac{1}{2}$
n	+1	$-\dfrac{1}{2}$
Λ	0	0
Σ^+	0	+1
Σ^-	0	−1
Σ^0	1	0
Ξ^-	−1	$-\dfrac{1}{2}$
Ξ^0		$+\dfrac{1}{2}$

Suppose that instead of simply listing these baryons in a table, as we have just done, we decided to make a graph. Suppose that on the vertical axis of the graph we had hypercharge, and on the horizontal axis of the graph we had isospin. Each particle, then, would correspond to a point on this graph. This is because each particle has one value of hypercharge and one value of isospin, and hence each particle corresponds to going up or down a certain distance on the vertical axis, and to the right or to the left a certain distance on the horizontal axis. For example, the proton is a particle that has hypercharge 1 and isospin +1/2. Consequently, it would appear on a graph of the type we are talking about as a point that is up one unit in the vertical axis and 1/2 unit to the right on the horizontal axis. The proton is shown in Fig. 15.4 on a hypercharge-isospin graph.

Just as we can plot a point that represents a proton on this sort of graph, we can also plot each of the other baryons in the table above as points on the graph. When we do so, we come up with a set of points that looks like that pictured in Fig. 15.5.

This way of representing the elementary particles turns out to be very significant. Because there are eight particles in this particular grouping, the grouping is sometimes given the name "the eight-fold way." Like the periodic table of

439

the elements, this particular arrangement of elementary particles allows us to read off properties of the particles. For
example, if we read down a diagonal in the above figure
going from the upper left to lower right, we have particles
that have the same electrical charge. These particles, then,
have the same electromagnetic interaction. If we read across
the rows, we have particles that have the same strangeness.
Consequently, particles in the same rows will have similar
weak interactions. The diagonal from upper right to lower
left has a similar significance, although we have not discussed
"u-spin," the properties which particles in this particular
diagonal share. Finally, all eight particles in this particular
grouping are baryons and have about the same mass. Thus,
the particles in this particular group have a certain similarity
to each other.

It turns out that all of the hadrons that have been discovered can be fit into groups like the one pictured above.
There is, for example, a group of eight mesons, including the
π mesons, which has an interpretation similar to the one we
have just discussed for the baryons.

Up to this point, we do not understand why it is that
when we arrange the particles in these groups of eight as
shown above, the rows and columns and diagonals of the
arrangement have particles with similar properties. This is
exactly analogous to the fact that when we first worked out
the periodic table of the elements, we knew that elements in
the same column in the periodic table had similar chemical
properties, but we did not understand why this was so. Just
as one of the great triumphs of the periodic table was the

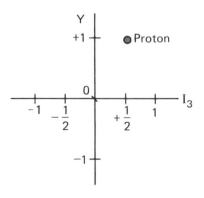

Figure 15.4.

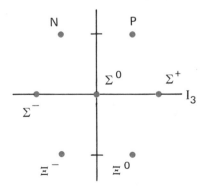

Figure 15.5.

discovery of new chemical elements, which appeared as gaps, or missing places in the periodic table, the discovery in the early 1960s of the eight-fold way led, in turn, to the discovery of new particles.

Actually, the "eight-fold way" does not always predict that particles will be arranged in groups of eight. In some cases the mathematical predictions involve particles arranged in groups of ten. This is analogous to atomic physics where different orbits contain different numbers of electrons. The point is that the number of particles in the group and the properties of these particles are predicted by a definite mathematical formula. The least number of particles that can appear in any group is eight, and most of the more "common" particles do appear in groups of eight. At the time of the discovery of the eight-fold way, however, there was a well-known group of particles that was supposed to fall into a grouping of ten. The Σ^* (which we discussed earlier) was one of these particles. If we make a plot of hypercharge versus isospin, as we did for the proton and its associated baryons, we would find something like that pictured in Fig. 15.6. The Σ^* and the Ξ^* are baryons which decay into the Σ and the Ξ, respectively. The gap in the grouping is obvious. The particle that would form the apex of the inverted triangle in the above figure was not known. However, from its position, we can make a number of predictions about it. For example, we know that it must have hypercharge minus two and isospin zero. This, together with the fact that it must have baryon number B=1, means that the particle must have S=-3 and must have a charge Q=-1. The reader should be familiar enough with the manipulation of quantum numbers by this time to understand why the second set of statements follows from the first (see Problem 2).

In 1964, a massive search was undertaken at several laboratories to find this missing particle, which was called the Ω^-. When the particle was actually seen in a bubble chamber (a device something like the cloud chamber we discussed earlier) at Brookhaven National Laboratories on Long Island, most physicists began to accept the idea that there was a way of ordering the elementary particles, and that the type of mathematical formulation which lay behind the eight-fold way—a branch of mathematics called "group

441

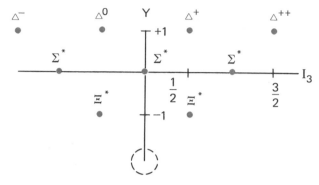

Figure 15.6.

theory"—would lead to a better understanding of the myriad elementary particles that have been discovered.

When Mendeleev discovered the periodic table, chemists had no idea as to why the elements arranged in that particular way had such a strong correlation to what was actually observed in nature. In the same way, arranging the particles according to the "eight-fold way" does not give us any deep insight into why nature has chosen to make particles that can be arranged in that way. At this point in the development, we are at the same stage as the 19th-century chemist. We know that order prevails among the elementary particles, and we know how to group them so that that order becomes manifest. We do not, however, understand why that order exists. This corresponds to the 19th-century chemist's lack of knowledge of quantum mechanics, the electronic structure of the atom, and the Pauli exclusion principle.

F. The Elementary Particle Anti-Zoo

As if the discovery of all of these elementary particles was not confusing enough, a parallel course of events was taking place that effectively doubled the number of particles that scientists had to deal with. In 1928 the British physicist P.A.M. Dirac suggested that certain rather abstract theoretical problems in quantum mechanics could be resolved if, for each particle that was known at that time, there was also an "anti-particle." The anti-particle would have the same mass as the particle, but all the other quantum numbers would be exactly the opposite. For example, the anti-particle to the electron (which we now call the positron) would have a mass of .5 MeV, but would have a positive (rather than negative) electrical charge.

Furthermore, it was predicted that if a particle and an anti-particle ever came together, they would annihilate each other. In the case of an electron and positron, for example, both the electron and positron would disappear and their energy would appear in the form of photons. Such an event, of course, would be a direct confirmation of the mass-energy equivalence we discussed in the context of the theory of relativity.

By the same token, it should be possible to convert the high energy of a cosmic ray particle directly into matter in the form of an electron-positron pair. For example, a photon striking a nucleus could, in addition to whatever else it did, convert some of its energy into making these two particles. Since the particles have opposite charges, the total charge before and after the reaction would be the same, so there is no reason why such an event could not occur.

In 1933, a young Cal. Tech. physicist named Carl Anderson detected such "pair production" events in a cloud chamber exposed to cosmic rays. Thus, in a single experiment both the existence of anti-particles and the direct conversion of mass and energy were established. Anderson was awarded the Nobel Prize for his work in 1936. (It is amusing by present-day standards that he was not promoted to full professor until three years later.)

In the time since Anderson's experiment, anti-particles have been discovered for every other elementary particle known. For example, there is an anti-proton that has a negative electrical charge, and baryon number –1. There is an anti Σ^+ that has negative charge of one unit, strangeness +1, and baryon number –1. All of the other particles discussed in Section E have their anti-particles, and each particle, if it should encounter its anti-particle, will annihilate.

The annihilation process follows the same rules of conservation of quantum numbers we have discussed. However, because of the way that anti-particles and particles are related, the total quantum number for any quantity (charge, strangeness, baryon number, etc.) will be zero in any reaction of the type

$$\text{particle} + \text{anti-particle} \rightarrow \text{anything}.$$

443

At the present time, anti-particles for all of the more common known particles have been seen in the debris from accelerator experiments, but annihilations have been observed directly only for the positron and anti-proton. The reason for this is the relative difficulty of making enough of both the particle and its anti-particle to have collisions between them.

G. The Frontier of Knowledge

Throughout the course of the development of our knowledge about the atomic, the nuclear, and the subnuclear world, we have always had a situation in which new information of a rather puzzling nature was encountered and then answers were found to the questions that this new information raised. Sooner or later, of course, this process has to end. Sooner or later, we are simply going to run out of answers. Unfortunately, we have come to that point now.

The knowledge of elementary particles that live for a short time inside of the atomic nucleus and that display such a wide variety of properties and characteristics is not complete. We do *not* understand the interactions that these particles undergo, nor do we really understand, on any fundamental level, why it is that they are the way they are. Just as the problems of the structure of the atom were dealt with by the scientists in the 1920s, and the problems of the structure of the nucleus by the scientists in the 1930s, the present generation of scientists is involved with trying to understand what the elementary particles are, how they interact with each other, and what part, if any, this knowledge will play in our picture of the world.

In this section we shall discuss some of the ideas and some of the theories about elementary particles that are popular at the present time. Unlike the theories we have discussed about the structure of the atom and the structure of the nucleus, however, these theories could very well be wrong. In fact, they probably are. Nevertheless, they represent our best guess at the present time as to what the structure of the elementary particles is.

The student should not be disappointed that answers to questions about elementary particles cannot be given. After all, we have repeatedly stressed that a scientist is not a man with answers; he is a man with questions. The questions about the elementary particles that are being asked today will presumably be answered at some time in the future. In a way, it is very exciting to think that we are living during the time when men are able to ask such fundamental questions about the nature of the physical world.

When we discussed the structure of the atom, we talked about two rather important milestones in our thinking. The first of these was the discovery of the periodic table of the chemical elements by Mendeleev, and the second was the experimental discovery of the nucleus by Rutherford. We have seen how the discovery in the 1960s of the "eight-fold way" parallels the discovery of the periodic table of elements.

The late 1960s also saw the carrying out of an experiment that is very similar to the Rutherford experiments. This experiment was done at the Stanford linear accelerator center (SLAC), a machine some two miles in length, whose function is to accelerate electrons up to an energy of 20 billion electron volts. These high energy electrons were then directed against proton targets. It was observed that, although most of the electrons came off in a forward direction, a rather large number—many more than would have been expected— came off at large angles. This is an exact analogy to what Rutherford found when he directed his alpha particles at the atom. And, just as Rutherford concluded that his experiment indicated the existence of a tiny, massive, point-like structure within the atom, which he called a nucleus, the Stanford experimenters suggested that the proton itself must have within it some tiny, massive, point-like objects. In other words, the proton (and other "elementary" particles) must have some sort of composite structure, just as the atom (which, you will recall, means something that cannot be split) was discovered to have a rather complicated internal structure. It is now accepted by most elementary particle physicists that there is some sort of structure to the elementary particles. In other words, we now believe that the elementary particles are probably not really "elementary" but are themselves made up of other kinds of particles.

The question is, what are these other kinds of particles like? This problem has engaged the attention of a large number of physicists since it became obvious that these new kinds of particles must exist. There have been many different kinds of theories about them, and we shall discuss only one here. This is the so-called "quark" theory. This theory was proposed in the late 1960s by several physicists simultaneously. The basic point of the theory is that the "elementary" particles that we see are not really "elementary" in the sense that they cannot be broken down further. The "elementary" particles that we see are, in fact, different arrangements of sub-elementary particles, which are called "quarks."*

Just as in the periodic table of the elements each chemical element is thought of as being made up of protons, neutrons, and electrons, in the quark scheme of the elementary particles, each of them is thought of as being some different arrangement of some still more elementary particles, which are given the name quarks.

From the facts that the elementary particles can be arranged according to the eight-fold way, the properties of the quarks can be deduced. As a matter of fact, the mathematical theories that led to the classification of the elementary particles into the eight-fold way suggest strongly that there are more basic particles than the particles that are being classified. The properties of the quarks that come out from mathematical analysis of this type are rather unusual. There are three kinds of quarks. They are called, respectively, the u, d, and s quarks. The letters stand for "up," "down," and "strange." Their properties (by which we mean their quantum numbers) are given in the table below.

quark	B	Q	S	Y	I_3
u	$\frac{1}{3}$	$\frac{2}{3}$	0	$\frac{1}{3}$	$\frac{1}{2}$
d	$\frac{1}{3}$	$-\frac{1}{3}$	0	$\frac{1}{3}$	$-\frac{1}{2}$
s	$\frac{1}{3}$	$-\frac{1}{3}$	-1	$-\frac{2}{3}$	0

It will become immediately obvious that these particles are not like any other kind of particles that we have seen. The most striking difference is the fact that the quarks have fractional charge quantum numbers. This means that the

*The name comes from a quote from *Finnigan's Wake*, "Three quarks for Master Mark." The significance of this will become obvious from the discussion.

quarks have charges that are, respectively, 1/3 and 2/3 the size of the charge on the electron. This would mean that quarks, if they could be created, would be very easy to recognize.

How would the existence of quarks make our understanding of elementary particles and their structures simpler? To answer this question, let us see how the various elementary particles that we have discussed so far would be constructed from quarks. Each quark has a baryon number equal to 1/3. This means that if we want to make a particle that has a baryon number 1, we must make it out of three quarks. Consequently, we have our first rule: all the baryons are composed of three quarks.

Once we have established how many quarks there are by looking at the baryon number, we then look at the charge quantum numbers and try to arrange them in such a way that the particle we are putting together has the correct charge. For example, we know that the proton has a charge of +1 and strangeness 0. The only way that we could make this particle from three quarks is by taking two u quarks and one d quark and putting them together. The total charge of such a system would be 2/3 + 2/3 - 1/3 = 1, while the total strangeness of such a system would be 0 + 0 + 0 = 0. In fact, in the quark picture, the proton would be made up of three quarks arranged as shown in Fig. 15.7. The quark spins are all 1/2, and the proton spin is 1/2, so the spins of the quarks must be arranged as we have shown—that is, we must have two spins in one direction and the third spin must be in the opposite direction. The total spin of such a system is 1/2 + 1/2 - 1/2 = 1/2, which is precisely the spin of the proton. What we have in Fig. 15.7, then, is a picture of the proton as it would be if it were made of three quarks.

How would the neutron be made? The only difference between the neutron and the proton is the electrical charge. We would have to choose our three quarks in such a way that the total charge added up to 0. This is simple to do. Rather than taking two u quarks and one d quark as we did in the case of the proton, we would simply take one u and two d quarks. The total charge of this new system would then be 2/3 - 1/3 - 1/3 = 0, and if we arrange the quarks as shown in Fig. 15.8, we would have a particle of charge 0, strangeness 0, baryon number 1, and spin 1/2. This particle is, of course, the neutron.

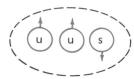

Figure 15.7.

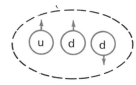

Figure 15.8.

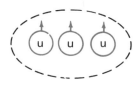

Figure 15.9.

When we come to discussing the structure of the other elementary particles, such as the Δ, the quark model works just as well. The Δ is a particle whose charge can be +2, +1, 0 or –1, has spin 3/2, baryon number 1, and strangeness 0. Suppose we were going to talk about the Δ^{++} which has a charge +2. How could we construct charge quantum number of +2 from the quarks listed in the above table? If we took three u quarks, we would have a particle whose charge was 2/3 + 2/3 + 2/3 = 2. In fact, to match the spin of the Δ, we would have something like that pictured in Fig. 15.9. This arrangement of quarks would correspond to a particle that has all the quantum numbers of the Δ(Q = +2). The only difference between this particle and the proton is that the d quark in the proton has been changed to a u quark, and its spin has been reversed, so that it lies in the same direction as the spin of the other quark. Higher resonances than the Δ would be constructed out of three quarks in the same way as the ones we have discussed so far, except that for these higher resonances the quarks would be moving in orbits around each other, rather than just sitting still. By simple processes like the ones we have discussed above, a representation of any non-strange elementary particles could be obtained.

What about the strange particles? If we look again at our table of quarks, we see that one of them—the s—has strangeness –1. Obviously, this quark would have to be included in any baryon whose strangeness was to be non-zero. Let us take as an example the Σ^+. The Σ^+ has a charge of +1, spin of 1/2, baryon number 1, and strangeness –1. Proceeding just as we did with the proton and neutron, we see that a choice of two u quarks and one s quark would give us all these quantum numbers, and that the Σ^+ made up of quarks must look like what we have pictured in Fig. 15.10. This arrangement of quarks has charge 2/3 + 2/3 - 1/3 = +1, strangeness 0 + 0 - 1 = –1, baryon number 1, and spin 1/2. Consequently, it has all the quantum numbers of the Σ^+. The other strange baryons could be made in the same way, and the higher lying strange resonances, like the higher lying non-strange resonances, would simply have the quarks moving in orbits rather than staying still in respect to each other.

Thus we see that all of the known baryons can be built up by picking the right combination of just three quarks. In the same way, all of the known anti-baryons could be built

448

up by having the appropriate combination of anti-quarks. In addition, as we have said above, the mathematical properties of the quarks are such that if the baryons are made up as we have discussed above, then they must arrange themselves in the groupings suggested by the eight-fold way. In this way, the existence of quarks could be taken as an "explanation" of the eight-fold way.

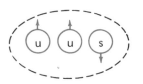

Figure 15.10.

What about the mesons? By definition, the mesons have baryon number zero. There is no way we can combine the quarks given in the above table to give a baryon number 0. All the quarks have baryon numbers +1/3. How, then, are we to make the mesons?

The answer to this question, of course, lies in the fact that in addition to the quarks, there are anti-quarks. They are listed and their quantum number given in the table below.

anti-quark	B	Q	S
$\bar{u}$	$-\dfrac{1}{3}$	$-\dfrac{2}{3}$	0
$\bar{d}$	$-\dfrac{1}{3}$	$+\dfrac{1}{3}$	0
$\bar{s}$	$-\dfrac{1}{3}$	$+\dfrac{1}{3}$	+1

We see instantly that one way to construct a particle out of quarks and anti-quarks and insure that the particle has baryon number 0 is to construct the particle out of a quark and an anti-quark. In this way, the baryon number will be +1/3 − 1/3 = 0. Although we could just as well construct the meson out of two quarks and two anti-quarks, the choice of the quark-anti-quark pair is simpler, and therefore is the one that is to be preferred.

Let us take the π^+ meson as an example. This is a particle with spin 0, baryon number 0, strangeness 0, and charge +1. How can we construct the π meson from the quarks and anti-quarks in our above tables? Suppose we took the combination of quarks and anti-quarks which is pictured in Fig. 15.11. This combination has baryon number zero, as discussed above. It has charge equal to 2/3 + 1/3 = 1, and since the spins of the quarks are anti-aligned, it has spin zero. Consequently, it has all of the quantum numbers of the π^+ meson.

449

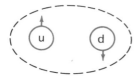

Figure 15.11.

In this way, we see that we can build all of the observed baryons out of combinations of three quarks and we can build all of the observed mesons out of combinations of quarks and anti-quarks. The remarks that were made above about the mathematics of the quarks implying that the baryons will fit themselves into representations consonant with the eightfold way hold for the mesons as well. Thus, *if quarks existed*, we would have an understanding of the observed properties and regularities among the elementary particles in terms of the three simple building blocks of nature—the three quarks. In this scheme of things, the quarks would be combined to make elementary particles, the elementary particles would be combined to make nuclei, electrons would be added to the nuclei to make atoms, and the atoms would be combined chemically to make all the substances we see around us. Once again, we would have a simple picture of the world.

Do quarks exist? Obviously, this is a question of paramount importance in modern science. Unfortunately, it is not a question that is easy to answer. Many experimenters have tried to find quarks in an almost unbelievable variety of ways. As of this writing, no one has produced evidence that he has seen a quark in nature. It would be as if we had a knowledge that there was such a thing as an atomic nucleus, but we had no way of producing protons and neutrons. We could say that the nuclei behave as if they were made up of protons and neutrons, but unless someone could actually produce a proton or a neutron in a lab, we would never be sure that these particles actually existed.

In the same way, we can say that elementary particles behave as if they were made up of quarks. However, since none has ever been seen in the laboratory, we remain unsure as to whether quarks exist in the sense that protons and neutrons exist. The absence of experimental confirmation of quarks has spawned a large number of competing theories about elementary particles, all of which assume that elementary particles are in some sense composite—that is, that they are made up of other kinds of particles—but involve "quarks" that are different than the ones we have discussed above. The picture remains clouded at the present time.

A further complication of the quark picture took place recently. Two scientists, Burton Richter of Stanford and

Samuel Ting of MIT and Brookhaven Laboratories, made an independent discovery of a new type of particle. In the language that we have been using, this new particle (called the ψ by Richter and the J by Ting) has a strangeness zero, but nevertheless decays slowly. Using the same sort of reasoning that went into defining strangeness in the first place, physicists have taken this discovery as evidence for yet a new quantum number called "Charm". It is denoted by "C."

By this line of reasoning, there should be many particles bearing this new quantum number, just as there are many particles carrying strangeness. In the language of quarks, this means that there must be a fourth quark (like the s) which, instead of carrying strangeness, carries charm. The quantum numbers of the new quark would be

name	B	Q	S	C
c	1/3	2/3	0	+1

The C quantum number of the "old" quarks would be zero.

Actually, the ψ/J particle for whose discovery Richter and Ting were awarded the Nobel Prize in 1976 was not a "charmed" particle, but was made from a c and an anti-c quark. Since that time, however, genuine charmed particles (by which we mean particles having only one c quark in them) have been seen in experiments. As of this writing some theorists are suggesting that still more quantum numbers (and hence more quarks) will be needed to explain recent experiments. And all this before even one quark has been seen in the laboratory.

The reader should have grasped by this time the fact that our knowledge of elementary particle physics is not complete. We know some things about elementary particles. We know that they can be grouped according to their presentations of the eight-fold way. We know that there is experimental evidence indicating that elementary particles are not really elementary, but that they contain some kind of structure. The precise kind of structure they contain is not clear at the present time because no one has been able to isolate or find one of these units of substructure in the laboratory.

Consequently, we shall have to wait for a while to find out whether the age-old dream of scientists—the dream of being able to discover a few basic building blocks which account for all of the infinite variety of matter that we see in the universe—will ever be realized.

SUGGESTED READING

Gell-Mann, M. and Rosenbaum, E. P. "Elementary Particles." *Scientific American,* July 1957. An account of the introduction of strangeness by its inventor.

Greider, K. *Invitation to Physics.* New York: Harcourt Brace Jovanovich, 1973, Chapter 20. An excellent and readable account of the elementary particle material discussed in this chapter.

Pine, Jerome. *Contemporary Physics.* New York: McGraw-Hill, 1972, Chapter 14. A somewhat more technical discussion of the material, but still suitable for the non-specialist.

Trefil, J. S. "Physicists Try to Peel off the Latest, but not the Last, Layer of Matter," and "New Particles Jolt Established Theories of Matter," 1975. *Smithsonian Magazine,* August 1974 and July 1975. Two articles on elementary particles written for non-specialists. The second one deals with the discovery of the ψ at Brookhaven and Stanford.

QUESTIONS AND DISCUSSION IDEAS

1. A new particle is observed. Let us call it the "Q" particle. It is observed to decay by the reaction

$$Q \rightarrow p + \pi^+ + \pi^-$$

    in 10^{-10} sec.
    a. What is the strangeness of this particle?
    b. What is its baryon number?
    c. What is its hypercharge?
    d. What is its charge quantum number?
    e. Is this particle like any discussed in the text? If so, which one?

2. From the fact that the Ω^- has baryon number 1, hypercharge -2, show that it has charge -1, strangeness -3.

3. There are hundreds of mesons known, all of which are hadrons, and all of which are at least three times heavier than the pion. Based on the idea of exchange forces and the uncertainty principle, explain why the pion is the most important meson in developing the strong force.

4. Let us consider a virtual pion such as the ones discussed in Section B. The mass of the pion is approximately 15% that of the proton.
   a. From the energy-time uncertainty relation, calculate roughly how long a virtual pi meson could live
   b. If it were traveling very close to the speed of light, how far could the virtual pi meson travel in that time?
   c. Most nuclei have radii of from 1–5×10^{-13} cm. Can you explain this on the basis of your answers to parts a and b?

5. a. From what you know about elementary particle quantum numbers, would you expect the decay

$$\Omega^- \rightarrow \Xi^- + \pi^0$$

   to be fast or slow? Why?
   b. The reaction given in (a) is one of the main decay modes of the Ω^-. Let's continue the chain. The main decay mode of the Ξ^- is

$$\Xi^- \rightarrow \Lambda + \pi^-$$

   Is this fast or slow? Why?

   c. Hence show that the final decay products of the Ω^- will be a proton and three pions. Why doesn't it decay directly into this final state, rather than through the steps outlined above?

   The following facts about a set of elementary particles will be needed to work the next problems.

name of particle	possible charge states	strangeness
π	-1,0,1	0
η	0	0
K^+	1	1
K^-	-1	-1
K^0	0	1
K^0	0	-1

6. All of these particles decay into final states that contain no protons or neutrons. What baryon number do they have?

7. Calculate the isospin component and hypercharge for each of the particles.

8. Hence, enter these particles on the type of I-Y plot shown in the text. Do you recognize the pattern?

9. Give the quark structure of the following particles discussed in the text:

   a. Δ^+

   b. Δ^0

   c. Δ^-

   d. Σ^-

APPENDIXES

Throughout this book we have used two different systems of units. We have used the English system, with which the reader is probably more familiar, and the metric system, which is widely used outside of North America. In this appendix, we shall summarize the names and definitions in these two systems of quantities we have discussed in the text.

The unit of length in the English system is the inch or the foot. The corresponding units of length in the metric system are the centimeter (one inch = 2.54 centimeters) and the meter (which is a little over a yard long). The meter is 100 times as long as the centimeter. Both systems use the second as the unit of time.

The two systems differ in their treatment of mass and weight. In the English system, the standard unit of force is the pound, which, as was discussed in the text, is simply the weight of THE pound at sea level. In this system, the mass of an object that weighs one pound at sea level is the pound-mass (not a commonly used unit).

In the metric system, the unit of mass is either the gram or the kilogram (1000 grams). The unit of force is then either the dyne (the force needed to accelerate one gram by one cm/sec^2) or the Newton (the force needed to accelerate one kilogram by one meter/sec^2). Thus, the weight of an object in the metric system will not be the same number as its mass. At sea level, for instance, a one-kilogram mass will weigh 9.8 Newtons.

Actually, although we have talked about "the metric system," there are two metric systems in common use. One of these uses the centimeter as the unit of length, the gram as the unit of mass, and the second as the unit of time. This is called the "cgs" system. The corresponding system in which the meter, kilogram, and second play the important roles goes by the name "mks."

In the following table, we summarize the definitions in the English, cgs, and mks system of a number of important quantities.

Quantity	English	cgs	mks
Force	pound—weight of standard pound at sea level	dyne—force which accelerates one gram one cm/sec^2	Newton—force which accelerates one Kg one m/sec^2
Work (Energy)	foot-pound—one pound acting through one foot	erg—one dyne acting through one cm	joule—one Newton acting through one meter
Power	horsepower—55 foot-pounds/sec		watt—one joule/sec

Units of Force, Work, and Power in the English, cgs, and mks systems.

APPENDIX B:
ARITHMETIC WITH LARGE AND SMALL NUMBERS

In this appendix we shall introduce and discuss the "powers of ten" notation that was used in the text whenever very large or very small numbers had to be manipulated. It is simply a way of dealing with such numbers without writing down a lot of zeros.

Consider a number like two million. Normally, we would write this as 2,000,000. Sometimes it just gets too bothersome to keep writing down all of those zeros. Therefore, we introduce a new notation, in which two million is written

$$2,000,000 = 2 \times 10^6$$

The second way of writing it is called the "powers of ten" notation. The six in the exponent of the 10 just tells us how many zeros there are, so writing something like 2×10^6 is equivalent to saying "the number is a two followed by six zeros."

There is another way to think of the powers of ten notation, which we could see most easily by rewriting the above equation as

$$2,000,000 = 2.0 \times 10^6$$

We can think of the exponent in this case as saying "move the decimal place six places to the right." This way of looking at things is, of course, equivalent to the way given above.

We can introduce a powers of ten notation for very small numbers as well. Consider the fraction one-millionth. We could write this as

$$1/1,000,000 = .000001 = 1.0 \times 10^{-6}.$$

In other words, when we want to talk about numbers less than one, we use a negative exponent with the ten, and we interpret this as saying "move the decimal point six places to the left."

In this way, any number, no matter how large or how small, can be written in the form

$$A \times 10^N,$$

where A is a number between one and ten, and N is either positive or negative. Of course, writing the number in this way does not affect its value in any way; it simply gives us a handy way of writing it down.

There are a few simple rules for multiplying and dividing two numbers in the power of ten notation. Suppose we have two such numbers, the one given above and the other written

$$B \times 10^M$$

Then the following two formulae tell how to multiply and divide:

i. $(A \times 10^N) \times (B \times 10^M) = (A \times B) \times 10^{(N+M)}$

ii. $\dfrac{A \times 10^N}{B \times 10^M} = (A/B) \times 10^{(N-M)}$

These formulae hold whether N and M are positive or negative.

Let's work out a few examples to illustrate these rules. Consider the numbers 2×10^3 (2000) and 3×10^2 (300). Then

$$(2 \times 10^3) \times (3 \times 10^2) = 6 \times 10^5,$$

whereas

$$(3 \times 10^2)/(2 \times 10^3) = 1.5 \times 10^{-1}.$$

The reader can verify these two examples by doing them without the powers of ten notation.

In addition to the powers of ten notation, there are a couple of approximations that come in very handy when we have to deal with very small numbers. For example, the expression $\sqrt{1 - v^2/c^2}$ occurs often in special relativity. If v is much less than c, then this expression has the form $\sqrt{1 - \epsilon}$, where $\epsilon = v^2/c^2$ and we require that ϵ be much less than one.

It turns out that the following expressions are approximately (but not exactly) true under these circumstances. These approximations are very useful in dealing with problems like the relativistic effects in slowly moving objects.

$$\sqrt{1 - \epsilon} = 1 - \frac{1}{2}\epsilon;$$

$$\frac{1}{\sqrt{1 - \epsilon}} = 1 + \frac{1}{2}\epsilon.$$

One more approximate equality is given here for the sake of completeness. If δ is a small number, then

$$\frac{1}{1 + \delta} = 1 - \delta.$$

Again, a few examples will help to fix these ideas in our minds. Consider an object moving at one-tenth the speed of light. Then $\epsilon = 0.01 = 10^{-2}$. We would have

$$\sqrt{1 - \epsilon} = 1 - \frac{1}{2} \times 10^{-2} = .995;$$

$$\frac{1}{\sqrt{1 - \epsilon}} = 1 + \frac{1}{2} \times 10^{-2} = 1.005;$$

$$\frac{1}{1 - \epsilon} = 1 + 10^{-2} = 1.01$$

APPENDIX C: SOME USEFUL CONSTANTS

For convenience, we present here some of the physical constants introduced in the book.

Newton's gravitational constant:

$$G = 6.67 \times 10^{-8} \text{ dyne cm}^2/\text{gm}^2$$

$$= 6.67 \times 10^{-11} \text{ Newton m}^2/\text{kg}^2.$$

Planck's constant:

$$h = 6.6 \times 10^{-34} \text{ joule sec.}$$

Speed of light:

$$c = 3 \times 10^8 \text{ m/sec.}$$

Mass of the electron:

$$m_e = 9.1 \times 10^{-28} \text{ gm.}$$

Mass of the proton:

$$m_p = 1.67 \times 10^{-24} \text{ gm.}$$

Charge on the electron:

$$e = 1.6 \times 10^{-20} \text{ emu.}$$

INDEX

Epistomology, 3
Erastosthenes, 56
Ether, 154, 189, 230
Euclid, 57 ff., 66, 113
Eudoxus, 49, 62
Experimental error, 81 ff.

Falling bodies, 103
Faraday, Michael, 181 ff.
Fermi, Enrico, 423, 425
Fields, 173
Fission, 378, 413
Fitzgerald contraction, 219, 224
Fluxions, 111
Force, 128, 456
Fourier, Jean, 269
Franklin, Benjamin, 163 ff.
Frequency, 254
Fuel processing, 408
Fundamental mode of vibration, 268

g, 105
G, 120
Galilei, Galileo, 2, 93 ff., 102 ff., 110, 122
Galvani, L., 167
Gamma rays, 344, 357
Geiger counter, 37
Geocentrism, 63
GeV, 414
Gilamesh, Epic of, 27, 28
Gilbert, William, 161, 170, 178
Goeppert Mayer, Maria, 354
Gravitational constant, 114, 120, 121
Gravitational field, 173
Graviton, 420
Gravity, 140
Gray, Stephen, 162

Hadron, 427
Half life, 365
Hammurabi, 27

Harmonics, 277
Heat, 247
Heisenberg, Werner, 318
   uncertainty principle, 309, 415
Heliocentric universe, 74, 89, 94
Heraclitus, 44
Hertz, Heinrich R., 199 ff.
High energy physics, 425
Histogram, 80
Horizon line, 23 ff.
Horsepower, 145
Hypercharge, 437
Hypatia of Alexander, 67

Induction, electromagnetic, 183
Integral calculus, 153
Interference, 255 ff., 343
Inverse square law, 119
Invisible college, 172
Ionia, 36 ff., 44, 45, 49, 66
   technical tradition, 36
Ionic bond, 331
Irrational numbers, 43
Iso spin, 437
Isotope, 364
Isotopic spin, 437

Jensen, J. H. D., 354

Kepler, Johann, 77, 88 ff., 97, 108 ff.
Kilogram, 118
Kilowatt, 195
Kinetic energy, 130, 143
Knowledge
   cultural, 3 ff.
   individual, 3, 6 ff.

Laser, 332 ff.
Lavoisier, Antoine, 241
Leibnitz, Gottfried Wilhelm, 112
Length contraction, 217
Lepton, 427